INTRODUCTION TO ABSTRACT ALGEBRA

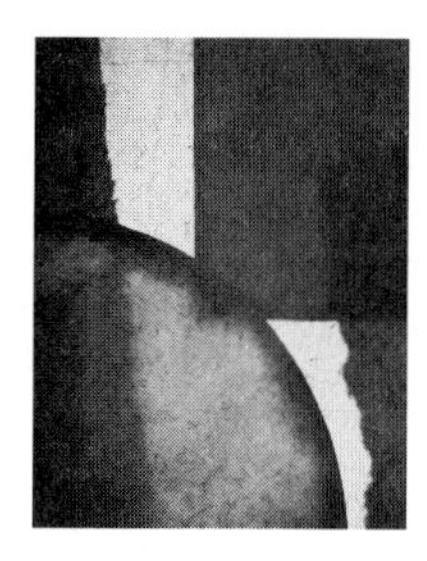

INTRODUCTION TO ABSTRACT ALGEBRA

Sixth Edition

Neal H. McCoy
Professor Emeritus of Mathematics
Smith College

Gerald J. Janusz
Professor of Mathematics
University of Illinois at Urbana-Champaign

A Harcourt Science and Technology Company

San Diego San Francisco New York Boston
London Toronto Sydney Tokyo

Sponsoring Editor	Barbara Holland
Production Editor	Julie Bolduc
Editorial Coordinator	Karen Frost
Marketing Manager	Marianne Rutter
Cover Design	Gary Ragaglia, Metro Design
Copyeditor	Dan Hays
Composition	MacroTEX
Printer	The Maple-Vail Book Manufacturing Group

This book is printed on acid-free paper. ∞

ACADEMIC PRESS
A Harcourt Science and Technology Company
525 B Street, Suite 1900, San Diego, CA 92101-4495, USA
http://www.academicpress.com

Academic Press
Harcourt Place, 32 Jamestown Road, London, NW1 7BY, UK
http://www.academicpress.com

Harcourt/Academic Press
200 Wheeler Road, Burlington, MA 01803, USA
http://www.harcourt-ap.com

Library of Congress Catalog Number: 00-108489
ISBN: 0-12-380392-6

Printed in the United States of America
00 01 02 03 04 05 MB 9 8 7 6 5 4 3 2 1

Dedicated to the memory of my son, Paul

NEAL H. MCCOY

To Madeleine and Annie –
whose smiles and love of everything in life
bring constant joy to their parents and grandparents

GERALD J. JANUSZ

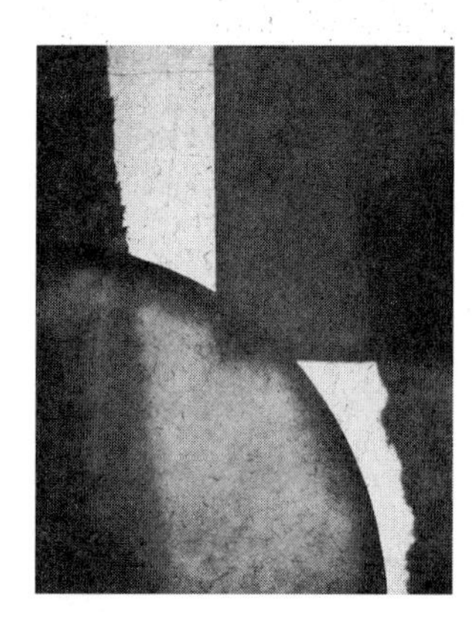

CONTENTS

III. INTEGRAL DOMAINS AND FIELDS

IV. FACTORIZATION

V. POLYNOMIALS

VI. GROUPS

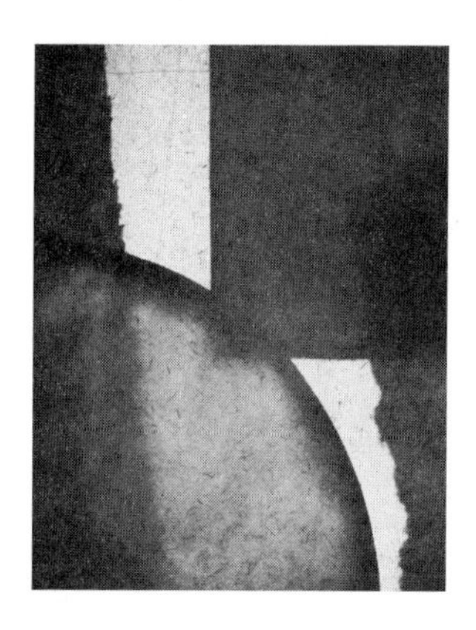

ACKNOWLEDGMENTS

The early editions of this book were written by Neal McCoy. His dedication to his students and his consideration of their needs led him to produce a classic text—one of the first in abstract algebra. I am grateful for his permission to do the revisions on the later editions and I express my heartfelt thanks to him for laying the solid foundations on which this edition is built.

G. Janusz

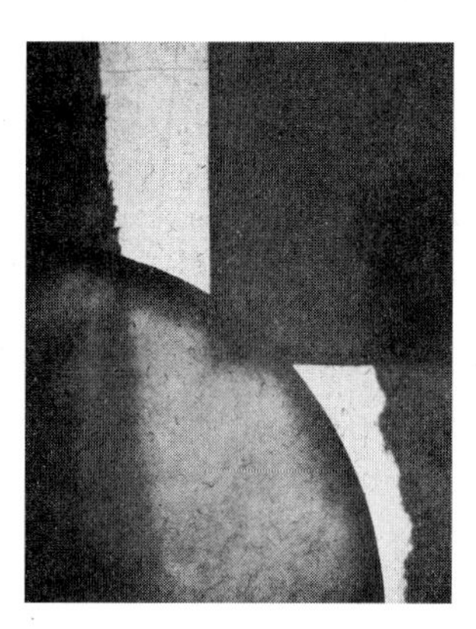

PREFACE

This edition is intended as a text for a first course in abstract algebra. As with the earlier editions, the goal continues to be that of making the exposition as simple and clear as possible while retaining the precision required to give the student an introduction to the ideas, methods, and results of abstract algebra.

In the present undergraduate curriculum, the first course in abstract algebra often serves several purposes. The student is expected to learn a collection of facts, to learn some methods and ideas, and to learn to read and write proofs of mathematical statements. All of these requirements have been carefully considered in the preparation of this text. The basic ideas of rings, fields, and groups are presented from an axiomatic point of view, and careful proofs, based on the axioms, are given starting with very elementary theorems and gradually building toward rather sophisticated proofs later in the text. The student is asked to provide some steps in proofs and also to provide complete proofs in the exercises.

Rings are presented before groups because the ongoing example of the ring of integers is familiar to the students and helps to make the abstract concepts more concrete. After a brief study of abstract rings, the ring of integers and the ring of integers modulo n are covered in detail. The notion of an equivalence relation is used early (Chapter II), and throughout the later chapters. Ideals are introduced, as are homomorphisms of rings also in Chapter II.

Integral domains and fields are studied in Chapters III–V. The study of unique factorization for the integers appears first and is imitated for the ring of polynomials over a field. The unique factorization theorem for principal ideal domains is given in Chapter IX. The study of polynomials with coefficients in a field is carried far enough to prove that every polynomial can be factored as a product of linear factors with coefficients in some extension field. This idea provides the key step in the study of finite fields; a proof of the existence of fields with p^n elements for every prime p and positive integer n is given. The uniqueness of such fields is proved in Chapter VII after the necessary preliminaries about finite abelian groups have been completed. The material on Euler's totient function and the application to RSA public key cryptology are new to this edition.

The three chapters on group theory can be used in different ways depending on the needs of the course. Chapter VI contains an introduction to groups in general, topics on cyclic groups, symmetric groups, and Lagrange's theorem. Many one-term courses will not have time for any additional material. For courses with additional time, either of Chapters VII, which covers the structure of finite abelian groups, or Chapter VIII, which covers finite groups through Sylow's theorem, can be presented. Chapters VII and VIII have been kept independent of each other so that either may be covered without reference to the other.

There is considerable information in the exercises. Some exercises contain important results that go beyond the scope of the text material, whereas others simply reinforce the material presented in the particular section. The students should be encouraged to read the exercises even though it is not reasonable to assign all of them to be worked. Some exercises are scattered throughout the text that deal with two-by-two matrices. Even though no formal matrix theory is presented or assumed, the student should have no serious difficulty with these if the exercises on this topic are worked from the beginning. Many new exercises have been added to this edition. A few new exercises are included to present more of a challenge to the strong student.

We thank the many readers who have sent comments and suggestions since the release of the fifth edition. We also thank Burton Fein, Oregon State University, and Frank DeMeyer, Colorado State University, for their comments and suggestions made after reviewing the manuscript, and Josh Mullet, University of Illinois, and James Keesling, University of Florida, for their careful proofreading of the first draft of the manuscript. Most of their suggestions have been incorporated into this edition.

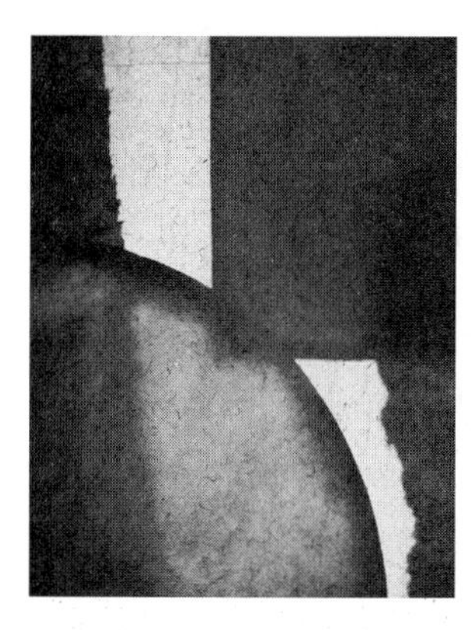

1 RINGS

In this chapter we will study the integers and prove a variety of facts about them. The integers may very well be the most commonly known algebraic system and readers may have widely differing knowledge of properties of the integers. The idea of *proving* facts about such familiar objects requires some discussion. We give some familiar properties, which we list as axioms, and show how conclusions may be drawn from these axioms. The power of abstract algebra is illustrated by applying the arguments to other algebraic systems that satisfy the same axioms. The integers are an example of an algebraic system called a *ring*; some of the properties of the integers are shared by all rings.

1 FORMAL PROPERTIES OF THE INTEGERS

The simplest numbers are the numbers 1, 2, 3,. . ., used in counting. These are called the "natural numbers." Addition and multiplication of natural numbers have simple interpretations when we consider a natural number as indicating the number of elements in a set. For example, suppose we have two jars of coins, one containing m coins and the second one containing n coins. Combine the two jars of coins by pouring the first jar into the second to obtain $m + n$ coins in the second jar. If, instead, we poured the second jar into the first, there results $n + m$ coins in the first jar. It seems quite obvious that

$$m + n = n + m$$

for every choice of the natural numbers m and n. The formal language is that the operation of addition of natural numbers is commutative. This is an example of a *law* or *formal property*. Another example is the associative law for addition which asserts

$$(a+b)+c=a+(b+c)$$

for every choice of natural numbers a, b, c.

Multiplication of natural numbers may be introduced as follows. If one has m jars of coins, each of which has n coins, and all the jars are poured into a single jar, then there are mn coins in the jar. It is also a familiar fact that multiplication of natural numbers is commutative (i.e., $mn = nm$) and associative [i.e., $a(bc) = (ab)c$]. Moreover, addition and multiplication are related by the distributive law, which reads

$$m(n+k)=mn+mk,$$

for all natural numbers m, n, k.

Historically, the natural numbers were used for centuries before there was any consideration of the formal properties. However, in modern algebra it is precisely such formal properties that are of central interest. Some reasons for this changed viewpoint will become evident later on in this chapter as well as in succeeding chapters.

If m and n are natural numbers, there need not be a natural number x such that $m + x = n$. In order to be able to solve all equations of this kind, we need to have available the negative numbers and zero as well as the positive integers. The properties we study in this and later sections are suggested by the well-known properties of the set of all integers (positive, negative, and zero). Later we consider a list of properties that characterize the system of integers, although for the most part we assume familiarity with the simpler properties of this system. In later chapters, other number systems of elementary algebra will be discussed in detail.

2 DEFINITION OF A RING

The concepts presented in this section are of fundamental importance, although a full realization of their generality will probably not become apparent until many examples are studied.

The addition of natural numbers is an example of a binary operation; that is, two natural numbers are "combined" to produce a third natural number. The abstraction of this concept is discussed next.

2.1 Binary Operations

Let R be a nonempty set of elements. A *binary operation* on R is a function that assigns to each ordered pair (r, s) of elements in $R \times R$ an element of R. It is safe to think of an operation as a rule for combining two elements of R to obtain another element of R. We may sometimes write the operation using the function notation, so $f(r, s)$ is the element assigned to the pair (r, s), and then say f is the operation. An element $e \in R$ is an identity for the operation f if

$$f(e, r) = f(r, e) = r$$

for every $r \in R$.

If R is the set of natural numbers, addition is the binary operation that assigns to the pair (m, n) the element which we write as $m + n$. No natural number is an identity for addition. The element 0 is an identity for the operation of addition of the set of all integers since

$$0 + m = m + 0 = m$$

for all integers m. Multiplication is the operation which assigns to (m, n) the element mn. The natural number 1 is an identity for multiplication on natural numbers.

Now we turn to the definition of a ring. We begin with a nonempty set R on which there are defined two binary operations, which we call "addition" and "multiplication" and for which we use the familiar notation. Accordingly, if $a, b \in R$, then $a + b$ and ab (or sometimes $a \cdot b$) are elements of R. By way of emphasis, we may state this fact in another way by saying that R is *closed* under the binary operations which we are calling addition and multiplication.

Definition 2.1 The set R together with two binary operations + and $\cdot$ (called addition and multiplication) is called a *ring* if the following properties hold for every selection of elements $a, b, c \in R$:

$\mathbf{P}_1$: $a + b = b + a$ *(commutative law of addition).*

$\mathbf{P}_2$: $(a + b) + c = a + (b + c)$ *(associative law of addition).*

$\mathbf{P}_3$: There is an element $0 \in R$ such that $a + 0 = a$ for every $a \in R$ *(existence of a zero).*

$\mathbf{P}_4$: If $a \in R$ there exists an $x \in R$ such that $a + x = 0$ *(existence of additive inverses).*

$\mathbf{P}_5$: $(ab)c = a(bc)$ *(associative law of multiplication).*

$\mathbf{P}_6$: $a(b + c) = ab + ac$ and $(b + c)a = ba + ca$ *(distributive laws).*

Here, we make a few remarks about the defining properties of a ring. First, the elements should necessarily not be thought of as *numbers*. We will see examples later in which the elements of R are polynomials, matrices, subsets of a set, as well as other possible objects. Moreover, addition and multiplication are not assumed to have any properties other than those specified by the listed properties. The element "0," whose existence is asserted in $\mathbf{P}_3$ and which we call *zero*, is an identity for the operation of addition since by $\mathbf{P}_1$ we have

$a + 0 = 0 + a = a$ for all $a \in R$. We do not assume that there is only one identity for addition, but later on we will prove that this is true. Again we should not think of 0 as the familiar number zero; it is merely an element of R having a special property–namely, it is an identity for the operation of addition. Finally, we point out that $\mathbf{P}_4$ does not assert that there is *only one* $x \in R$ such that $a + x = 0$, but this fact will also be proved eventually.

The properties used to define a ring are familiar properties of the integers. Hence, with the usual definition of addition and multiplication, the set of integers is a ring. Henceforth this ring will be denoted by $\mathbb{Z}$. For this ring, the zero whose existence is asserted by $\mathbf{P}_3$ is the familiar number 0.

Now we give another example of a ring closely associated to $\mathbb{Z}$. Let E be the set of all even integers (positive, negative, and zero). So E is a subset of $\mathbb{Z}$ and the addition and multiplication that are defined for elements of $\mathbb{Z}$ are also defined for elements of E. Moreover, if x and y are in E then $x + y$ and xy as the sum or product of even integers is again even. That is, E is closed under the operations of addition and multiplication. We claim that E is a ring with these operations. To verify this one must check that properties $\mathbf{P}_1$–$\mathbf{P}_6$ hold for E. It is immediate that properties $\mathbf{P}_1$, $\mathbf{P}_2$, $\mathbf{P}_5$, and $\mathbf{P}_6$ hold because they hold for all elements of $\mathbb{Z}$. To check that $\mathbf{P}_3$ holds, the only question is whether the 0 for $\mathbb{Z}$ is an element of E–which it is by definition of E. To check that $\mathbf{P}_4$ holds we start with an arbitrary element $a \in E$ so that a is an even integer. Then the element x in $\mathbb{Z}$ that is asserted by $\mathbf{P}_4$ for $\mathbb{Z}$ is $x = -a$. Since $-a$ is an even integer when a is even, it follows that $-a \in E$ and so $\mathbf{P}_4$ holds for E. Thus, E is a ring. We say that E is a *subring* of $\mathbb{Z}$. Here is the definition of subring in a more general context.

Let R be a ring with operations $+$ and $\cdot$ and let S be a subset of R. Then S is a *subring* of R if S is closed under the addition and multiplication that are defined for all elements of R and if S is a ring with respect to these two operations.

Consider the set of all odd integers. Although the set of odd integers is closed under multiplication, it is not a subring of $\mathbb{Z}$ because it is not closed under the operation of addition. That is, if x and y are odd integers, then $x + y$ is not always an odd integer (in fact, it never is an odd integer).

The definition of a ring does not require that multiplication be commutative. However, we will frequently want to consider rings that have this additional property so we give it a number for easy reference:

$\mathbf{P}_7$: If $a, b \in R$ then $ab = ba$ *(commutative law of multiplication).*

A ring which has property $\mathbf{P}_7$ is called a *commutative* ring. When $\mathbf{P}_7$ does not hold for a ring R, then there exist elements $c, d \in R$ such that $cd \neq dc$ and we say R is a *noncommutative* ring.

In a ring, there need not be an identity for the operation of multiplication. The ring of even integers is an example in which there is no identity for multiplication. Many of the rings we consider will have such an element and we indicate this by saying R is a ring *with identity*. We give this property a number:

P$_8$: There is an element $e \in R$ such that $ea = ae = a$ for all $a \in R$ *(existence of an identity for multiplication).*

Note that the terminology "ring with identity" should not cause the reader to wonder if identity refers to addition or multiplication. Every ring has an identity with respect to addition; the qualifier "with identity" will always mean an identity with respect to multiplication.

We emphasize that a ring need not have either of the properties **P**$_7$ or **P**$_8$. However, most of the rings we study will have both properties. One significant exception will be the study of rings of matrices which will fail to be commutative.

3 EXAMPLES OF RINGS

In order to give an example of a ring R, it is necessary to specify the elements of R and to define the operations of addition and multiplication on R so that the properties **P**$_1$–**P**$_6$ hold. The reader is aware of several examples of rings although perhaps not aware in the formal sense that the required properties have previously been considered. In addition to the ring $\mathbb{Z}$ of integers, there is the ring of all *rational numbers* described as

$$\mathbb{Q} = \left\{ \frac{a}{b} : a, b \in \mathbb{Z}, b \neq 0 \right\};$$

this is the set quotients of integers with addition and multiplication defined in the familiar way:

$$\frac{a}{b} + \frac{c}{d} = \frac{ad + bc}{bd}, \quad \frac{a}{b} \cdot \frac{c}{d} = \frac{ac}{bd}.$$

The ring $\mathbb{Z}$ is a subring of $\mathbb{Q}$ as $\mathbb{Q}$ is a subring of the ring of all real numbers which in turn is a subring of the ring of all complex numbers. All these number systems will be considered in detail in later chapters.

We proceed to give some other, less familiar, examples of rings. For the most part, we will not write out the verification of the properties **P**$_1$–**P**$_6$. Some will be given as exercises. The main purpose of these examples is to clarify the concept of a ring and to show that there are rings of many different kinds.

Example 3.1 Let $S = \{x + y\sqrt{2} : x, y \in \mathbb{Z}\}$ with the expected definition of addition and multiplication given by:

$$\begin{aligned}(a+b\sqrt{2})+(c+d\sqrt{2}) &= (a+c)+(b+d)\sqrt{2},\\ (a+b\sqrt{2})(c+d\sqrt{2}) &= (ac+2bd)+(ad+bc)\sqrt{2}.\end{aligned}$$

These definitions show that S is closed under addition and multiplication. It may be verified that the conditions $\mathbf{P}_1$–$\mathbf{P}_6$ hold; in fact, $\mathbf{P}_7$ and $\mathbf{P}_8$ also hold so S is a commutative ring with identity. Of course, S is a subring of the ring of all real numbers.

Example 3.2 Let $T = \{u + v\sqrt[3]{2} + w\sqrt[3]{4} : u, v, w \in \mathbb{Q}\}$ with the usual definition of addition and multiplication. This is similar to the previous example but with two differences: The coefficients are rational numbers in place of integers and the radicals are cube roots rather than square roots. We cannot use the smaller set of elements of the form $u + v\sqrt[3]{2}$ because the set must be closed under multiplication; it is necessary that $(\sqrt[3]{2})(\sqrt[3]{2}) = \sqrt[3]{4}$ also be used. One may check that T is a commutative ring with identity.

Example 3.3 Let C be the set of all real-valued functions that are defined and continuous on the interval $[0, 1]$. For two functions f and g in C, define their sum and product by the rules

$$\begin{aligned}(f+g)(x) &= f(x)+g(x)\\ (fg)(x) &= f(x)g(x).\end{aligned}$$

From calculus, we know that the sum and product of two continuous functions are continuous so C is closed under these operations. This definition of sum and product is called *pointwise addition* and *pointwise multiplication*. The validity of the axioms of a ring may be checked. The zero of the ring is the function $z(x)$ defined by $z(x) = 0$ for every x. The ring is a commutative ring with identity. What is the identity element?

In the examples so far given, the rings have had infinitely many elements. If a ring has only a finite number of elements, one may describe the rules for addition and multiplication by simply giving all possible sums and products in a table. We present two such examples.

Example 3.4 Let $R = \{u, v, w, x\}$ be a set with four elements. Define addition and multiplication by the data in the following tables:

$(+)$	u	v	w	x
u	u	v	w	x
v	v	u	x	w
w	w	x	u	v
x	x	w	v	u

$(\cdot)$	u	v	w	x
u	u	u	u	u
v	u	v	w	x
w	u	w	w	u
x	u	x	u	x

Information is read from the tables as follows. If we wish to determine $v + x$, look at the intersection of the row having v at the left and the column having x at the top. Since w appears at this position, $v + x = w$. In the analogous way we discover that $x + w = v$, $w + w = u$, $vw = w$, and $xx = x$. It would take a serious amount of time to verify that the commutative law for addition, the associative law of addition and multiplication, and the distributive laws all hold for these operations and that indeed R is a ring. We can see by inspection that u is the zero of the ring, as $u + y = y + u = y$ for every $y \in R$. (Just look at the row to the right of u and the column below u in the addition table.) Similarly, v is the multiplicative identity. It can be checked that this is a commutative ring with identity.

Example 3.5 Let $\mathbb{Z}_2$ be a set with two elements which we label as z and i. The rules for addition and multiplication are described in the following tables:

$(+)$	z	i
z	z	i
i	i	z

$(\cdot)$	z	i
z	z	z
i	z	i

Since $\mathbb{Z}_2$ has only two elements, it is possible to actually verify the validity of the ring axioms. Visibly z is the zero and i is the multiplicative identity and $\mathbb{Z}_2$ is a commutative ring with identity. This ring will make an appearance later in a somewhat different form. We give a preview of that now. Suppose we let z stand for the set of all even integers and let i stand for the set of all odd integers so that $\mathbb{Z}_2$ is a collection of two sets of integers. We attempt to define addition and multiplication of these sets of integers by the following rule. Select an integer from each of the two sets to be added and add the integers. This sum lies in one of the two sets, so declare that set to be the sum of the two sets. For example, take any even integer, e.g., 4, from z and an odd integer, e.g., 7, from i; then $4 + 7 = 11$ is odd and lies in i so we declare that $z + i = i$. As one may easily check, this definition does not depend on the particular elements selected; for example, if we select 8 in z in place of 4 and select -11 in i in place of 7, the sum $8 + (-11) = -3$ is still odd. The analogous rules for multiplication also hold. By using properties of integers, it is fairly easy to verify that $\mathbb{Z}_2$ is a commutative ring with identity.

Example 3.6

The following example will appear in several places later in the text. Let $M_2(\mathbb{Z})$ be the set of all symbols of the form

$$\begin{bmatrix} a & b \\ c & d \end{bmatrix},$$

where a, b, c, d are arbitrary elements of $\mathbb{Z}$. This symbol is called a two-by-two *matrix*. Addition and multiplication of these symbols are defined as follows:

$$\begin{bmatrix} a & b \\ c & d \end{bmatrix} + \begin{bmatrix} e & f \\ g & h \end{bmatrix} = \begin{bmatrix} a+e & b+f \\ c+g & d+h \end{bmatrix}$$

$$\begin{bmatrix} a & b \\ c & d \end{bmatrix} \cdot \begin{bmatrix} e & f \\ g & h \end{bmatrix} = \begin{bmatrix} ae+bg & af+bh \\ ce+dg & cf+dh \end{bmatrix}.$$

These operations make $M_2(\mathbb{Z})$ into a ring. The multiplication rule may seem a bit unusual and is most easily remembered using the "row by column" multiplication; each entry of the product of two matrices is obtained by the following row-by-column rule:

$$[x \quad y] \cdot \begin{bmatrix} p \\ q \end{bmatrix} = xp + yq.$$

With these operations, $M_2(\mathbb{Z})$ is a ring. It is called the *ring of all two-by-two matrices over the integers*. By looking a specific examples, the reader may verify that this is a noncommutative ring. It has an identity element—what is it?

If we modify this example by letting a, b, c, d be rational numbers in place of integers, we obtain the ring $M_2(\mathbb{Q})$ of two-by-two matrices over the rational numbers. In fact, if R is any ring and we let a, b, c, d be arbitrary elements of R we obtain the ring $M_2(R)$ of two-by-two matrices over R.

The final example of this section is quite different from any of the previous examples; before describing it we present a quick review of some notions of sets.

3.1 Review of Set Operations

Let A be any set; the collection of all subsets of A, including the empty set, $\emptyset$, and the entire set A is called the *power set of* A and is denoted by R. The empty set $\emptyset$ is the subset of A having no elements; it is regarded as a subset of every subset of A. We will denote elements of R by lowercase letters a, b, and so on; keep in mind that they are subsets of A. If u is an element of A and if a is a subset of A then $u \in a$ means that u is *an element of* a. If a and b are subsets of A then $a \subseteq b$ means every element of a is also an element of b (read "a is contained in b"). This notation allows the possibility that $a = b$. If there is an element of b that is not in a (so that $a \subseteq b$ and $a \neq b$), we may write $a \subset b$.

We use the following familiar definitions of union, intersection, and difference of subsets:

$$\begin{aligned} a \cup b &= \{x : \text{ either } x \in a \text{ or } x \in b\} && \textit{(union)} \\ a \cap b &= \{x : x \in a \text{ and } x \in b\} && \textit{(intersection)} \\ a \backslash b &= \{x : x \in a \text{ and } x \notin b\} && \textit{(difference)}. \end{aligned}$$

Using words in place of these symbolic definitions we see that the union of a and b is the set of elements that lie in *either a or b or both*; the intersection of a and b is the set of elements that lie in *both a and b*; the difference $a \backslash b$ is the set of elements that are in a but not in b. This set is denoted as $a - b$ by some authors, but we use the notation $a \backslash b$ to avoid confusion with other uses of the minus sign. If $A = \{1, 2, 3, 4, 5, 6, 7, 8\}$ and $a = \{1, 2, 3\}$ and $b = \{2, 4, 6, 8\}$ then

$$a \cup b = \{1, 2, 3, 4, 6, 8\}, \quad a \cap b = \{2\}, \quad a \backslash b = \{1, 3\}.$$

Example 3.7 Let A be any set and let R be the power set of A. We define operations of addition and multiplication that make R into a ring. For elements $a, b \in R$ let

$$a + b = (a \cup b) \backslash (a \cap b) \quad \text{and} \quad ab = a \cap b.$$

In words, $a + b$ is the set of elements that lie in the subset a or the subset b but not in both. The ring axioms may be verified with a bit of effort. The zero element (additive identity) is the empty set $\emptyset$ and the multiplicative identity is the set A. With these operations, R is a commutative ring with identity.

This ring has several rather unusual properties. For example, if $a \in R$ then $a + a = \emptyset = 0$ so that each element is its own additive inverse. Moreover, $a \cdot a = a \cap a = a$ so each element equals its own square.

In case A is a finite set, R is a finite ring. The number of elements in R is the number of subsets of the finite set A. For example, if $A = \{1, 2\}$ is a set with two elements, then R is the ring with the four elements

$$\emptyset, \quad \{1\}, \quad \{2\}, \quad \{1, 2\}.$$

More generally, if A has n elements, then R has 2^n elements.

We will call R the *ring of all subsets of the set A*. Later, when we mention the ring of all subsets of a set it is always understood that addition and multiplication are defined as in this example.

3.2 Direct Sums

We conclude this section not by giving still another example but by presenting a simple, but quite useful, way to construct new rings from given rings. Suppose R and S are rings, either distinct or identical, and let us consider the Cartesian product $R \times S$ whose elements are the ordered pairs (r, s) with $r \in R$ and $s \in S$. On the set $R \times S$ we define addition and multiplication as follows:

$$\begin{aligned}(r_1, s_1) + (r_2, s_2) &= (r_1 + r_2, s_1 + s_2),\\ (r_1, s_1) \cdot (r_2, s_2) &= (r_1 r_2, s_1 s_2).\end{aligned}$$

It is understood, of course, that $r_1, r_2 \in R$, and $s_1, s_2 \in S$. Moreover, although the same symbol is used for addition in R and in S, $r_1 + r_2$ means the addition of elements defined for the ring R and $s_1 + s_2$ means the addition defined for elements in the ring S (and similarly for products). We leave as an exercise the proof that with respect to the previous definitions the set $R \times S$ becomes a ring. It is convenient to have a name for the ring obtained by this construction. Accordingly, we make the following definition:

Definition 3.1

If R and S are given rings, the ring whose elements are the elements of the product set $R \times S$, with addition and multiplication as defined previously, is called the **direct sum** of the rings R and S, and it is usually denoted by $R \oplus S$.

What conditions on R and S will ensure that $R \oplus S$ is a commutative ring? Under what conditions does $R \oplus S$ have an identity?

EXERCISES

In these exercises, it is to be assumed that the real numbers (in particular, the rational numbers and the integers) have all the properties that are freely used in elementary algebra.

1. Which of the following are rings with respect to the usual definition of addition and multiplication? In this exercise, E denotes the ring of all even integers.
(a) The set of all positive integers.
(b) The set of all integers (positive, negative, and zero) that are divisible by 3.
(c) The set of all real numbers of the form $x + y\sqrt{2}$, where $x, y \in E$.

(d) The set of all real numbers of the form $x + y\sqrt[3]{2}$, where $x, y \in \mathbb{Z}$.
(e) The set of all real numbers of the form $x + y\sqrt[3]{2} + z\sqrt[3]{4}$, where $x, y, z \in \mathbb{Z}$.
(f) The set of all real numbers of the form $x + y\sqrt{3}$, where $x \in E$ and $y \in \mathbb{Z}$.
(g) The set of all rational numbers that can be expressed in the form m/n, where $m \in \mathbb{Z}$ and n is a positive odd integer.

2. What is the additive inverse of each element of the ring R described in Example 3.4?

3. Verify that the subset $\{u, v\}$ of the ring R in Example 3.4 is a subring. Show that after a change of notation this is the same ring as described in Example 3.5.

4. For the ring R described in Example 3.4, use the tables to verify each of the following:
(a) $(u + v) + w = u + (v + w)$
(b) $(v + w) + x = v + (w + x)$
(c) $w(v + x) = wv + wx$
(d) $(w + v)x = wx + vx$
(e) $(xv)w = x(vw)$

5. For the ring $M_2(\mathbb{Z})$ described in Example 3.6, verify the associative law of multiplication and the distributive laws. What is the zero of this ring? Verify that

$$\begin{bmatrix} 1 & 0 \\ 0 & 1 \end{bmatrix}$$

is the identity of $M_2(\mathbb{Z})$. Give examples to show that $M_2(\mathbb{Z})$ is a noncommutative ring.

6. For the ring R of all subsets of a given set (Example 3.7) give a formal proof of the associative law of addition by describing in words the elements of A that lie in each set $(a + b) + c$ and $a + (b + c)$.

7. For the ring R of all subsets of a given set (Example 3.7) give a formal proof of the distributive law by describing in words the elements of A that lie in each set $(a + b)c$ and $ac + bc$. How does the commutative law for multiplication ensure that the other distributive law must hold?

8. On the set $S = \mathbb{Z} \times \mathbb{Z}$, define addition and multiplication by the rules

$$\begin{aligned} (a, b) + (c, d) &= (a + c, b + d) \\ (a, b) \cdot (c, d) &= (ac + 2bd, ad + bc). \end{aligned}$$

Prove that S is a commutative ring with identity.

9. It can be shown that the set $\{a, b, c, d\}$ is a ring if addition and multiplication are defined by the following tables.

$(+)$	a	b	c	d
a	a	b	c	d
b	b	c	d	a
c	c	d	a	b
d	d	a	b	c

$(\cdot)$	a	b	c	d
a	a	a	a	a
b	a	c	a	c
c	a	a	a	a
d	a	c	a	c

Is this a commutative ring? What is the zero of the ring? What is the additive inverse of each element? Does the ring have a multiplicative identity?

10. Show that neither of the following can possibly be the addition table for a ring consisting of the set $\{a, b, c, d\}$ of four elements:

$(+)$	a	b	c	d
a	a	b	c	a
b	b	c	d	a
c	c	d	a	b
d	a	a	b	c

$(+)$	a	b	c	d
a	a	b	c	d
b	b	c	a	d
c	c	a	d	b
d	d	c	d	a

11. If a ring has exactly n elements, then its addition table is a square array of n rows and n columns having the ring elements as indices for the rows and columns. Based on an examination of the tables in this section, one might guess that each ring element appears exactly once in each row and column of the addition table (of course, not counting the element that serves as an index for the row or column). Prove that this is indeed the case.

12. On the set of integers let addition be defined in the usual way but define the "product" of any two integers to be 0. Is the set of all integers a ring with respect to addition and this new "multiplication"?

13. If R is any ring, verify that the definitions of addition and multiplication of two-by-two matrices over R as given by the rules in Example 3.6 (with a, b, c, d arbitrary elements in R) make $M_2(R)$ into a ring.

14. Prove that if a ring R contains elements s and t such that $st \neq 0$, then $M_2(R)$ is a noncommutative ring.

15. Make addition and multiplication tables for the ring of all subsets of $A = \{1, 2\}$. Verify that the four subsets can be assigned the names

u, v, w, x in such a way that the addition and multiplication tables are the ones given in Example 3.4.

16. The following is an addition table and part of the multiplication table for a ring with three elements. Make use of the distributive laws to fill in the rest of the table.

$(+)$	a	b	c
a	a	b	c
b	b	c	a
c	c	a	b

$(\cdot)$	a	b	c
a	a	a	a
b	a	b	$*$
c	a	$*$	$*$

Is this ring a commutative ring? Does it have an identity?

17. The following is an addition table and part of the multiplication table for a ring with four elements. Make use of the distributive laws to fill in the rest of the table.

$(+)$	a	b	c	d
a	a	b	c	d
b	b	a	d	c
c	c	d	a	b
d	d	c	b	a

$(\cdot)$	a	b	c	d
a	a	a	a	a
b	a	$*$	$*$	a
c	a	$*$	c	$*$
d	a	b	c	$*$

Is this ring a commutative ring? Does it have an identity?

18. On the set of all integers, we define a new addition and multiplication indicated by $\oplus$ and $\odot$:

$$a \oplus b = a + b - 1, \qquad a \odot b = a + b - a \cdot b,$$

where the "+," "−," and "·" on the right side of the equal signs are the usual operations defined for integers. Verify that with respect to these new operations of addition and multiplication the set of all integers is a commutative ring with identity. Identify the zero and the multiplicative identity.

19. If R and S are rings, give a detailed proof that the direct sum $R \oplus S$ is a ring.

20. Let m and n be positive integers and R a ring with m elements and S a ring with n elements.

(a) How many elements are in the direct sum $R \oplus S$?

(b) Give an example of a commutative ring with 16 elements and an example of a noncommutative ring with 16 elements.

(c) Give an example of a ring with 32 elements that does not have an identity.

4 SOME PROPERTIES OF ADDITION AND MULTIPLICATION

So far we have given the definition of a ring and presented examples of rings of many different kinds. By now it should be clear that when we think of an arbitrary ring, we should not necessarily think of one of our familiar number systems. Accordingly, we cannot consider properties of a ring as being obvious, except those actually used in the definition. In this section we give proofs of many properties of any ring. At first we consider only properties of addition, and hence we use only the axioms $\mathbf{P}_1$–$\mathbf{P}_4$.

Before proceeding, it may be well to recall that if a and b are elements of a set (in particular of a ring) then the statement $a = b$ means that a and b are identical elements of the set or, looked at another way, a and b are two symbols for the same element. As a consequence of this usage of "equality," we may use arguments that assert if $a = b$ and $c = d$, then $a + c = b + d$ and $ac = bd$. We freely use these facts without further mention.

We begin by showing the uniqueness of the zero element.

Theorem 4.1 *The zero of a ring R, whose existence is asserted by $\mathbf{P}_3$, is unique.*

Proof: The uniqueness assertion means that if $0, 0' \in R$ and if these two elements both have the property

$$a + 0 = a \quad \text{and} \quad a + 0' = a$$

for every $a \in R$, then $0 = 0'$. Here is the argument. Since $a + 0 = a$ for every $a \in R$, we apply this with $a = 0'$ to conclude

$$0' + 0 = 0'. \tag{1}$$

In like manner, from the equation $a + 0' = a$ we use $a = 0$ to conclude

$$0 + 0' = 0. \tag{2}$$

Since the commutative law holds for addition we know $0 + 0' = 0' + 0$ and hence we can conclude from Eqs. (1) and (2) that $0 = 0'$, and the proof is complete. ∎

In view of this result, we are justified in speaking of *the* zero of a ring. We may observe that this is the first result proved for an arbitrary ring. It would have been easy to verify the truth of the theorem in all the examples of ring so far presented but that would not constitute a proof that it is always true. Now we know it is true in every ring. Many other results will be obtained for arbitrary rings as we proceed.

Next we consider cancelation laws.

Theorem 4.2 CANCELATION LAWS OF ADDITION. *If a, b, and c are elements of a ring R, the following hold:*

(i) *If $a + c = b + c$, then $a = b$.*
(ii) *If $c + a = c + b$, then $a = b$.*

Proof: By property $\mathbf{P}_4$, there is an element $t \in R$ such that $c + t = 0$. Starting from the equation

$$a + c = b + c, \tag{3}$$

we add t to obtain

$$(a + c) + t = (b + c) + t.$$

However, now we have

$$\begin{aligned}(a + c) + t &= a + (c + t) \quad \textit{(assoc. law)}\\ &= a + 0 \quad \textit{(def. of } t\textit{)}\\ &= a \quad \textit{(def. of } 0\textit{)}.\end{aligned}$$

Similarly,

$$(b + c) + t = b + (c + t) = b + 0 = b.$$

From these calculations and Eq. (3) we see that $a = b$.

In view of the commutative law of addition, statement (ii) follows from statement (i). ∎

Although the definition of the zero element requires that $a + 0 = a$ for *every* element a of the ring R, we can now observe that the zero of the ring is completely determined by any *one* element; that is, if there is just one element $a \in R$ such that $a + z = a$, then z is the zero element. This is a consequence of the cancelation law just proved because $a + z = a$ and $a + 0 = a$ imply $a + z = a + 0$ and so $z = 0$.

The preceding theorem can be used to give a short proof of the uniqueness of the additive inverse of an element.

Corollary 4.1 *The additive inverse of an element a of a ring R, whose existence is asserted by property $\mathbf{P}_4$, is unique.*

Proof: To prove this statement suppose that $a + x = 0$ and $a + y = 0$. Then $a + x = a + y$ and $x = y$ by the cancelation law. ∎

4.1 Notation for Additive Inverses

Since each element a of R has exactly one additive inverse, we will denote this additive inverse by $-a$. Thus, we have the condition that $-a$ is the unique element such that $a + (-a) = 0$. We will often write an expression $b + (-a)$ in the shorter form $b - a$. The verbal definition of the symbol $-a$ is that $-a$ is the unique element which gives 0 when added to a.

Since $a + (-a) = 0$ it follows that a is the additive inverse of $-a$; in symbols we write $-(-a) = a$. We have established the first of the following properties, where a, b, and c are arbitrary elements of a ring R:

Properties 4.1

(i) $-(-a) = a$,

(ii) $-(a + b) = -a - b$,

(iii) $-(a - b) = -a + b$,

(iv) $(a - b) - c = a - (b + c)$.

Let us prove the second of these statements. By definition, $-(a + b)$ is the additive inverse of $(a + b)$. Let us verify that $-a - b$ is also an additive inverse of $(a + b)$:

$$
\begin{aligned}
(a+b)+(-a-b) &= (a+b)+((-a)+(-b)) && \textit{(notation)}\\
&= [(a+b)+(-a)]+(-b) && \textit{(assoc. law)}\\
&= [(-a)+(a+b)]+(-b) && \textit{(comm. law)}\\
&= [(-a+a)+b]+(-b) && \textit{(assoc. law)}\\
&= [0+b]+(-b) && \textit{(def. of } -a)\\
&= b+(-b) && \textit{(def. of } 0)\\
&= 0 && \textit{(def. of } -b).
\end{aligned}
$$

We see therefore that both $-(a+b)$ and $-a-b$ are additive inverses of $(a+b)$. Hence, the uniqueness of additive inverses implies that

$$-(a + b) = -a - b,$$

and the proof is complete. The proofs of the other statements can be carried out in a similar manner and will be given as exercises.

The next theorem shows that certain equations have unique solutions in any ring.

Theorem 4.3 *If a and b are elements of a ring R, there is a unique element x in R that satisfies the equation $a + x = b$.*

Proof: It is easy to verify that $x = b - a$ is a solution. For we have

$$\begin{aligned} a + (b - a) &= a + (-a + b) \quad \textit{(comm. law)} \\ &= (a + (-a)) + b \quad \textit{(assoc. law)} \\ &= 0 + b = b. \end{aligned}$$

The uniqueness of the solution follows from the cancelation laws. For if we have $a + x = b$ and $a + y = b$ then $a + x = a + y$ and hence, $x = y$. ■

Next we establish some properties of a ring that involve multiplication only and some that involve both addition and multiplication.

Theorem 4.4 *A ring R can have at most one (multiplicative) identity element.*

Proof: The proof is much like the proof of the uniqueness of zero. Suppose e and e' are elements of R such that for every $a \in R$ we have

$$ea = ae = a \tag{1}$$

and also

$$e'a = ae' = a. \tag{2}$$

In particular, Eq. (1) must hold for $a = e'$; that is,

$$ee' = e'e = e'.$$

Similarly, using $a = e$ in Eq. (2) we obtain

$$e'e = ee' = e.$$

Thus, ee' is equal to both e and e' and we conclude $e = e'$. If a ring has an identity, we may therefore properly speak of *the* identity of the ring. ■

4.2 Multiplicative Inverses

The defining properties of a ring ensure that every element has an additive inverse. If the ring has an identity one could also ask about the inverse of an element with respect to the operation of multiplication. We make the following definition:

Definition 4.1 Let R be a ring with identity e and let $a \in R$. If there exists an element $s \in R$ such that

$$as = sa = e,$$

then s is called **a multiplicative inverse** of a.

In the ring of all real numbers, every nonzero element has a multiplicative inverse. In the ring $\mathbb{Z}$ of all integers, exactly two elements have multiplicative inverses, namely, 1 and -1. In the ring of all subsets of a given set S (see Section 1.3, Example 7), the only element with a multiplicative inverse is the identity of the ring, the subset A. Recall that for two subsets a and b their product is the intersection $ab = a \cap b$, and the identity, A, element is equal to this intersection only when $a = b = A$. Hence, $ab = e$ only if $a = e$ and $b = e$.

In view of these examples, it is clear that we must never take it for granted that an element of a ring necessarily has a multiplicative inverse. However, the following result which asserts the uniqueness of an inverse is easy to establish:

Theorem 4.5 *If an element a of a ring with identity e has a multiplicative inverse, then it is unique.*

Proof: Suppose that both s and t are multiplicative inverses of the element a. Then, using the fact that $sa = e$ and the associative law, we see

$$s(at) = (sa)t = et = t.$$

However, since $at = e$, it is also true that

$$s(at) = se = s$$

and it follows that $s = t$ since both are equal to $s(at)$. ∎

In case a has a multiplicative inverse, we denote it by a^{-1}.

The zero of a ring is defined as an element having a special property with respect to addition. We now show it has a familiar property with respect to multiplication.

Theorem 4.6 *For each element a of a ring R, we have*

$$a \cdot 0 = 0 \cdot a = 0.$$

Proof: Start with the equation $0 = 0 + 0$ and multiply by a on the left and use the distributive law:

$$a \cdot 0 = a \cdot (0 + 0) = a \cdot 0 + a \cdot 0.$$

If we give the name z to the element $a \cdot 0$, this equation reads $z = z + z$ and so the cancelation law implies $z = 0$, that is, $a \cdot 0 = 0$. In a similar way we may argue that $0 \cdot a = 0$. ∎

The following properties relating products and additive inverses will be used frequently:

Properties 4.2 *For arbitrary elements a, b, and c of a ring, the following hold:*

$$\begin{array}{lrcl} \text{(i)} & a(-b) & = & -(ab), \\ \text{(ii)} & (-a)b & = & -(ab), \\ \text{(iii)} & (-a)(-b) & = & ab, \\ \text{(iv)} & a(b-c) & = & ab-(ac), \\ \text{(v)} & (b-c)a & = & ba-(bc). \end{array}$$

Proof: The proof of (i) is carried out by showing that $a(-b)$ is an additive inverse of ab and so, by uniqueness of additive inverses, it equals $-(ab)$. We have

$$a(b+(-b)) = a \cdot 0 = 0.$$

The distributive law allows us to rewrite the left-hand side to conclude

$$a(b+(-b)) = ab + a(-b) = 0.$$

The last equality implies that $a(-b)$ is the additive inverse of ab and so $a(-b) = -(ab)$ as required. The proofs of (ii) and (iii) follow by using similar arguments and will be left as exercises. To prove (iv) we apply the distributive law to obtain

$$a(b-c) = ab + a(-c).$$

By property (i) this may be written as

$$a(b-c) = ab + (-(ac)) = ab - ac,$$

as required. The proof of (v) is accomplished in a similar way. ∎

Since the elements $-(ab)$, $(-a)b$, and $a(-b)$ are all the same, we will write $-ab$ for any one of these equal expressions.

4.3 Cancelation in Products

The cancelation laws given for addition in Theorem 1.2 do not have a general analog for products. In a general ring the equation $ab = ac$ need not imply that $b = c$. This will be the case, however, when a has a multiplicative inverse (and in many other cases as well).

Theorem 4.7 *If R is a ring with identity $a, b, c \in R$ and a is an element having a multiplicative inverse, the equation $ab = ac$ implies $b = c$; similarly, $ba = ca$ implies $b = c$.*

Proof: If a has a multiplicative inverse, a^{-1} (so that $a^{-1}a = e$ is the identity element), then we may multiply both sides of the given equation $ab = ac$ by a^{-1} to obtain

$$a^{-1}(ab) = a^{-1}(ac), \tag{1}$$

but then

$$a^{-1}(ac) = (a^{-1}a)c = ec = c$$

and also $a^{-1}(ab) = b$. It follows that the left side of Eq. (1) equals b and the right side equals c; thus, $b = c$. The obvious modification of this proof gives the cancelation of a from the equation $ba = ca$. ■

4.4 Subrings

Here we discuss the notion of a *subring* of a given ring. Let R be a ring and S a subset of R (written $S \subseteq R$). We assume S is not the empty set. Since R is a ring with addition and multiplication defined, it follows that addition and multiplication are defined for elements of S. We say *S is a subring of R* if S is a ring with respect to the addition and multiplication that are defined for elements of R. We are assured by the definition of the ring R that the sum of two elements of S is an element of R; similarly, the product of two elements of S is an element of R. For S to be a subring, it is necessary that the sum and the product of two elements of S be an element of S (not just an element of R). The zero of R must be the zero of S and the additive inverse of an element of S must be the same as the additive inverse of the element when it is viewed as an element of R (by uniqueness of additive inverses). Several of the defining properties of a ring hold in a subring S simply because they hold in R. For example, the associative laws for addition and multiplication hold in S because $S \subseteq R$.

The following theorem furnishes a convenient way to determine whether a set of elements of a ring R is actually a subring of R:

Theorem 4.8

Let R be a ring and S a nonempty subset of R. Then S is a subring of R if and only if the following condition holds: Whenever a, b are in S then $a - b$ and ab are in S.

Proof: It is first necessary to show that S is closed under addition. We know that for any elements of a and b of S that $a - b$ is in S. We show 0 is in S. Take any $c \in S$; then $c - c = 0$ is in S. For any b in S use the hypothesis to conclude $0 - b = -b$ is in S. Now for any a and b in S we have seen that $-b \in S$, so the hypothesis yields $a - (-b) = a + b$ in S. Thus, S is closed under addition, and it is closed under multiplication by hypothesis. Now we check the defining properties $\mathbf{P}_1$– $\mathbf{P}_6$ to show that S is a ring. $\mathbf{P}_1$, $\mathbf{P}_2$, $\mathbf{P}_5$, and $\mathbf{P}_6$ hold because they hold for all elements of R and thus for the elements of S in particular. Property $\mathbf{P}_3$ requires $0 \in S$. We previously verified this to be true. Property $\mathbf{P}_4$ requires that the additive inverse of each element $b \in S$ also be in S. We saw previously that $b \in S$ implies $-b \in S$, so all the defining properties are satisfied and S is a ring. ■

Example 4.1 Consider the ring R consisting of all real numbers of the form $x + y\sqrt{2}$ with $x, y \in \mathbb{Z}$, and let S be the subset of R consisting of all elements of the form $x + y\sqrt{2}$, where y is an even integer and x is an arbitrary integer. Clearly, $S \subseteq R$. To verify that S is a subring we check the hypothesis of the previous theorem: Take $a = x + y\sqrt{2}$ and $b = x' + y'\sqrt{2}$, where x, x' are integers and y, y' are even integers. Then

$$\begin{aligned} a - b &= (x - x') + (y - y')\sqrt{2} \\ ab &= (xx' + 2y'y) + (xy' + x'y)\sqrt{2}. \end{aligned}$$

Since y and y' are even integers, $y - y'$ and $xy' + x'y$ are also even integers and thus $a - b$ and ab lie in S. It follows that S is a subring of R.

EXERCISES

1. Let R be the four-element ring described in Example 3.4.
(a) Verify the equation $a = -a$ for each element of R.
(b) Determine which elements of R have a multiplicative inverse.

2. Complete the proof of Properties 4.1 by verifying the properties (iii) $-(a - b) = -a + b$, and (iv) $(a - b) - c = a - (b + c)$ for elements a, b, c of a ring.

3. Complete the proof of statements (ii), (iii), and (v) of Proposition 4.2.

4. Prove the following identities for elements in any ring:
(a) $(a + b)(c + d) = (ac + ad) + (bc + bd)$,
(b) $(a + b)(c + d) = (ac + bc) + (ad + bd)$,
(c) $(a - b)(c + d) = (ac + ad) - (bc + bd)$,
(d) $(a(-b))(-c) = a(bc)$.

5. Let R be the ring consisting of all elements of the form $x + y\sqrt{5}$, with x and y arbitrary rational numbers. Prove that every nonzero element has a multiplicative inverse. [Hint: Rationalize the denominators by using $(x + y\sqrt{5})(x - y\sqrt{5}) = x^2 - 5y^2$.]

6. Let R be the ring of two-by-two matrices over $\mathbb{Z}$ as described in Example 3.6.

(a) Find the multiplicative inverse of the element $x = \begin{bmatrix} 2 & 5 \\ 1 & 3 \end{bmatrix}$.

(b) Show that the element $\begin{bmatrix} 1 & 2 \\ 0 & 3 \end{bmatrix}$ does not have a multiplicative inverse.

(c) Use the relation

$$\begin{bmatrix} a & b \\ c & d \end{bmatrix} \begin{bmatrix} d & -b \\ -c & a \end{bmatrix} = \begin{bmatrix} ad - bc & 0 \\ 0 & ad - bc \end{bmatrix}$$

to show that $\begin{bmatrix} a & b \\ c & d \end{bmatrix}$ has an inverse in $M_2(\mathbb{Z})$ if and only if $ad - bc = \pm 1$.

7. In the ring $M_2(\mathbb{Q})$ of two-by-two matrices over the rational numbers, show that $\begin{bmatrix} a & b \\ c & d \end{bmatrix}$ has an inverse if and only if $ad - bc \neq 0$. [Hint: Use the relation in the previous exercise which also holds when the entries in the matrix are rational numbers.]

8. Let R be a ring containing an identity and two elements $a \neq 0$ and $b \neq 0$ such that $ab = 0$. Prove that neither a nor b has a multiplicative inverse.

9. If a ring has an identity e and has more than one element, show $e \neq 0$.

10. Prove that the additive identity 0 of a ring has a multiplicative inverse only if the ring consists of exactly one element, namely, 0.

11. Let R be a ring with identity. If a and b are elements of R that have multiplicative inverses, prove that the product ab has a multiplicative inverse by showing that $(ab)^{-1} = b^{-1}a^{-1}$.

12. Let a be a fixed element of a ring R and let $S = \{x : x \in R, xa = 0\}$. Prove that S is a subring of R.

13. Let A be any set and let R be the ring of all subsets of A with operations as described in Example 3.7. Let p be some element of A and let S be the collection of all subsets of A that do not contain p. Show that S is a subring of R.

14. Let R be the ring of continuous functions on the interval $[0, 1]$ with operations described in Example 3.3. Select any set of n real numbers, $x_1, x_2, \ldots, x_n$ between 0 and 1 and let S be the set of all functions $f(x)$ in R that satisfy $f(x_i) = 0$ for $i = 1, 2, \ldots, n$. Prove that S is a subring of R.

5 DEFINING PROPERTIES OF THE INTEGERS

We began the discussion of rings by examining some properties well-known for the ring of integers and then discovered that there are many different rings that share these properties. One may ask if there are additional properties possessed by the integers that are not shared by other rings. One obvious condition is that we should consider only those rings for which multiplication is commutative; this assumption will be in force throughout the next two sections. We will also consider the "order" properties of the integers and eventually show how the integers may be characterized by restricting these properties.

5.1 Order Properties

If we think of the integers as being exhibited in the following way:

$$\cdots, -4,\ -3,\ -2,\ -1,\ 0,\ 1,\ 2,\ 3,\ 4,\ \cdots$$

then we say "a is greater than b" if the integer a appears in the list to the right of the integer b. This is equivalent to the assertion that $a - b$ is a positive integer. This observation suggests that the concept of an "ordering" can be defined in terms of the concept of "positive." Accordingly, we make the following definition:

Definition 5.1

Let D be a commutative ring. We say D is **ordered** if there is a subset D^+ of D with the following properties:

(i) If $a, b \in D^+$, then $a + b \in D^+$ *(closed under addition).*

(ii) If $a, b \in D^+$, then $ab \in D^+$ *(closed under multiplication).*

(iii) For each element $a \in D$ exactly **one** of the following is true:

$$a = 0, \quad a \in D^+, \quad -a \in D^+ \qquad \textit{(trichotomy law).}$$

The elements of D^+ are called the *positive* elements. Note that 0 is not in D^+ by the trichotomy law. The nonzero elements of D that are not in D^+ are called the *negative* elements.

We may emphasize that D^+ is simply the notation used to designate a particular subset of D. Obviously, the ring $\mathbb{Z}$ of integers is ordered with $\mathbb{Z}^+$, the set of positive integers. There are other commutative rings that are ordered. For example, the ring of all rational numbers and the ring of all real numbers are ordered commutative rings. After some facts about orderings have been proved, it will be possible to show that the ring of all complex numbers is not ordered; that is, there is no subset of the complex numbers that has the three properties in the definition.

The definition of an ordered commutative ring implies the following property:

Theorem 5.1

If D is an ordered commutative ring, then the product of two nonzero elements is nonzero.

Proof: Suppose that $a, b \in D$ and neither a nor b equals 0. Then, by the trichotomy law, either a or $-a$ is in D^+ and, similarly, either b or $-b$ is in D^+. Consider first the case with both a and b in D^+. Then $ab \in D^+$ so, in particular, $ab \neq 0$ since 0 is not in D^+. Consider the case $-a \in D^+$ and $b \in D^+$. Then $(-a)b = -(ab) \in D^+$ and so again it is impossible to have $ab = 0$. In the remaining two cases, we find either $a(-b)$ or $(-a)(-b)$ are in D^+ and the same conclusion that $ab \neq 0$ holds. ∎

Rings with the property stated in the conclusion of this theorem appear frequently so we give them a name in the following definition:

Definition 5.2

Let D be a commutative ring with identity $e \neq 0$. Then D is called an **integral domain** if the equation $ab = 0$ with $a, b \in D$ implies either $a = 0$ or $b = 0$.

Using this terminology, we have proved that an ordered commutative ring with identity is an integral domain. However, as mentioned previously, there do exist integral domains that are not ordered.

Now let D be any ordered integral domain, and let D^+ be the set of positive elements of D, that is, the set having the three properties stated in the preceding definition. We now show how D^+ is used to define an order relation on D. If $c, d \in D$ we define the $c > d$ (read as "c greater than d") to mean that $c - d \in D^+$. Thus, the statement $a > 0$ means that $a \in D^+$, that is, a is a positive element of D. Similarly, $a < 0$ means $0 - a = -a \in D^+$ or that a is a negative element of D. The three properties in the definition of D^+ can be restated as follows:

Properties 5.1

The order relation $c > d$ defined for elements of D has the following properties:

(i) *If $a > 0$ and $b > 0$, then $a + b > 0$.*
(ii) *If $a > 0$ and $b > 0$, then $ab > 0$.*
(iii) *If $a \in D$, then exactly one of the following holds:*

$$a = 0, \quad a < 0, \quad a > 0.$$

The relation "$c > d$" will sometimes be written as "$d < c$" (d less than c), as is customary when dealing with the familiar relation defined for integers or real numbers.

Since it is possible (see Exercise 1) for a given integral domain to have more than one set satisfying the properties listed in the definition of D^+, and therefore it is possible to have more than one order relation defined on D, we will use the notation $(D, >)$ to denote the ordered integral domain D with a particular order relation $>$ defined for it.

We now list some properties of the order relation that are familiar for the integers but require proof for a general ordered integral domain.

Properties 5.2

Let $(D, >)$ be an ordered integral domain and a, b, c arbitrary elements of D. The following hold:

(i) *If $a > b$, then $a + c > b + c$.*
(ii) *If $a > b$ and $c > 0$, then $ac > bc$.*
(iii) *If $a > b$ and $c < 0$, then $ac < bc$.*
(iv) *If $a > b$ and $b > c$, then $a > c$.*
(v) *If $a \neq 0$, then $a^2 > 0$.*

Proof: Here is a proof of statement (i). If $a > b$, then by definition $a - b \in D^+$. For any element $c \in D$,

$$(a + c) - (b + c) = a - b + c - c = a - b \in D^+.$$

Thus, $a + c > b + c$. For a proof of (ii) we argue that $(a - b)$ and c in D^+ imply $(a - b)c \in D^+$ and so $ac - bc \in D^+$. Thus, $ac > bc$. Statement (iii) is proved analogously. Statement (iv) follows from the equation

$$(a - b) + (b - c) = a - c$$

and the fact that D^+ is closed under addition. For statement (v), we must have either $a \in D^+$ or $-a \in D^+$ because $a \neq 0$. In the first case, $a^2 \in D^+$ because the product of two elements of D^+ lies in D^+. In the second case, $-a \in D^+$ implies $(-a)(-a) = a^2$ is in D^+. All the properties have been proved. ■

Notice the following consequence of the trichotomy law. If $a, b \in D$, then exactly one of the following is true:

$$a < b, \qquad a = b, \qquad a > b.$$

To see this apply the trichotomy law to the element $a-b$: $a-b < 0, a-b = 0$, or $a - b > 0$. Then add b to both sides of the inequality or equation as in Property 5.4(i). It is sometimes simpler to use some slight variations of the notation just introduced. The symbols $a < b$ and $b > a$ mean exactly the same thing. Also, we define $a \geq b$ (or $b \leq a$) to mean either $a = b$ or $a > b$, without specifying which. If $a \geq 0$, we say a is **nonnegative**. By writing $a < b < c$ we mean that $a < b$ and also $b < c$.

If e is the identity of D, then $e^2 = e$ and $e \neq 0$. Thus, $e = e^2 > 0$ so the *identity element of an ordered integral domain is a positive element.*

In any ordered integral domain it is possible to introduce the concept of absolute value in the usual way.

Definition 5.3 Let D be any ordered integral domain and $a \in D$. The **absolute value** of a, written as $|a|$, is defined as follows:

$$|a| = \begin{cases} a & \text{if } a \geq 0, \\ -a & \text{if } a < 0. \end{cases}$$

From this definition, it follows that either $a = 0$ or $|a| > 0$.

A few properties of the absolute value which follow from the definition and the properties of the positive elements are given in the exercises.

EXERCISES

1. Let $D = \{x+y\sqrt{2} : x, y \in \mathbb{Z}\}$, so that D is an integral domain contained in the ring of all real numbers, which is also an ordered integral domain. Let $a > b$ have the usual meaning for real numbers a and b.

(a) Let A be the subset consisting of all $x+y\sqrt{2}$ in D with $x+y\sqrt{2} > 0$. Show that A has the properties required of D^+ in the definition of an ordered commutative ring. Thus, $(D, >)$ is an ordered integral domain.

(b) Let B be the set consisting of all $x + y\sqrt{2}$ in D with $x - y\sqrt{2} > 0$. Show that B also has the properties required of D^+ in the definition of an ordered commutative ring.

(c) Let $>_A$ denote the ordering of D for which A is the set of positive elements and let $>_B$ denote the ordering for which B is the set of positive elements. Show that $1 >_A 1 - \sqrt{2}$ but $1 <_B 1 - \sqrt{2}$.

This example shows that an ordered integral domain may have more than one ordering.

2. Do you think there is more than one ordering of the set of integers? Make a guess and then try to prove it correct. [Hint: You may wish to start the argument using the fact that in any ordering, 1 must be positive.]

In Exercises 3–9, the symbols a, b, and c represent elements of an ordered integral domain.

3. Prove that if $a > b$, then $-a < -b$.

4. Prove that if $a > b$ and $c > d$, then $a + c > b + d$.

5. Prove that if $a > b > 0$ and $c > d > 0$, then $ac > bd$.

6. Prove that if $a > 0$ and $ab > ac$, then $b > c$.

7. Prove that $|ab| = |a| \cdot |b|$.

8. Prove that $-|a| \leq a \leq |a|$.

9. Prove that $|a + b| \leq |a| + |b|$. (This is called the *triangle inequality*).

10. Prove that an ordered integral domain D cannot have a largest element; that is, for any $c \in D$ there is some $x \in D$ with $x > c$.

11. Use the result of the preceding exercise to give a convincing argument that an integral domain with a finite number of elements cannot be an ordered integral domain.

12. Let R be the ring of all real numbers which we view as an ordered integral domain with $a > b$ having the usual meaning for real numbers a and b. Let S be any subring of R such that $1 \in S$. Prove that S is an ordered integral domain.

13. Let C be any integral domain with identity $e \neq 0$. Suppose C contains an element i such that $i^2 = -e$. Then C is not an ordered integral domain. [Hint: If there is a set of positive elements, then either i or $-i$ is positive. See where that leads you.]

6 WELL-ORDERING AND MATHEMATICAL INDUCTION

We need one further condition to characterize the ring of integers among the ordered integral domains. Let us first make the following general definition:

Definition 6.1

A set S of an ordered integral domain is said to be **well-ordered** if each nonempty subset U of S contains a least element; that is, each nonempty subset U contains an element u such that $u \leq w$ for every $w \in U$.

It is almost obvious that the set of all positive integers is well-ordered. An argument might be fashioned as follows. Let U be some nonempty subset of the set of positive integers. Since U is nonempty, pick an element $n \in U$. Then $n > 0$ since U consists of only positive integers. Consider each integer $1, 2, 3, \ldots, n$ in turn and determine the smallest one, call it w, that lies in U. Then w is the smallest element of U and $w \leq u$ for every $u \in U$.

The set of positive rational numbers in not well-ordered. In fact, the set of all positive rational numbers does not have a least element. Here is a proof. If r is a positive rational number, then $r/2$ is also a positive rational number and $r/2 < r$ since $r > 0$. Hence, for any positive rational number r, there is a smaller positive rational number; thus, there cannot be a smallest positive rational number.

6.1 Defining Properties of the Integers

Let us pause to clarify our point of view about the ring of integers. We have not *proved* any property of the integers; instead, occasionally we have asked the reader to recall some familiar properties that are used in elementary mathematics and assumed they were true of the integers. We are now able to state precisely just what properties of the integers we will assume as known. Namely, *the ring of integers is an ordered integral domain in which the set of positive elements is well-ordered.* Accordingly, when we speak of a proof of any property of the integers, we mean a proof based on these properties (and any property logically derived from these properties). The main theorem of the next section will indicate why no other property is required.

We have already seen that the identity e of an ordered integral domain D lies in D^+ (because $e = e^2 > 0$). If D^+ is well-ordered, we now assert that *e is the smallest element of D^+*. Let c denote the smallest element of D^+ and suppose that $c < e$. Of course, then $0 < c$ so $0 < c < e$. Multiply each term of this inequality by the positive element c to obtain $0 < c^2 < c$ since

$ce = c$. Thus, we have $c^2 \in D^+$ and c^2 is less than the smallest element c of D^+. This is an impossible situation arising from the assumption that $c < e$. Hence, it must be that $e \leq c$ and it follows that e is the least element in D^+. In particular, our assumption about the ring of integers assures us that 1 is the least positive integer. No surprise!

The following theorem, which is the basis of proofs by mathematical induction, is just as "obvious" as the fact that the set of positive integers is well-ordered. However, in accordance with our chosen point of view, we give a proof.

Theorem 6.1

INDUCTION PRINCIPLE. *Let K be a set of positive integers with the following two properties:*

(i) $1 \in K$.

(ii) *If k is any positive integer such that $k \in K$, then also $k + 1 \in K$.*

Then K consists of the set of all positive integers.

Proof: To prove this theorem, let us assume that there is a positive integer not in K and obtain a contradiction. Let U be the set of all positive integers not in K, so therefore by assumption, U is not empty. By the well-ordering property, U must contain a least element m. Since by (i), $1 \in K$, we cannot have $m = 1$ and so $m > 1$. Thus, in particular $m - 1 > 0$. Moreover, $m - 1$ is in K because $m - 1$ is smaller than the least element that is not in K. Now apply property (ii) which asserts that because $k = m - 1$ is in K, then so is $k + 1 = m$. However, $m \in U$, and we have reached a contradiction. Thus, U must be empty after all. ∎

The most frequent application of this theorem is to a proof of the following kind. Suppose there is assigned to each positive integer n a *statement* (or proposition) S_n which is either true or false, and suppose we want to prove that S_n is true for every positive integer n. Let K be the set of all positive integers n such that S_n is true. If we can show $1 \in K$ (i.e., S_1 is true) and that whenever $k \in K$ then $k + 1 \in K$ (i.e., on the supposition that S_k is true we can show S_{k+1} is true), then it follows from the previous theorem that K is the set of all positive integers and so S_n is true for every positive integer. We reformulate these remarks in the following form:

Statement 6.1

INDUCTION PRINCIPLE. *Suppose there is associated with each positive integer n a statement S_n such that the following hold:*

(i) *S_1 is true.*

(ii) *If k is any positive integer such that S_k is true, then S_{k+1} is true.*

Then S_n is true for every positive integer n.

A proof making use of the induction principle is usually called a proof by mathematical induction.

As a first illustration of the language and notation just introduced, we consider a simple example from elementary algebra. If n is a positive integer, let S_n be the statement that *the sum of the first* n *even positive integers equals* $n(n+1)$. In symbols, we write

$$S_n : \qquad 2+4+6+\cdots+2n = n(n+1).$$

We now prove S_n is true for every positive integer n by verifying (i) and (ii) of the induction principle. S_1 is the statement $2 = 1(1+1)$ which is true. Now suppose k is a positive integer for which S_k is true. Then we have the equation

$$2+4+6+\cdots+2k = k(k+1).$$

By adding the next even integer, $2(k+1)$, to both sides of this equation we obtain

$$2+4+6+\cdots+2k+2(k+1) = k(k+1)+2(k+1) = (k+2)(k+1).$$

This shows that the statement S_{k+1} is true, and hence we have verified both (i) and (ii) of the induction principle; thus, S_n is true for every positive integer n.

There is a slightly more general form of the induction principle that is sometimes useful. In Statement 6.1, the induction begins with $n = 1$, that is, we first determine that S_1 is true. The following version permits us to begin at any integer n_0 (positive, negative, or zero) and gives us the conclusion that the statement is true for every $n \geq n_0$:

Statement 6.2

INDUCTION PRINCIPLE: SECOND FORM. *Let n_0 be any integer. Suppose there is associated with each integer n greater than or equal to n_0 a statement S_n such that the following hold:*

(i) *S_{n_0} is true.*

(ii) *If k is any integer such that $k \geq n_0$ and S_k is true, then S_{k+1} is true.*

Then S_n is true for every $n \geq n_0$.

Proof: The proof is by direct appeal to the induction principle and amounts to a "renumbering" of the statements. Let T_n be the statement S_{n_0-1+n}. Then $T_1 = S_{n_0}$ is true by hypothesis. If $T_k = S_{n_0+k-1}$ is true, then $T_{k+1} = S_{n_0+k}$ is true by statement (ii). Hence, T_n is true for all $n \geq 1$. It follows that S_n is true for all $n \geq n_0$. ∎

Here is an application of this form of the induction principle. Consider the statement

$$S_n : \qquad 2^n \geq 50n.$$

This inequality is not true for $n = 1, 2, \ldots, 8$ but it is true for $n = 9$. Let us prove it is true for all $n \geq 9$. Apply Statement 6.2 with $n_0 = 9$. Then S_9 is true because $2^9 = 512 \geq 50 \cdot 9 = 450$. Assume $k \geq 9$ and S_k is true. Then we have the true statement

$$2^k \geq 50k \tag{1}$$

and we want to prove that

$$2^{k+1} \geq 50(k+1) \tag{2}$$

is true. Multiply both sides of Eq. (1) by 2 to obtain

$$2^{k+1} \geq 2 \cdot 50k = 100k.$$

If we show that $100k \geq 50(k+1)$ then Eq. (2) will be true. This latter inequality is easily seen to be true as $50k \geq 50$ (since $k \geq 9$) and by adding $50k$ to each side. This shows that S_k is true for all $k \geq 9$.

6.2 Multiples and Powers

We now show how induction can be used not only to prove theorems but also to give definitions. Let R be a ring and $a \in R$. We will define the *multiples* ma for an integer m and *powers* a^n for a nonnegative integer n using induction.

Definition 6.2

Multiples. Let a be an element of a ring R. Define $1a = a$. If k is a positive integer such that ka has been defined, then $(k+1)a = ka + a$. For a negative integer m define $ma = -(-m)a)$. For $m = 0$, define $ma = 0$.

Notice that the definition of ma includes the case $m = 1$, and whenever ka is defined it also defines $(k+1)a$. Thus, the induction principle implies that ma is defined for all positive integers m.

For negative integers m, note that $-m$ is a positive integer so $(-m)a$ is already defined and then $ma = -((-m)a)$ is the additive inverse of $(-m)a$. Finally, we set $0a = 0$, where the 0 on the left is the integer zero while the 0 on the right is the zero of R.

It should be emphasized that ma is a convenient way of indicating a certain sum of elements of R. Since m is an integer and a is in R, it is not correct to regard ma as the product of two elements of a ring.

Next we define the *powers* a^n of a.

Definition 6.3

Powers. Let a be an element of a ring R. Define $a^1 = a$. If k is a positive integer for which a^k is defined, we define $a^{k+1} = a^k \cdot a$. Then a^n is defined for all positive integers n.

Just as for multiples, the induction principle implies that a^n is defined for all positive integers n. The definitions of multiples and powers are examples of *recursive* definitions.

Let us give an illustration of how certain properties may be proved using induction. We prove the familiar *law of exponents* which asserts that if m and n are arbitrary positive integers then

$$a^m \cdot a^n = a^{m+n}. \tag{1}$$

Let S_n be the statement that for the positive integer n, Eq. (1) is true for every positive integer m. Then by definition $a^{m+1} = a^m \cdot a^1$, and hence S_1 is true. Let us now assume that k is a positive integer such that S_k is true; that is,

$$a^m \cdot a^k = a^{m+k}$$

for every positive integer m. We now verify that S_{k+1} is true as follows:

$$\begin{aligned} a^m \cdot a^{k+1} &= a^m \cdot a^k \cdot a && (\text{by def. of } a^{k+1}) \\ &= a^{m+k} \cdot a && (\text{because } S_k \text{ is true}) \\ &= a^{m+k+1} && (\text{by def. of } a^{(m+k)+1}). \end{aligned}$$

We have proved S_{k+1} is true and the induction principle then assures us that S_n is true for every positive integer n.

A slightly more complicated argument can be given to prove the analogous formula for multiples, which asserts that for any integers m and n,

$$ma + na = (m + n)a.$$

The proof is similar to the previous statement about powers but has the added complication that m and n can be nonpositive integers. See the hint given in the exercises following this section.

6.3 General Sums and Products

The operations of addition and multiplication are *binary* operations; that is, they apply to *two* elements only. We use induction to give meaning to sums or products of three or more elements of a ring.

Let $a_1, a_2, \ldots, a_n$ be elements of a ring R and assume $n \geq 2$. We wish to define what is meant by the sum $a_1 + a_2 + \cdots + a_n$ of n elements. Of course, if $n = 2$ the sum $a_1 + a_2$ is already defined because the addition in R is defined. For a positive integer k such that

$$a_1 + a_2 + \cdots + a_k$$

is defined, we set

$$a_1 + a_2 + \cdots + a_k + a_{k+1} = (a_1 + a_2 + \cdots + a_k) + a_{k+1}. \tag{2}$$

This is a recursive definition of the sum of n elements, and by the induction principle (second form) the sum is defined for every positive integer $n \geq 2$. Note that the definition of the sum of n elements is given by grouping the elements in a certain way to produce the sum of two elements, one of which is the sum of $n-1$ elements. There are many ways to groups the n elements of the sum to produce a sum of fewer than n elements and the point of these maneuvers is that the same sum is obtained no matter how the parentheses are inserted. Let us give a formal proof of this.

Let S_n be the statement: If $a_1, a_2, \ldots, a_n$ are arbitrary elements of a ring and for each positive integer r such that $1 \leq r < n$, we have

$$(a_1 + \cdots + a_r) + (a_{r+1} + \cdots + a_n) = a_1 + a_2 + \cdots + a_n.$$

The statement S_2 is true because it says only that $(a_1) + (a_2) = a_1 + a_2$. Assume k is a positive integer $k \geq 2$ such that S_k is true. We will show that S_{k+1} is true. Otherwise expressed, we must prove that if r is an integer such that $1 \leq r < k+1$, then

$$(a_1 + \cdots + a_r) + (a_{r+1} + \cdots + a_{k+1}) = a_1 + a_2 + \cdots + a_{k+1}. \tag{3}$$

The case $r = k$ is true by Eq. (2) of the right side of Eq. (3). Suppose then that $1 \leq r < k$. As a special case of Eq. (2), we have

$$a_{r+1} + a_2 + \cdots + a_k + a_{k+1} = (a_{r+1} + a_2 + \cdots + a_k) + a_{k+1}.$$

This will be used in the first step of the following chain of equalities:

$$\begin{aligned}
&(a_1 + \cdots + a_r) + (a_{r+1} + \cdots + a_{k+1}) \\
&\quad = (a_1 + \cdots + a_r) + \big((a_{r+1} + \cdots + a_k) + a_{k+1}\big) \\
&\quad = \big((a_1 + \cdots + a_r) + (a_{r+1} + \cdots + a_k)\big) + a_{k+1} && \text{(by assoc. law)} \\
&\quad = (a_1 + \cdots + a_k) + a_{k+1} && \text{(by } S_k\text{)} \\
&\quad = a_1 + \cdots + a_{k+1} && \text{(by def.).}
\end{aligned}$$

This completes the proof that S_k implies S_{k+1} and so S_n holds for all integers $n \geq 2$. This property of addition is called the *generalized associative law for addition.*

With only the substitution of $\cdot$ for $+$, one may define products of n ring elements recursively and prove the generalized associative law for products; namely,

$$a_1 \cdot a_2 \cdots a_n = (a_1 \cdot a_2 \cdots a_{n-1}) \cdot a_n$$

is the definition and the generalized associative law states

$$(a_1 \cdot a_2 \cdots + a_r) \cdot (a_{r+1} \cdots a_{k+1}) = a_1 \cdot a_2 \cdots a_{k+1}.$$

Some other properties of multiples, powers, and generalized laws are given in the exercises following this section.

6.4 The Binomial Theorem

Here, we discuss important application of the induction principle that has arises frequently in many topics of algebra and other areas of mathematics. Before stating it, we introduce the *factorial* notation to enable economical statement of the results.

For a positive integer n define the symbol $n!$ to stand for the product of all the positive integers between 1 and n; that is,

$$n! = 1 \cdot 2 \cdots (n-1) \cdot n \quad \text{read as "}n\text{ factorial."}$$

We also define $0! = 1$. Thus, for example, $4! = 24$ and $7! = 7 \cdot 6!$. More generally, we have

$$(m+1)! = (m+1) \cdot m!$$

for any nonnegative integer m.

Theorem 6.2

THE BINOMIAL THEOREM. *Let a and b be elements of a commutative ring and n a positive integer. Then*

$$(a+b)^n = a^n + C(n,1)a^{n-1}b + \cdots + C(n,j)a^{n-j}b^j + \cdots + b^n, \quad (1)$$

where $C(n,j)$ is defined for all positive integers n and j with $0 \leq j \leq n$ by the formulas

$$C(n,j) = \frac{n!}{j!(n-j)!}, \qquad 1 \leq j \leq n.$$

Proof: Let S_n be the statement of the conclusion in Eq. (1). Then S_1 is true because the $C(1,j)$ do not even enter into the statement. Statement S_2 is true because $C(2,1) = (2 \cdot 1)/(1)(1) = 2$ and

$$(a+b)^2 = a^2 + 2ab + b^2.$$

Let us suppose that $k \geq 2$ is a positive integer for which S_k is true. We must verify that S_{k+1} is true. Using Eq. (1) for $n = k$, we multiply both sides of the equation by $(a+b)$ to obtain

$$\begin{aligned}(a+b)^{k+1} &= \\ &a[a^k + C(k,1)a^{k-1}b + \cdots + C(k,j)a^{k-j}b^j + \cdots + b^k] \\ &\quad +b[a^k + C(k,1)a^{k-1}b + \cdots + C(k,j)a^{k-j}b^j + \cdots + b^k] \\ &= a^{k+1} + [k+1]a^k b + \cdots + [C(k,r) + C(k,r-1)]a^{k+1-r}b^r + \cdots + b^{k+1}.\end{aligned}$$

In the last line of the display, we collect the terms with equal powers so r takes the values $1, 2, \ldots, k$. To prove the formula correct, it is necessary to verify that

$$C(k,r) + C(k,r-1) = C(k+1,r).$$

Using the definition of the coefficients $C(n, j)$, we have

$$\begin{aligned} C(k,r)+C(k,r-1) &= \frac{k!}{r!(k-r)!}+\frac{k!}{(r-1)!(k-r+1)!} \\ &= \frac{k!}{r\cdot(r-1)!(k-r)!}+\frac{k!}{(r-1)!(k-r+1)(k-r)!} \\ &= \frac{k!}{(r-1)!(k-r)!}\left(\frac{1}{r}+\frac{1}{k-r+1}\right) \\ &= \frac{k!}{(r-1)!(k-r)!}\left(\frac{k+1}{r(k-r+1)}\right) \\ &= C(k+1,r). \end{aligned}$$

This verifies the induction step so the statement S_n is true for all positive integers n. ■

The coefficients $C(n, j)$ are called the *binomial coefficients*. These numbers are visibly rational numbers but, in fact, they are integers. Of course, we deduce they are integers from the expansion of $(a + b)^n$ since only integer coefficients appear in the product. An alternative argument can be given by proving that the number $C(n, j)$ is the number of subsets with j elements that can be made from a set with n elements. A proof of this would automatically give a proof that the binomial coefficients are integers. We do not do this here.

EXERCISES

1. Prove the distributive law:

$$b(a_1 + a_2 + \cdots + a_n) = ba_1 + ba_2 + \cdots + ba_n.$$

2. Prove, for positive integers m and n and any element a of a ring, that

$$(a^m)^n = a^{mn}.$$

3. If a and b are elements of a *commutative* ring, prove $(ab)^n = a^n b^n$.

4. If $a_1, a_2, \ldots, a_n$ are elements of an integral domain such that the product $a_1 a_2 \cdots a_n = 0$, show that at least one of the a_i must be zero.

5. Prove that if a and b are elements of a ring, for every integer m (positive, negative, or zero),

$$m(a + b) = ma + mb.$$

6. Prove that if a and b are elements of a ring, for all integers m and n,

$$(m + n)a = ma + na.$$

[Hint: Treat a number of cases: either m or n is zero; both m and n are positive; either m or n is negative; both m and n are negative.]

7. If a and b are elements of a commutative ring R, verify that $(ab)^3 = a^3b^3$.

8. In the ring $M_2(\mathbb{Z})$, let

$$A = \begin{bmatrix} 1 & 2 \\ 0 & 0 \end{bmatrix}, \quad \text{and} \quad B = \begin{bmatrix} 0 & 1 \\ 0 & 1 \end{bmatrix}.$$

Verify that $(AB)^2 \neq A^2B^2$.

9. Let A be a nonempty set and R the ring of all subsets of A (Example 3.7). Show that $x^2 = x$ and $2x = 0$ for all $x \in R$.

10. An element e_1 of a ring R is called a *left* (multiplicative) identity if $e_1x = x$ for every $x \in R$. Similarly, $e_2 \in R$ is called a *right* identity if $xe_2 = x$ for every $x \in R$. Verify that the set S of all two-by-two matrices of the form

$$\begin{bmatrix} a & b \\ 0 & 0 \end{bmatrix} \qquad a, b \in \mathbb{Z},$$

is a subring of the ring $M_2(\mathbb{Z})$ of two-by-two matrices over $\mathbb{Z}$. Then prove:

(a) S has a left identity but no right identity.

(b) S has infinitely many left identities.

11. Using the definitions in the preceding exercise, prove that if a ring R has a *unique* left identity, then it is also a right identity and hence is an identity. [Hint: If e_1 is a left identity and $c \in R$, show that $e_1 + ce_1 - c$ is also a left identity.]

12. Use induction to prove the following two rules for products of elements in $M_2(\mathbb{Z})$:

$$(a) \quad \begin{bmatrix} 1 & 1 \\ 0 & 1 \end{bmatrix}^n = \begin{bmatrix} 1 & n \\ 0 & 1 \end{bmatrix} \quad \text{and} \quad (b) \quad \begin{bmatrix} 2 & 0 \\ 0 & -1 \end{bmatrix}^n = \begin{bmatrix} 2^n & 0 \\ 0 & (-1)^n \end{bmatrix}.$$

13. Find a suitable integer n_0 and then use the second form of the principle of induction to prove $3^n \geq 1000n$ for all $n \geq n_0$.

14. Suppose x is an element of a ring with identity e such that $x^3 = e$. Give a proof using induction that $x^n \in \{e, x, x^2\}$ for all positive integers n. [Hint: For $n > 3$, $x^n = x^{n-3}x^3$.]

15. Verify that $C(n, j)$ is an integer in the special cases $C(10, 4)$, $C(12, 7)$, and $C(14, 6)$.

16. Verify the general rule $C(n, j) = C(n, n - j)$.

17. Use the equation $(1+1)^n = 2^n$ to verify that the sum

$$C(n,0) + C(n,1) + C(n,2) + \cdots + C(n,n) = 2^n.$$

18. Why is the analog of the binomial theorem not true for noncommutative rings? Show it is false by example.

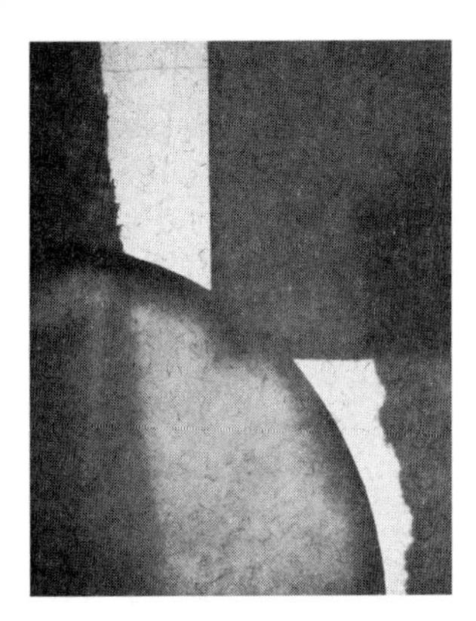

II CONGRUENCES

We examine a method of constructing new rings from a given ring by using congruences. Starting with the ring of integers, we use the idea to produce finite rings called the integers modulo n; for more general rings the idea leads us to study ideals and congruences modulo an ideal.

1 EQUIVALENCE RELATIONS

We have already studied "order relations"–$a < b$ defined for real numbers or certain integral domains. There is a more general concept of a *relation on a set* A. In rough terms, a relation is some rule by which we may assert that two elements "are related" or "are not related." We can make this notion precise.

If A is any nonempty set, then $A \times A$ is the set of all ordered pairs (a, b) with $a, b \in A$. A *relation on* A is a subset $\mathcal{R}$ of $A \times A$. Thus, for each element $(a, b) \in A \times A$ either (a, b) is in the subset $\mathcal{R}$ or it is not; accordingly, we could say that either a *is related to* b or not. It is customary to write $a\mathcal{R}b$ and to say that a is in the relation $\mathcal{R}$ to b, when $(a, b) \in \mathcal{R}$. We may also say that "a is related to b" to indicate that $(a, b) \in \mathcal{R}$.

As a concrete example consider the real numbers $\mathbb{R}$ and the familiar relation, less than, $a < b$. The definition of this relation as the subset of $\mathbb{R} \times \mathbb{R}$ is the set of all elements (x, y) in the Cartesian plane which lie below straight line $y = x$ through the origin. For any pair (x, y), it is either true or false that $x < y$; that is, either $(x, y) \in \mathcal{R}$ or $(x, y) \notin \mathcal{R}$ according as $x < y$ or $x \not< y$.

We will not usually be concerned with relations in general, but primarily with those relations having the particular properties stated in the following definition.

Definition 1.1

A relation $\mathcal{R}$ defined on a set A is called an **equivalence relation** if it has the following three properties, where a, b, and c denote arbitrary elements of A:

(i)	$a\mathcal{R}a$	(reflexive property)
(ii)	If $a\mathcal{R}b$ then $b\mathcal{R}a$.	(symmetric property)
(iii)	If $a\mathcal{R}b$ and $b\mathcal{R}c$ then $a\mathcal{R}c$.	(transitive property)

In the future, we will use a more suggestive notation such as $a \sim b$ to denote an equivalence relation. One may then read $a \sim b$ as "a is equivalent to b." We sometimes write $a \not\sim b$ to indicate that a is not equivalent to b.

Using the definition of a relation as a subset of $A \times A$, we see that a subset $\mathcal{R}$ is an equivalence relation if and only if

(i) $(a, a) \in \mathcal{R}$ for every $a \in A$
(ii) If $(a, b) \in \mathcal{R}$, then $(b, a) \in \mathcal{R}$
(iii) If $(a, b) \in \mathcal{R}$ and if $(b, c) \in \mathcal{R}$, then $(a, c) \in \mathcal{R}$.

Let us emphasize how property (i) differs in an essential way from the other two properties. If we have an equivalence relation "$\sim$" defined on a set A then property (i) asserts $a \sim a$ for every $a \in A$. On the other hand, if a and b are given, property (ii) does not assert anything about the truth of the statement $a \sim b$; it asserts only that *if* $a \sim b$ is true, then $b \sim a$ must also be true. Property (iii) states that *if* $a \sim b$ and *if* $b \sim c$ are true, then $a \sim c$ is true.

The simplest example of an equivalence relation on a set A is the relation of *equality*; that is, $a \sim b$ is defined to mean $a = b$. This is easily seen to satisfy the three properties in the definition of equivalence relation.

Here is a less trivial example of an equivalence relation. Let $\mathbb{R}$ denote the set of all real numbers; for any two elements $a, b \in \mathbb{R}$, let $a \sim b$ be true if and only if $a - b$ is an integer. Thus, $3.21 \sim 19.21$ and $22/7 \sim 1/7$ because $3.21 - 19.21 = -16$ and $22/7 - 1/7 = 3$ are integers. Also, $1.5 \not\sim 4.2$ because $1.5 - 4.2 = -2.7$ is not an integer.

1.1 Congruence Modulo n

Here is an important example that will be studied extensively. Let n be a nonzero integer. We define a relation on the set $\mathbb{Z}$ of integers called *congruence modulo* n as follows: For integers a and b we write $a \equiv b \pmod{n}$ to mean $a - b$ is a multiple of n [or equivalently that $(a - b)/n$ is an integer].

Let us prove that congruence modulo n is an equivalence relation. For any integer a, $a - a = 0 = 0 \cdot n$, so $a \equiv a \pmod{n}$ and the reflexive property holds. If $a \equiv b \pmod{n}$, then $a - b = nt$ for some integer t. Then $b - a = n(-t)$ and so $b \equiv a \pmod{n}$ and the symmetric property holds. Now suppose $a \equiv b \pmod{n}$ and $b \equiv c \pmod{n}$. Then $a - b = nt$ and $b - c = ns$

for some integers t and s, and we have

$$a - c = (a - b) + (b - c) = nt + ns = n(t + s).$$

This shows that $a \equiv c \pmod{n}$ and that the transitive property holds.

For example, $27 \equiv 12 \pmod{5}$ because $27 - 12 = 15$ is a multiple of 5; $103 \equiv -27 \pmod{10}$ because $103 - (-27) = 130$ is a multiple of 10.

1.2 Partitions and Equivalence Classes

A collection of subsets of a set A is called a *partition* if A is the union of the subsets and any two of the subsets have empty intersection. Partitions of a set A are very closely related to equivalence relations on A in the following way. Suppose $A_1, A_2, \ldots, A_m$ are subsets of A (so $A_i \subseteq A$ for each i, $1 \leq i \leq m$) and

$$A_i \cap A_j = \emptyset, \quad \text{if } i \neq j \quad \text{and} \quad A_1 \cup A_2 \cup \cdots \cup A_m = A.$$

Then the collection $\{A_1, A_2, \ldots, A_m\}$ is a partition of A. We use the partition to define a relation $\sim$ by the rule $a \sim b$ if a and b lie in the same subset A_j. Thus, an element a in A_j is equivalent to all the other elements of A_j (and to no others). Then it is quite easy to verify that this is an equivalence relation. Just check the three conditions!

We may reverse this situation since we may define a partition by starting with an equivalence relation. Suppose $\sim$ is an equivalence relation on A. For each element $b \in A$, define a subset

$$[b] = \{x \in A : x \sim b\}.$$

Such subsets occur frequently enough to warrant giving them a name.

Definition 1.2 Let A be a set and $\sim$ an equivalence relation on A. If $a \in A$, the subset of elements of A which consists of all x such that $x \sim a$ is called an **equivalence class** or, more precisely, the **equivalence class containing** a.

It can happen that $[b] = [c]$ even if $b \neq c$. In fact, we can prove that $[b] = [c]$ if and only if $b \sim c$. This is part of the following theorem:

Theorem 1.1 *Let $\sim$ be an equivalence relation defined on A, and for each $b \in A$ let $[b] = \{x : x \sim b\}$.*

(i) *If $[b] \cap [c] \neq \emptyset$, then $[b] = [c]$.*

(ii) *The distinct sets $[b]$ for $b \in A$ form a partition of A.*

Proof: Let b and c be elements of A and suppose there is an element x that lies in both $[b]$ and $[c]$. Then by definition $x \sim b$ and $x \sim c$. Since we are dealing with an equivalence relation, it follows that $b \sim x$ (symmetric property) and so $b \sim c$ by transitivity. Now let $z \in [b]$. Then $z \sim b$ and $b \sim c$ implies $z \sim c$ and so $z \in [c]$. Thus, $[b] \subseteq [c]$. By reversing the roles of b and c and applying the same reasoning, we also get $[c] \subseteq [b]$. It follows that $[b] = [c]$. Now to prove statement (ii) we must show that the distinct sets $[b]$, $b \in A$ have union equal to A and pairwise have empty intersection. We have just seen that two distinct sets must have no common element since the existence of some x in both $[b]$ and $[c]$ implies $[b] = [c]$. For any element $a \in A$ we have $a \in [a]$ since $a \sim a$. Thus, every element of A is in some subset and the distinct equivalence classes $[b]$ give a partition of A. ∎

It follows that there is a one-to-one correspondence between partitions and equivalence relations.

Consider the equivalence relation of congruence modulo 3. Let us describe the equivalence classes of elements of $\mathbb{Z}$. The equivalence class $[0]$ contains an element x if and only if $x \equiv 0 \pmod 3$; that is, $x - 0 = x$ is divisible by 3. It follows that

$$[0] = \{\cdots, -9, -6, -3, 0, 3, 6, 9, \cdots\}$$

is the set of integer multiples of 3. The equivalence class that contains 1 is the set of integers x such that $x \equiv 1 \pmod 3$, which means $x - 1$ is divisible by 3. Thus, $x - 1$ is in the equivalence class of 0 so the element x lies in the set we obtain by adding 1 to each element of $[0]$; thus,

$$[1] = \{\cdots - 8, -5, -2, 1, 4, 7, 10, \cdots\}.$$

The equivalence class containing 2 is the set of integers x such that $x - 2$ is divisible by 3; in other words, x is obtained by adding 2 to an element of $[0]$:

$$[2] = \{\cdots, -7, -4, -1, 2, 5, 8, 11, \cdots\}.$$

These are all of the equivalence classes modulo 3, as we can see from the following argument: Let b be any integer and let k be the largest integer less than or equal to the (possibly rational) number $b/3$. Then $k \leq b/3$ implies $0 \leq b - 3k$ and k is the largest integer with this property. We now show that $b - 3k \leq 2$. If this were false, then $b - 3k \geq 3$ and so $b - 3(k+1) \geq 0$, which contradicts the defining property of k. It follows that the integer $t = b - 3k$ equals $0, 1$, or 2. In other words, the arbitrary integer that we selected must be in the equivalence class of 0,1, or 2.

A more general argument along these lines is presented in the next section.

EXERCISES

1. Let $\mathcal{R}$ be the relation on the set $\{1, 2\}$ defined by $2\mathcal{R}2$, and the relation $\mathcal{R}$ holds for no other ordered pair except the pair $(2, 2)$. Show that $\mathcal{R}$ has exactly two of the three defining properties of an equivalence relation.
2. Let $\mathcal{R}$ be the relation on $\{1, 2, 3\}$ defined by $2\mathcal{R}3$ and $1\mathcal{R}2$, but $\mathcal{R}$ holds for no other pairs. Which of the three properties of an equivalence relation hold for $\mathcal{R}$?
3. Let A be the set $\{1, 2, 3\}$. How many partitions of A are possible? (Here we mean partitions into *nonempty* subsets. Two partitions are different if at least one subset from the first partition is not one of the subsets in the second partition.) How many equivalence relations can be defined on A? How many relations (not necessarily equivalence relations) can be defined on A?
4. Let $\sim$ be an equivalence relation on the set A. Give a careful proof based only on the definition of an equivalence relation of the following: If $a, b, c, d \in A$ such that $c \in [a]$, $d \in [b]$, and $[a] \neq [b]$, then $c \not\sim d$.
5. Let $A = \{1, 2, 3, 4\}$. Give an example of an equivalence relation on A in which two of the equivalence classes have a different number of elements.
6. Let $\mathbb{R}$ denote the set of all real numbers; for any two elements $a, b \in \mathbb{R}$ let $a \sim b$ be true if and only if $a - b$ is an integer. Prove that $\sim$ is an equivalence relation and that each equivalence class $[x]$ contains one, and only one, number y with $0 \leq y < 1$.
7. Let $\mathbb{R}$ be the set of real numbers with the equivalence relation $\sim$ defined as in the preceding exercise. Show that for any real numbers a, b, x, y with $x \in [a]$ and $y \in [b]$ it follows that $x + y \in [a + b]$. Show also that it need **not** be true that $xy \in [ab]$ for certain choices of a, b, x, y.

2 INTEGERS MODULO n

In this section we describe a class of rings called the *integers modulo* n using the equivalence classes introduced in the previous section.

Let n be a fixed positive integer; the equivalence class of an integer k modulo n is denoted by $[k]$ and is defined as

$$[k] = \{x : x \in \mathbb{Z}, x \equiv k \pmod{n}\}.$$

If m is an element in $[k]$, then $m - k = nt$ for some integer t. Thus, $m = k + nt$ and so we know something about the form of elements in the equivalence class $[k]$. On the other hand, if m is an integer and $m = k + nt$ for some integer t, then $m \equiv k \pmod{n}$ and m is in $[k]$. Thus, we have an exact description of the elements in $[k]$ and we may write

$$[k] = \{k + nt : t \in \mathbb{Z}\}.$$

We saw in the previous section that the number of equivalence classes modulo 3 was exactly 3. The equivalence classes modulo 4 can be exhibited easily:

$$\begin{aligned}
[0] &= \{\cdots, -12, -8, -4, 0, 4, 8, 12, \cdots\} \\
[1] &= \{\cdots, -11, -7, -3, 1, 5, 9, 13, \cdots\} \\
[2] &= \{\cdots, -10, -6, -2, 2, 6, 10, 14, \cdots\} \\
[3] &= \{\cdots, -9, -5, -1, 3, 7, 11, 15, \cdots\}.
\end{aligned}$$

It follows from Theorem 1.1 that if $a \in [2]$, then $[a] = [2]$. Hence, $[2] = [-6] = [-14]$ and so on.

For general n, there are exactly n equivalence classes. A complete proof requires the division algorithm for integers.

2.1 The Division Algorithm for Integers

The familiar process of long division taught in elementary school says that an integer can be divided by any positive integer to obtain a quotient and a remainder that is smaller than the number by which we are dividing. Here is a more formal statement using symbols:

Theorem 2.1

THE DIVISION ALGORITHM. *If $a, b \in \mathbb{Z}$ with $b \neq 0$, then there exists integers q and r such that*

$$a = bq + r \quad \textit{and} \quad 0 \leq r < |b|.$$

Proof: Let us first give the proof using an additional assumption that $b \geq 1$. If a is a multiple of b, then $a = bq$ for some integer q and the conclusion holds with $r = 0$. Assume from now on that a is not a multiple of b. Consider the set $U = \{a - bx : x \in \mathbb{Z}, \text{ and } a - bx > 0\}$. We first show that U is not the empty set. Since $b \geq 1$ it follows that $-b \leq -1$, and after multiplying by the nonnegative number $|a|$ we have

$$-b|a| \leq -|a| \leq a, \quad \text{and so} \quad 0 \leq a + |a|b = a - b(-|a|).$$

We cannot have $a - b(-|a|) = 0$ since a is not a multiple of b (by assumption). It follows that $a - bx \in U$ with $x = -|a|$ and so U is not the empty set. By the well-ordering property, there is a smallest positive integer in U; call it s and let p be the integer such that $s = a - bp$. Suppose we had $b < s$; this would imply

$$0 < s - b = a - bp - b = a - b(p + 1),$$

which in turn would imply that $s - b$ is a positive integer in U smaller than the smallest positive integer in U. This is an impossible situation and so it must be true that $0 < s \leq b$. If $s = b$, then $b = a - bp$ implies $a = b(p + 1)$. We have excluded the case in which a is a multiple of b, so this cannot hold. Thus, $0 < s < b$ and the conclusion holds with $r = s$ and $q = p$.

We still have to do the case with $b < 0$. In this case, $-b > 0$, so by the part already proved there exist integers q and r with $a = (-b)q + r$ and $0 \leq r < (-b)$. Rewrite this as $a = b(-q) + r$ and $0 \leq r < |b|$ so the conclusion also holds in this case. ∎

The number q is the *quotient* and r is the *remainder* of division of a by b.

We use the division algorithm to describe the equivalence classes modulo n.

Theorem 2.2 *Let n be a positive integer. There are exactly n equivalence classes of integers modulo n. They are the classes $[0]$, $[1]$, ..., $[n - 1]$. If a is any integer and if $a = nq + r$ with $0 \leq r < n$, then $[a] = [r]$.*

Proof: If a is any integer, then there are integers q and r such that $a = nq + r$ with $0 \leq r < n$ by the division algorithm. Then $[a] = [r]$ since $a \equiv r \pmod{n}$. Hence, every equivalence class $[a]$ equals one of the classes $[0], [1], \ldots, [n - 1]$. Next we argue that no two of the classes are equal. If $[i] = [j]$ with $0 \leq i, j \leq n - 1$, then $i - j$ is a multiple of n; because of the restriction on the size of i and j we know that $i - j$ lies between $-(n - 1)$ and $(n - 1)$. The only multiple of n on this interval is $n \cdot 0 = 0$, so $i - j = 0$ and $i = j$. Hence, the n equivalence classes $[0], [1], \ldots, [n-1]$ are all different. ∎

For any positive integer n, we denote the set of equivalence classes modulo n by $\mathbb{Z}_n$. Thus, we have

$$\mathbb{Z}_n = \{[0], [1], \ldots, [n - 1]\}.$$

2.2 The Ring of Integers Modulo n

Next we want to make $\mathbb{Z}_n$ into a ring by defining addition and multiplication that satisfy the ring axioms. Let us illustrate the procedure to be used in the general case with the case $n = 5$. We have

$$\mathbb{Z}_5 = \{[0], [1], [2], [3], [4]\}.$$

How should we define the sum $[2] + [4]$? A first thought would be that $[2] + [4] = [6]$ because the sum of the integers 2 and 4 is 6. Keeping in mind that equivalence classes have many different elements, for example, $[2] = [7]$ and $[4] = [-6]$, we should consider the necessity that

$$[2] + [4] = [7] + [-6].$$

Since $7 - 6 = 1 \equiv 6 \pmod 5$, we see that the equivalence class of $2 + 4 = 6$ is the same as the equivalence class of $7 - 6 = 1$. In fact, it is not difficult to show that if $x \in [2]$ and $y \in [4]$, then $x + y \in [6]$. Here is the proof. Since $x \in [2]$ it follows that $x = 2 + 5t$ for some integer t. Similarly, $y \in [4]$ implies $y = 4 + 5s$ for some integer s. Thus,

$$x + y = (2 + 5t) + (4 + 5s) = 6 + 5(t + s).$$

It follows that $x + y \in [6]$. Thus, the sum $[2] + [4]$ is the equivalence class that contains $x + y$, where x is any element of $[2]$ and y is any element of $[4]$.

We face the analogous question for products. How should we define the product of $[2]$ and $[4]$? The logical guess is $[2][4] = [8] = [3]$. Again considering just one of the many other elements in the equivalence classes, $[2] = [7]$ and $[4] = [-6]$, we will want $[7][-6] = [3]$. Since $(7)(-6) = -42 \equiv 3 \pmod 5$, the equality in this case is verified. More generally, for any $x \in [2]$ and any $y \in [4]$, the product xy lies in $[3]$. Thus, the obvious guess for the rules of addition and multiplication are suitable but one should understand that there are several details that must be checked. Here is a statement of the general case:

Theorem 2.3 *Let n be a positive integer and let $[k]$ and $[h]$ be any two equivalence classes in $\mathbb{Z}_n$. Then the equivalence class $[k + h]$ contains the sum of any integer in $[k]$ and any integer in $[h]$. Similarly, the equivalence class $[kh]$ contains the product of any integer in $[k]$ and any integer in $[h]$.*

Proof: Let $x \in [k]$ and $y \in [h]$. Then

$$x = k + nt \quad \text{and} \quad y = h + ns$$

for some integers t and s. Then $x + y = (k + h) + n(t + s)$, so

$$x + y \equiv k + h \pmod n.$$

Thus, $x + y \in [k + h]$ and $[x + y] = [k + h]$. For the product of x and y we have

$$xy = (k + nt)(h + ns) = kh + n(ks + ht + nts) \equiv kh \pmod n.$$

Thus, $[xy] = [kh]$. ■

This theorem permits us to define addition and multiplication of elements in $\mathbb{Z}_n$ as follows: Let U and V be any equivalence classes in $\mathbb{Z}_n$. Define the sum and product of U and V by the rules

$$\begin{aligned} U + V &= [k + h] \quad \text{for any} \quad k \in U, \quad h \in V \\ UV &= [kh] \quad \text{for any} \quad k \in U, \quad h \in V. \end{aligned}$$

The point of Theorem 2.3 is that in making the computation of $U+V$ or UV, *any* elements of U and V may be used for the computation and the result will be the same. A more succinct description of addition and multiplication may be given by writing

$$[k]+[h]=[k+h], \qquad [k][h]=[kh].$$

Here are some examples of computations in $\mathbb{Z}_5$:

$$\begin{aligned}
&[18]+[7]=[25]=[0]\\
&[13]+[-19]=[-6]=[-1]=[4]\\
&[3][3]=[9]=[4]\\
&[-7]^3=[-2]^3=[-8]=[2]\\
&[2]^{100}=([2]^4)^{25}=[16]^{25}=[1]^{25}=[1].
\end{aligned}$$

Here are some computations in $\mathbb{Z}_{12}$:

$$\begin{aligned}
&[4]+[4]+[4]=[12]=[0]\\
&[3][4]=[12]=[0]\\
&[10]([8]+[6])=[10][14]=[10][2]=[20]=[8]\\
&[5]^2=[25]=[1]\\
&[5]^{80}=([5]^2)^{40}=[1]^{40}=[1].
\end{aligned}$$

For any positive integer n we have constructed a ring $\mathbb{Z}_n$ that contains exactly n elements.

2.3 Solution of Congruences

Congruences modulo n may be treated as equations in $\mathbb{Z}_n$. For example, suppose we wish to find an integer x that satisfies a congruence $3x \equiv 4 \pmod{11}$. If x is such an integer, then $[3x]=[4]$ in $\mathbb{Z}_{11}$. It happens that we have selected an equation that can be easily solved by observing that $[3][4]=[12]=[1]$; that is, $[3]$ has a multiplicative inverse. So we have

$$[4][3x]=[4][4] \quad \text{or} \quad [4][3][x]=[x]=[16]=[5].$$

Thus, $x=5$ is one solution to the congruence. Every solution has the form $x=5+11t$ for some integer t.

The key step in this example relies on the fact that the class $[3]$ has a multiplicative inverse. We will return to this idea later and determine which elements of $\mathbb{Z}_n$ have multiplicative inverses. To solve a congruence involving elements that do not have inverses, some other technique must be used. Suppose we wish to find all integers x that satisfy $6x \equiv 14 \pmod{20}$. This is equivalent to the equation $[6x] = [14]$ in $\mathbb{Z}_{20}$. There is a secondary question that must be answered: Is there any solution to the congruence? A brute-force method of solution would be to compute all products $[6][x]$ as $[x]$ runs through the 20 equivalence classes in $\mathbb{Z}_{20}$ and determine if any satisfy the given condition. A slightly better way is to recognize that the congruence is equivalent to the equation $6x = 14 + 20t$ for some integer t. We cancel 2 to get the equation $3x = 7 + 10t$, which is equivalent to $[3]_{10}[x]_{10} = [7]_{10}$ in $\mathbb{Z}_{10}$. In this case $[3]_{10}[-3]_{10} = [1]_{10}$, so it follows that $[x]_{10} = [-21]_{10} = [-1]_{10}$. Thus, every solution is an integer of the form $x = -1 + 10t$ for an integer t. These solutions fall into two equivalence classes in $\mathbb{Z}_{20}$; namely, $[x]_{20} = [-1]_{20}$ and $[x]_{20} = [9]_{20}$ obtained by taking $t = 0$ or 1. So this congruence has solutions and we have found all of them.

2.4 Divisibility by Nine

Here is an application of congruences to verify the rule, often taught in elementary school, to test an integer for divisibility by 9: An integer m is divisible by 9 if the sum of its digits (written in base 10) is divisible by 9. For example, the integer 2943 is divisible by 9 because the sum of its digits $2+9+4+3 = 18$ is divisible by 9. Why does this work?

We use the ring $\mathbb{Z}_9$ to help verify the test. The proof is based on the idea that an integer m is divisible by 9 precisely when $[m] = [0]$ in $\mathbb{Z}_9$.

Now incorporate the digits of m. Suppose d_i stands for a digit (an integer between 0 and 9) and $m = d_k d_{k-1} \cdots d_1 d_0$ is the decimal representation of m in base 10. This means

$$m = (d_k \times 10^k) + (d_{k-1} \times 10^{k-1}) + \cdots + (d_2 \times 10^2) + (d_1 \times 10) + d_0.$$

For example,

$$2943 = 2 \times 10^3 + 9 \times 10^2 + 4 \times 10 + 3.$$

Now we rely on the equations

$$[1] = [10] = [10]^2 = [10]^3 = \cdots$$

in $\mathbb{Z}_9$ to conclude

$$\begin{aligned} [m] &= [(d_k \times 10^k) + (d_{k-1} \times 10^{k-1}) + \cdots + (d_1 \times 10) + d_0] \\ &= [d_k][10^k] + [d_{k-1}][10^{k-1}] + \cdots + [d_1][10] + [d_0] \end{aligned}$$

$$\begin{aligned} &= [d_k] + [d_{k-1}] + \cdots + [d_1] + [d_0] \\ &= [d_k + d_{k-1} + \cdots + d_1 + d_0]. \end{aligned}$$

This equality implies that the equivalence class modulo 9 of m is equal to the equivalence class modulo 9 of the sum of its digits. In other words, m is in the equivalence class of 0 (i.e., m is divisible by 9) if and only if the same is true of the sum of its digits.

EXERCISES

1. Find the integers x, if any, which satisfy the following congruences:

(a) $3x \equiv 2 \pmod 7$ (b) $12x \equiv 1 \pmod{13}$

(c) $12x \equiv 38 \pmod{12}$ (d) $6x \equiv 10 \pmod{14}$

(e) $10x \equiv 3 \pmod 5$ (f) $25x \equiv 30 \pmod{15}$

In each case interpret the result as an equation in $\mathbb{Z}_n$ for appropriate n.

2. Write out the multiplication table for $\mathbb{Z}_6$. Show that exactly two elements have a multiplicative inverse.

3. Find all solutions of the equation $x^2 = 0$ in the ring $\mathbb{Z}_4$. Do the same for the ring $\mathbb{Z}_5$.

4. Find all solutions of the equation $x^2 = x$ in each of the rings $\mathbb{Z}_2$, $\mathbb{Z}_4$, and $\mathbb{Z}_6$.

5. Which of the rings $\mathbb{Z}_3$, $\mathbb{Z}_4$,$\mathbb{Z}_5$, or $\mathbb{Z}_6$ contain an element x such that $[2]x = [1]$? Which contain an element $y \neq 0$ such that $[2]y = [0]$?

6. In the ring $\mathbb{Z}_{12}$, find all four solutions to the equation $x^2 = x$.

7. In the ring $\mathbb{Z}_{12}$, write the additive inverse of each class $[a]$ as $-[a] = [b]$ with $0 \leq b \leq 11$. What are all the elements that satisfy $-[a] = [a]$?

8. If $n = 2k$ is an even integer, prove that $x = [0]$ and $x = [k]$ are the only two elements of $\mathbb{Z}_n$ that satisfy $x = -x$.

9. If n is an odd integer, prove that $x = [0]$ is the only element in $\mathbb{Z}_n$ that satisfies $x = -x$.

10. Find the sum of all the elements in each of the rings $\mathbb{Z}_2$, $\mathbb{Z}_3$, $\mathbb{Z}_4$,$\mathbb{Z}_5$, $\mathbb{Z}_6$, and $\mathbb{Z}_7$. Guess the answer for the sum of all the elements of $\mathbb{Z}_n$ and try to prove your guess is correct. [Hint: The two previous exercises might be helpful.]

11. The following rule provides a check on hand calculations of products: For an integer m, form the sum of the digits of m; if the result has more than one digit, form the sum of the digits of that number and repeat until only one digit remains; call it $s(m)$. For two integers m and n, the product mn and the product $s(m)s(n)$ are congruent modulo 9. Explain why this works.

12. Use the rule in the previous exercise to quickly find the integer i with $0 \leq i \leq 8$ such that $(1234) \cdot (6789) \equiv i \pmod 9$.

13. If the integer a is obtained by rearranging the digits of an integer b in some order, prove $a \equiv b \pmod 9$. [For example, $379 \equiv 973 \pmod 9$.]

14. Let $m \in \mathbb{Z}$ be written in its "digit form" as $m = d_k d_{k-1} \cdots d_1 d_0$. Prove that m is divisible by 11 if and only if the alternating sum $d_0 - d_1 + d_2 - \cdots + (-1)^k d_k$ is divisible by 11.

3 HOMOMORPHISMS AND ISOMORPHISMS

If we want to compare two rings, one way to approach this is to consider mappings between them. The concept of mappings between rings or other algebraic objects is one of the most important tools for describing properties of these objects.

Definition 3.1

A **mapping** of a set A to a set B is a correspondence that associates with each element a of A a unique element b of B. The notation $a \to b$ is sometimes used to indicate that b is the element associated with the element a of A under the given mapping. We may say that a **maps to** b or b is **the image of** a under this mapping. The mapping is **onto** B if for every element b of B there is some $a \in A$ such that a is associated with b; that is, $a \to b$.

The reader who has studied calculus will recognize that a mapping from a set A to a set B is just a function f such that $f(a)$ is defined for every $a \in A$ and whose values lie in B; that is, $f(a) \in B$ for every $a \in A$. Thus, for example, the function $f(x) = x^2 + 4x^4$ maps an arbitrary real number x to the nonnegative real number $x^2 + 4x^4$; that is, $x \to x^2 + 4x^4$ under this mapping. This function maps the real numbers *onto* the nonnegative real numbers because, we learn from calculus, for any nonnegative real number b there is some x such that $x^2 + 4x^4 = b$.

For the study of rings, we are not concerned with arbitrary functions from one ring to another but rather with those functions that "preserve" the ring properties given in the definition of a ring. Such functions are called homomorphisms.

Definition 3.2

A mapping $\theta : R \to S$ of a ring R to a ring S is called a **homomorphism** if

$$\theta(a + b) = \theta(a) + \theta(b) \quad \text{and} \quad \theta(ab) = \theta(a)\theta(b) \tag{2.1}$$

for every $a, b \in R$.

Note that the symbols for addition and multiplication are those appropriate for the particular ring; that is, in the symbol $\theta(a + b)$ the plus sign is addition in

R, whereas in $\theta(a) + \theta(b)$ the plus sign is addition in S. Similar comments apply to multiplication.

If there is a homomorphism θ that maps R *onto* S, we say S is a **homomorphic image** of R.

Example 3.1

Let $\mathbb{Z}$ be the ring of integers and $\mathbb{Z}_n$ the ring of integers modulo n for some $n > 1$. Let θ be the map from $\mathbb{Z}$ to $\mathbb{Z}_n$ defined by

$$\theta : x \longrightarrow [x].$$

We verify that θ is a homomorphism using the definitions of the operations in $\mathbb{Z}_n$:

$$\begin{aligned} \theta(a+b) &= [a+b] = [a] + [b] = \theta(a) + \theta(b) \\ \theta(ab) &= [ab] = [a][b] = \theta(a)\theta(b). \end{aligned}$$

This verifies that θ is a homomorphism from $\mathbb{Z}$ to $\mathbb{Z}_n$. It is also a homomorphism of $\mathbb{Z}$ *onto* $\mathbb{Z}_n$ because for any equivalence class $[k]$ in $\mathbb{Z}_n$, $\theta(k) = [k]$.

When a map θ is defined on a ring and the conditions of Definition 3.2 are satisfied, we say θ *preserves addition and multiplication.* Thus, we have just shown that the map $x \rightarrow [x]$ preserves addition and multiplication.

Example 3.2

Consider the mapping $\theta : \mathbb{Z} \rightarrow \mathbb{Z}_{10}$ given by $\theta(x) = [5x]$. One easily checks that θ preserves addition. Let us verify that it also preserves multiplication:

$$\begin{aligned} \theta(xy) &= [5xy] \\ \theta(x)\theta(y) &= [5x][5y] = [25][xy] = [5][xy] = [5xy] \end{aligned}$$

because $[25] = [5]$ in $\mathbb{Z}_{10}$. The homomorphism is neither one-to-one nor onto. It is not one-to-one because two different integers may have the same image under the mapping. For example, $\theta(2) = [10] = [0] = \theta(0)$ but $0 \neq 2$ in $\mathbb{Z}$. It is not onto because there is no integer x with $\theta(x) = [1]$ (for example). The only elements in the image of θ are $[0]$ and $[5]$ because $[5x] = [0]$ if x is even and $[5x] = [5]$ if x is odd.

Example 3.3

Let $R = \{u, v, w, x\}$ be the ring with addition and multiplication given by the tables below. This is the ring that was described in Example 3.4 of Chapter I.

$(+)$	u	v	w	x
u	u	v	w	x
v	v	u	x	w
w	w	x	u	v
x	x	w	v	u

$(\cdot)$	u	v	w	x
u	u	u	u	u
v	u	v	w	x
w	u	w	w	u
x	u	x	u	x

Now consider the ring $S = \mathbb{Z}_2 \times \mathbb{Z}_2$ and a map from R to S defined as follows:

$$\begin{aligned} \theta(u) &= ([0],[0]) \\ \theta(v) &= ([1],[1]) \\ \theta(w) &= ([1],[0]) \\ \theta(x) &= ([0],[1]). \end{aligned}$$

We claim that θ is a homomorphism. Note that u, the zero of R, corresponds to the zero of S under the map θ and the identity v of R corresponds to the identity $([1],[1])$ of S. We check only a few of the necessary equations: Starting from $w + x = v$ and $wx = u$, we have

$$\begin{aligned} \theta(w+x) &= \theta(v) = ([1],[1]) \\ \theta(w)+\theta(x) &= ([1],[0]) + ([0],[1]) = ([1],[1]) \end{aligned}$$

and

$$\begin{aligned} \theta(wx) &= \theta(u) = ([0],[0]) \\ \theta(w)\theta(x) &= ([1],[0])([0],[1]) = ([0],[0]). \end{aligned}$$

This verifies that $\theta(w+x) = \theta(w)+\theta(x)$ and that $\theta(wx) = \theta(w)\theta(x)$. Since $([1],[1])$ is the identity of S and v is the identity of R, all products involving these elements are easily checked. We leave the remaining equations to the reader.

This example of a homomorphism has the additional property that it is a one-to-one mapping of R onto S; that is, for each element a in S there is one, and only one, element p of R such that $\theta(p) = a$. A homomorphism with this property is so important that we give it a special name.

Definition 3.3 A function $\theta : R \to S$ from a ring R to a ring S is called an **isomorphism** if

(i) θ is a homomorphism;
(ii) θ is onto; that is, for each $s \in S$ there is some $r \in R$ with $\theta(r) = s$;
(iii) θ is one-to-one; that is, for $a, b \in R$ the equation $\theta(a) = \theta(b)$ holds only if $a = b$.

When there is an isomorphism from R to S we say R is **isomorphic** to S.

Here, some examples to illustrate various properties that a homomorphism may have are provided. If $\theta : R \to S$ is an isomorphism of the ring R with the ring S, then there is a closely related isomorphism of S to R denoted as θ^{-1} and defined as follows: For each $s \in S$ there is one and only one element $r \in R$ such that $\theta(r) = s$; let the unique r be denoted by $\theta^{-1}(s)$. Thus, $\theta(r) = s$ if and only if $\theta^{-1}(s) = r$. Let us verify that θ^{-1} is a homomorphism.

To verify that $\theta^{-1}(u+v) = \theta^{-1}(u) + \theta^{-1}(v)$ we start with the equations $\theta(a) = u$ and $\theta(b) = v$ for uniquely determined elements $a, b \in R$. Then $\theta(a+b) = \theta(a) + \theta(b) = u + v$ and the definition of θ^{-1} implies $a + b = \theta^{-1}(u+v)$. Then $a = \theta^{-1}(u)$ and $b = \theta^{-1}(v)$ so θ^{-1} preserves addition. In a similar way one checks that θ^{-1} preserves multiplication.

Thus, if R is isomorphic to S, then S is isomorphic to R and we may say that R and S are *isomorphic rings*. If R and S are isomorphic rings, then any property possessed by R that is a consequence of the ring properties is also possessed by S. They will differ only in the notation that is used to indicate the elements of the rings. One says that isomorphic rings are *abstractly identical.*

Next we give an example of a nontrivial isomorphism. In it we will simultaneously be using the rings $\mathbb{Z}_2$, $\mathbb{Z}_3$, and $\mathbb{Z}_6$. In order to distinguish the elements of the different rings, we attach subscripts to the brackets so that elements $[a] \in \mathbb{Z}_2$, $[b] \in \mathbb{Z}_3$, and $[c] \in \mathbb{Z}_6$ are denoted by

$$[a]_2 \in \mathbb{Z}_2, \quad [b]_3 \in \mathbb{Z}_3, \quad [c]_6 \in \mathbb{Z}_6.$$

We streamline the notation by removing some redundant parenthesis, for example, by writing $\theta[u]_6$ in place of $\theta([u]_6)$.

Theorem 3.1 *There is a mapping*

$$\theta : \mathbb{Z}_6 \longrightarrow \mathbb{Z}_2 \oplus \mathbb{Z}_3$$

having the property $\theta[x]_6 = ([x]_2, [x]_3)$ for every integer x. The map θ is an isomorphism of $\mathbb{Z}_6$ with the direct sum of $\mathbb{Z}_2$ and $\mathbb{Z}_3$.

Proof: We attempt to define a function θ of $\mathbb{Z}_6$ by the following procedure: For any equivalence class U in $\mathbb{Z}_6$, select an integer $u \in U$ and set $\theta(U) = ([u]_2, [u]_3) \in \mathbb{Z}_2 \oplus \mathbb{Z}_3$. There is something that needs to be checked here. It is necessary to show that the definition of $\theta(U)$ does not depend on the particular choice of element in U. Suppose we select a different element $v \in U$. Will it follow that

$$([u]_2, [u]_3) = ([v]_2, [v]_3)?$$

Let us verify that this is the case. If $u \in U$, then $U = [u]_6$ and another element $v \in U$ must have the form $v = u + 6t$ for some integer t. Using the facts that $[6]_2 = [0]_2$ and $[6]_3 = [0]_3$ we have

$$([v]_2, [v]_3) = ([u+6t]_2, [u+6t]_3) = ([u]_2, [u]_3).$$

Thus, the definition of θ does not depend on the choice of element in the equivalence class. We use the phrase θ is *well-defined* by the equation

$$\theta[u]_6 = ([u]_2, [u]_3).$$

Next we verify that θ is a homomorphism. For any integers a and b we have

$$\begin{aligned}\theta([a]_6 + [b]_6) &= \theta[a+b]_6 = ([a+b]_2, [a+b]_3)\\ &= ([a]_2 + [b]_2, [a]_3 + [b]_3) = ([a]_2, [a]_3) + ([b]_2, [b]_3)\\ &= \theta[a]_6 + \theta[b]_6.\end{aligned}$$

$$\begin{aligned}\theta([a]_6[b]_6) &= \theta[ab]_6 = ([ab]_2, [ab]_3)\\ &= ([a]_2[b]_2, [a]_3[b]_3) = ([a]_2, [a]_3)([b]_2, [b]_3)\\ &= \theta[a]_6 \cdot \theta[b]_6.\end{aligned}$$

Thus, θ preserves addition and multiplication and consequently θ is a homomorphism.

Next we show that θ is one-to-one. Suppose x and y are integers such that $\theta[x]_6 = \theta[y]_6$. In order to prove θ is one-to-one, it is necessary to show that $[x]_6 = [y]_6$. The equality of the images under θ implies

$$([x]_2, [x]_3) = ([y]_2, [y]_3)$$

and so $[x]_2 = [y]_2$ and $[x]_3 = [y]_3$. The first of these implies $x = y + 2t$ for some integer t, while the second equality implies $x = y + 3s$ for some integer s. These last two equations imply that $2t = 3s$; therefore, in particular, s is even because $3s$ is equal to the even number $2t$. Because s is even, it can be written as $s = 2k$ for some integer k. Substituting $2k$ for s gives us the equation

$$x = y + 3s = y + 6k,$$

and it follows that $[x]_6 = [y]_6$. This shows that θ is one-to-one.

In order to complete the proof, it is necessary to show that θ is onto; that is, every element in $\mathbb{Z}_2 \oplus \mathbb{Z}_3$ equals $\theta(U)$ for some $U \in \mathbb{Z}_6$. Several more lines of computation could be given to prove this but instead we do some simple counting. The ring $\mathbb{Z}_6$ has six elements and θ must map them to six distinct elements of $\mathbb{Z}_2 \oplus \mathbb{Z}_3$. (We just proved θ is one-to-one.) However, the ring $\mathbb{Z}_2 \oplus \mathbb{Z}_3$ has exactly $2 \cdot 3 = 6$ elements, so each of them is in the image of θ and θ must be an onto map. This completes the proof that the map is an isomorphism. ∎

The most fundamental properties of homomorphisms are stated in the following theorem. It should be kept in mind that an isomorphism is a special case of a homomorphism, so isomorphisms certainly have the stated properties.

Theorem 3.2 *Let $\theta : R \to S$ be a homomorphism of the ring R into the ring S. Then each of the following is true:*

(i) *If 0 is the zero of R, then $\theta(0)$ is the zero of S.*

(ii) *If $a \in R$ then $\theta(-a) = -\theta(a)$.*

(iii) *The set $\theta(R) = \{\theta(a) : a \in R\}$ is a subring of S.*

(iv) *If R has an identity e and if θ is an onto map, then $\theta(e)$ is the identity of S.*

(v) *Suppose that R has an identity and that θ is an onto map. If $a \in R$ has a multiplicative inverse, then $\theta(a)$ has an inverse in S and $\theta(a^{-1}) = \theta(a)^{-1}$.*

(vi) *If R is a commutative ring and θ is an onto map, then S is a commutative ring.*

Proof: We prove only selected parts and leave the remaining as exercises. (i) The zero of a ring is the unique element z such that $z + z = z$; $z = \theta(0)$ satisfies this equation since $\theta(0) = \theta(0 + 0) = \theta(0) + \theta(0)$. (iii) The set, denoted by $\theta(R)$, is closed under addition and multiplication by the definition of homomorphism and under taking additive inverses by (ii). Thus, $\theta(R)$ is a subring of S by Theorem 4.8 of Chapter I.

The statements (iv)–(vi) all require that θ be onto; that is, $\theta(R) = S$. To prove (iv), it is necessary to show that $s\theta(e) = \theta(e)s = s$ for all $s \in S$. Since θ is onto, any $s \in S$ may be written as $s = \theta(r)$ for some $r \in R$. Since e is the identity of R we have $er = re = r$ and so $\theta(er) = \theta(re) = \theta(r)$. Since multiplication is preserved by θ, it follows that $\theta(r)\theta(e) = \theta(e)\theta(r) = \theta(r)$ or $s\theta(e) = \theta(e)s = s$, as required. ■

Let us use some of these properties to show that two rings are **not** isomorphic. The rings $\mathbb{Z}_4$ and $\mathbb{Z}_2 \oplus \mathbb{Z}_2$ are both commutative rings with exactly four elements. Let us show there is no isomorphism between them. Suppose there is a homomorphism $\theta : \mathbb{Z}_4 \to \mathbb{Z}_2 \oplus \mathbb{Z}_2$. We will prove that it cannot be an isomorphism.

Note that every element $x \in \mathbb{Z}_2 \oplus \mathbb{Z}_2$ satisfies the equation $2x = x + x = 0$ because this is obviously true in $\mathbb{Z}_2$. Thus, for any integer t, we must have $2\theta[t]_4 = ([0]_2, [0]_2)$ because $\theta[t]_4$ is an element of $\mathbb{Z}_2 \oplus \mathbb{Z}_2$. In particular,

$$\theta[2]_4 = \theta([1]_4 + [1]_4) = \theta[1]_4 + \theta[1]_4 = ([0]_2, [0]_2).$$

However, we also have $\theta[0]_4 = ([0]_2, [0]_2)$ by property (i) of the theorem just proved. Hence, we have $\theta[0]_4 = \theta[2]_4$ $[= ([0]_2, [0]_2)]$ even though $[0]_4 \neq [2]_4$. That is, θ is not one-to-one and hence is not an isomorphism.

We point out to the reader that we proved this map is not one-to-one by showing that two different elements were mapped to zero. The set of elements that are mapped to zero by a homomorphism is called the *kernel* of the homomorphism and will be studied in the next section.

EXERCISES

The first three exercises assume the context of Theorem 3.2.

1. Use only the definitions (of a homomorphism and additive inverses) to prove $\theta(-a) = -\theta(a)$.

2. Suppose $\theta(R) = S$ and that R has an identity. If $a \in R$ has a multiplicative inverse, then $\theta(a)$ has an inverse in S and $\theta(a^{-1}) = \theta(a)^{-1}$.

3. Suppose $\theta(R) = S$ and R is a commutative ring. Show that S is a commutative ring.

4. Show that there are exactly two homomorphisms from $\mathbb{Z}_4$ to $\mathbb{Z}_2$ and one of them sends every element of $\mathbb{Z}_4$ to $[0]_2$.

5. If R and S are any two rings, then the subset of $R \oplus S$ consisting of all elements of the form $(r, 0)$ is a subring isomorphic to R.

6. If R and S are rings, show that $R \oplus S$ is isomorphic to $S \oplus R$.

7. Let $A = \{x\}$ be a set with one element and S the ring of all subsets of A. Show that S is isomorphic to $\mathbb{Z}_2$.

8. Let $A = \{1, 2\}$ and $B = \{1, 2, 3\}$ be two sets. Let R be the ring of all subsets of A and S the ring of all subsets of B. Find a homomorphism of R to S that is one-to-one.

9. Keeping the same notation as in the previous exercise, find a homomorphism from S onto R. [Hint: You could try to get a function from S to R by assigning any elements of A to the elements of B and using that to transform subsets of B to subsets of A.]

10. Define a function θ from the ring $\mathbb{Z}[\sqrt{2}]$ of all elements $x + y\sqrt{2}$, $x, y \in \mathbb{Z}$ to the ring $\mathbb{Z}_2$ by the rule $\theta : x + y\sqrt{2} \to [x]_2$. Show that θ is a homomorphism of $\mathbb{Z}[\sqrt{2}]$ onto $\mathbb{Z}_2$.

11. Show that the function from $\mathbb{Z}[\sqrt{2}]$ to $\mathbb{Z}_3$ defined by the rule $x + y\sqrt{2} \to [x]_3$ is **not** a homomorphism.

12. Let U be the subring of $M_2(\mathbb{Z})$ consisting of all elements of the form

$$\begin{bmatrix} x & 0 \\ y & z \end{bmatrix}, \qquad x, y, z \in \mathbb{Z}.$$

Let L be the set of all triples (a, b, c) with $a, b, c \in \mathbb{Z}$. How should addition and multiplication be defined to make L a ring isomorphic to U. [Hint: What conditions are required to make

$$\begin{bmatrix} x & 0 \\ y & z \end{bmatrix} \to (x, y, z)$$

an isomorphism?]

13. Let U be the ring defined in the previous exercise. Define $\theta : U \to \mathbb{Z}$ by the rule

$$\theta : \begin{bmatrix} x & 0 \\ y & z \end{bmatrix} \longrightarrow x.$$

Show θ is a homomorphism. Is it onto? Is it one-to-one?

14. Show that there is a well-defined mapping $\mathbb{Z}_{10} \to \mathbb{Z}_5$ defined by $[x]_{10} \to [x]_5$ and verify that it is a homomorphism.

15. The mapping $\mathbb{Z}_6 \to \mathbb{Z}_2 \oplus \mathbb{Z}_3$ defined by $[x]_6 \to ([x]_2, [x]_3)$ was shown to be a well-defined isomorphism. Show that the inverse map θ^{-1} : $\mathbb{Z}_2 \oplus \mathbb{Z}_3 \to \mathbb{Z}_6$ is given by $([a]_2, [b]_3) \to [3a - 2b]_6$.

16. Prove there is an isomorphism of $\mathbb{Z}_{10}$ with $\mathbb{Z}_2 \oplus \mathbb{Z}_5$.

17. Show that the two rings $\mathbb{Z}_2 \oplus \mathbb{Z}_4$ and $\mathbb{Z}_8$ (each with eight elements) are **not** isomorphic.

18. Show that exactly one of the two rings $\mathbb{Z}_4 \oplus \mathbb{Z}_3$ and $\mathbb{Z}_2 \oplus \mathbb{Z}_2 \oplus \mathbb{Z}_3$ is isomorphic to $\mathbb{Z}_{12}$.

19. Let m and n be positive integers. Show that there is a well-defined function $\theta : \mathbb{Z}_m \to \mathbb{Z}_{mn}$ that satisfies $\theta[x]_m = [xn]_{mn}$. Furthermore, show that θ preserves addition but that θ preserves multiplication only when $n^2 \equiv n \pmod{mn}$. Conclude that θ is a homomorphism whenever $n \equiv 1 \pmod{m}$.

20. Let $\mathbb{R}$ denote the ring of all real numbers and let C denote the ring of all real-valued functions that are continuous on the interval $[0, 1]$. Let $\phi : C \to \mathbb{R}$ be the function defined by the rule $\phi(f) = f(0)$ for every function $f \in C$. Show that ϕ is a homomorphism from C to $\mathbb{R}$. Show that ϕ maps C *onto* $\mathbb{R}$ but that ϕ is not one-to-one. [Hint: The constant functions are continuous.]

21. Let $C[a, b]$ denote the set of all continuous, real-valued functions on the interval $[a, b]$ so that $C[a, b]$ is a ring for every pair of real numbers $a < b$. Show that the mapping $C[0, 2] \to C[0, 1]$, defined by sending the function $f(x)$ (defined for x on $[0, 2]$) to the function $f(2x)$ (defined for x on $[0, 1]$), is an isomorphism.

22. As a generalization of the preceding exercise, show that there is an isomorphism of the ring $C[a, b]$ with $C[0, 1]$ for any pair of real numbers $a < b$.

4 IDEALS

We will describe how to obtain homomorphic images of a ring. The answer is given in terms of objects in the ring, called **ideals**, which are usually easy to describe. Even though we are primarily interested in commutative rings, the definitions are given to allow for noncommutative rings as well.

Definition 4.1 Let A be a subset of the ring R.

(i) A is a **right ideal** of R if for every $a, b \in A$ and $r \in R$ the elements $a - b$ and ar lie in A.
(ii) A is a **left ideal** of R if for every $a, b \in A$ and $r \in R$ the elements $a - b$ and ra lie in A.
(iii) A is an **ideal** of R if for every $a, b \in A$ and $r \in R$ the elements $a - b$, ar, and ra lie in A.

To emphasize that A is an ideal rather than a left or right ideal, we will sometimes say A is a *two-sided ideal*. We are primarily interested in the study of ideals, although the study of right or left ideals plays an important role in more advanced topics. It is clear from the definitions that an ideal is both a left ideal and a right ideal. Moreover, if the ring is a commutative ring, then the three concepts of left ideal, right ideal, and ideal are identical.

In any ring R, the subset $A = \{0\}$ consisting of only the zero element is an ideal, and the entire ring $A = R$ is also an ideal. These two ideals are called the *trivial ideals*. Ideals are closely related to subrings. Here is the precise statement of the relationship:

Property 4.1 *An ideal A of a ring R is a subring that is closed under multiplication (on the left and right) by elements of R.*

Proof: Let A be an ideal and $a, b \in A$. By definition of ideal $a - b \in A$ and $ab \in A$ because we may apply the condition $ar \in A$ using $r = b$. Thus, A is a subring by Theorem 4.8 in Chapter I. The definition of ideal implies that A is closed under multiplication on the left or right by elements of R. Conversely, if A is a subring of R closed under multiplication by elements of R it is an ideal. ∎

Let us examine a few nontrivial examples.

Consider the set E of all even integers in the ring $\mathbb{Z}$ of integers. Then E is a subring, and if $a \in E$ and $n \in \mathbb{Z}$ then $an \in E$ because an is an even integer whenever a is an even integer.

More generally, let n be any integer and let A be the set of all integer multiples of n. Then A is an ideal. If $a = ns$ and $b = nt$ are elements of A, the $a - b = n(s - t)$ is in A, and if $z \in \mathbb{Z}$ then $az = nsz$ is in A. The ideal consisting of all multiples of n is denoted by (n).

The ring $\mathbb{Z}$ of integers is a subring of the rational numbers $\mathbb{Q}$ but not an ideal of $\mathbb{Q}$. For example, $3 \in \mathbb{Z}$ and $\frac{1}{2} \in \mathbb{Q}$ but $3 \cdot \frac{1}{2} \notin \mathbb{Z}$.

In order to give an example of a left ideal (or a right ideal) which is not an ideal we must deal with a noncommutative ring. Let $R = M_2(\mathbb{Z})$ be the ring of two-by-two matrices with integer entries and let A be the subset

$$A = \left\{ \begin{bmatrix} x & 0 \\ y & 0 \end{bmatrix} : x, y \in \mathbb{Z} \right\}.$$

Then one may check that A is a left ideal but not a right ideal. Similarly, the set

$$B = \left\{ \begin{bmatrix} x & y \\ 0 & 0 \end{bmatrix} : x, y \in \mathbb{Z} \right\}$$

is a right ideal but not a left ideal.

4.1 Kernels of Homomorphisms

Now let us indicate the connection between ideals and homomorphisms. Let R be any ring and $\theta : R \to S$ a homomorphism of R into some ring S.

Definition 4.2 The **kernel** of θ is the set of elements of R that are mapped to 0 by θ. We denote the kernel of θ by $ker\ \theta$.

In the example used earlier in which $\theta : \mathbb{Z} \to \mathbb{Z}_n$ by $\theta(x) = [x]_n$, the kernel of θ is the set of all integers x such that $\theta(x) = [0]_n$. This holds if and only if $[x]_n = [0]_n$ or equivalently if $x \equiv 0 \pmod{n}$. Thus, x is in the kernel of this map if and only if x is a multiple of n. We have seen that the set of all multiples of n is an ideal in $\mathbb{Z}$, so this example shows that the kernel of θ is an ideal. This illustrates a general phenomenon.

Theorem 4.1 *If θ is a homomorphism defined on a ring R, then the kernel of θ is an ideal of R.*

Proof: Let $A = \{x \in R : \theta(x) = 0\}$ be the kernel of θ. If $a, b \in A$ we show $a - b \in A$ as follows:

$$\theta(a - b) = \theta(a) - \theta(b) = 0 - 0 = 0$$

and so $a - b$ is in A, the kernel of θ. Suppose r is any element of R. We show $ra \in A$ using the equations

$$\theta(ra) = \theta(r)\theta(a) = \theta(r) \cdot 0 = 0.$$

Hence, $ra \in A$. Similarly, one checks that $ar \in A$ and so A is an ideal. ∎

We will see later that every ideal is the kernel of some homomorphism. Another reason to consider the kernel of a homomorphism is that it tells us when the homomorphism is one-to-one.

Theorem 4.2

A homomorphism θ is one-to-one if and only if $ker\ \theta = (0)$.

Proof: Suppose θ is one-to-one and $x \in ker\ \theta$. Then

$$\theta(x) = 0 = \theta(0).$$

Both x and 0 are mapped by θ to 0; the one-to-one property implies $x = 0$. Since x was an arbitrary element of the kernel, it follows that $ker\ \theta = (0)$. Now suppose that $ker\ \theta = (0)$. We must show the one-to-one property: namely, $\theta(a) = \theta(b)$ implies $a = b$. Since θ is a homomorphism, the equation $\theta(a) = \theta(b)$ implies

$$0 = \theta(a) - \theta(b) = \theta(a - b)$$

and so $a - b \in ker\ \theta$. However, $ker\ \theta = (0)$, so $a - b = 0$ and $a = b$ as required. ■

We may use this fact to see how isomorphisms may be identified.

Theorem 4.3

Let $\theta : R \to S$ be a homomorphism from a ring R to a ring S. Then θ is an isomorphism if and only if $\theta(R) = S$ and $ker\ \theta = (0)$.

Proof: The statement $\theta(R) = S$ is equivalent to saying that θ maps R onto S. The statement $ker\ \theta = (0)$ is equivalent to saying that θ is one-to-one. Since an isomorphism is a homomorphism that is one-to-one and onto, the conclusion follows. ■

4.2 Principal Ideals

There is a class of ideals that can be easily described.

Definition 4.3

Let R be a commutative ring with identity. For an element $a \in R$, the set

$$(a) = \{ar : r \in R\}$$

of all multiples of a by elements of R is an ideal called the **principal ideal** generated by a and a is called a **generator** of the ideal. Any ideal of the form (a) is called a **principal ideal**.

One could make the definition without the assumption that R has an identity but there could result a bit of awkwardness. With the current definition, the principal ideal (a) contains a because $a = ae$ is a multiple of a. If R has no identity, then a might not be an element of (a). Hence, we use the term only as given in the definition.

Principal ideals are the simplest to describe, and in some cases every ideal is principal. We prove this is the case for the ring of integers.

Theorem 4.4 *Every ideal in the ring $\mathbb{Z}$ of integers is a principal ideal.*

Proof: Let A be an ideal. If $A = (0)$, then A is the principal ideal generated by 0. Therefore, from now on we may assume A contains some nonzero elements. We assert that in fact A must contain some positive elements. To see this, select a nonzero $x \in A$; then $(-1)x = -x$ is also in A because A is an ideal and either x or $-x$ is positive. By the well-ordering property of $\mathbb{Z}$, the set of positive integers in A has a smallest element—call it n. We will prove that every element of A is a multiple of n. Let x be any element of A. We apply the division algorithm and divide x by n to get $x = nq + r$ for integers q and r with $0 \leq r < n$. Since $n \in A$ it follows that $nq \in A$. Since $x \in A$ it follows that $r = x - nq \in A$. However, n was selected as the smallest positive element in A and the number r is less than n and nonnegative. The only possibility is that $r = 0$ and hence $x = nq$. Thus, every element of A is a multiple of n and, of course, every multiple of n lies in A. Thus, $A = (n)$ is the set of all multiples of n. ∎

Later we will study some other classes of rings, namely, polynomial rings and Euclidean rings, for which all ideals are principal ideals. We will see that they share some familiar properties with the integers. Many additional properties of ideals are given in the following exercises.

EXERCISES

1. Let R be a ring with identity e and let A be an ideal (right, left, or two-sided) that contains e. Prove $A = R$.
2. Let R be a ring with identity e and let A be an ideal (right, left, or two-sided) that contains an element x that has a multiplicative inverse in R. Prove $A = R$.
3. Let R denote either the ring of all real numbers or the ring of all rational numbers. Prove that the only ideals of R are the trivial ones, (0) and R.
4. If A and B are ideals of R, prove $A \cap B$ is an ideal of R.
5. If A and B are ideals of R, let us define

$$A + B = \{a + b : a \in A, b \in B\}.$$

Prove $A + B$ is an ideal of R that contains both A and B.

6. Let A be an ideal of $\mathbb{Z}$ and suppose that A contains 2. Prove that either $A = (2)$ or $A = \mathbb{Z}$.
7. Let (m) and (n) be nonzero ideals of $\mathbb{Z}$. Show that $(m) \subseteq (n)$ if and only if m is a multiple of n.

8. Find generators for the ideals $(10) \cap (15)$ and $(10) + (15)$ in $\mathbb{Z}$. [See Exercise 5 for the definition of the sum of two ideals.]

9. Let d be an integer such that both m and n are multiples of d. Show $(m) + (n) \subseteq (d)$.

10. Let $\theta : \mathbb{Z} \oplus \mathbb{Z} \to \mathbb{Z}_n \oplus \mathbb{Z}_m$ be defined by $\theta(a, b) = ([a]_n, [b]_m)$. Show that θ is a homomorphism and describe its kernel explicitly.

11. Let R be a commutative ring with identity and $x \in R$. Give a complete proof that the set (x) defined by

$$(x) = \{rx : r \in R\}$$

is an ideal of R.

12. Let R be a commutative ring with identity such that the only ideals of R are the two trivial ideals (0) and R. Prove that every nonzero element $x \in R$ has a multiplicative inverse.

13. If R is the ring of all real numbers of the form $x + y\sqrt{2}$ with $x, y \in \mathbb{Z}$, show that the set $A = \{2a + b\sqrt{2} : a, b \in \mathbb{Z}\}$ is an ideal of R.

14. Let R be the ring described in the preceding exercise and let $\theta : R \to R$ be the mapping defined by $\theta(x + y\sqrt{2}) = x - y\sqrt{2}$. Show that θ is an isomorphism of R with itself.

15. Let R be any ring and θ the mapping from $R \oplus R$ to itself defined by the rule $\theta(r, s) = (s, r)$. Show that θ is an isomorphism.

5 A CHARACTERIZATION OF THE RING OF INTEGERS

Now that we have discussed isomorphisms, we can prove the statement made in Section 1.6 regarding the properties of the integers from which all other properties can be derived. Here is the precise statement.

Theorem 5.1 *Let D be an ordered integral domain for which the set of positive elements is well-ordered. Then D is isomorphic to the ring $\mathbb{Z}$ of integers.*

The consequence of this statement is that any ordered integral domain for which the set of positive elements is well-ordered is “the same” as the ring of integers so long as isomorphic rings are not considered as different.

As the first step of the proof we give a lemma that asserts some properties of the ordered integral domain D. The lemma requires the properties of multiples as given in Section 1.6 and the exercises following it; these will be used without comment.

Lemma 5.1 *Let D be an ordered integral domain in which the set D^+ of positive elements is well-ordered. If e is the multiplicative identity of D, then*

$$D^+ = \{me : m \in \mathbb{Z}^+\}$$

and

$$D = \{ne : n \in \mathbb{Z}\}.$$

Moreover, if m and n are elements of $\mathbb{Z}$ such that $me = ne$, then $m = n$.

Proof: Let the order relation in D be written as $a < b$ so that $0 < a$ is equivalent to $a \in D^+$. We have already observed that $e > 0$ (because $e = e^2$) and, in fact, that e is the least element of D^+. For each positive integer n let S_n be the statement that $ne > 0$. Since $1e = e \in D^+$ we know S_1 is true. Let k be a positive integer such that S_k is true. Then ke and e are in D^+ an so their sum $e + ke = (k+1)e$ is in D^+. Thus, S_{k+1} is true and by the induction principle $me > 0$ for every positive integer m. Next we show that every element of D^+ has this form. Select any $d \in D^+$. If $d = e$, then d is a multiple of e as we wish to prove. Suppose $d \neq e$. Then $d > e$ because e is the least element of D^+. Let U be the set of **positive** elements of the form $d - me$, with m a positive integer. Then U is not empty because $d - e \in U$. By the well-ordering property for subsets of D^+, U has a least element, u and $u = d - ne$ for some $n \in \mathbb{Z}^+$. Since u is positive we must have $u = d - ne \geq e$ because e is the least element in D^+. If we had $u = d - ne > e$, then $u - e = d - (n+1)e > 0$ is smaller than u and is still positive. This would conflict with our choice of u so in fact we must have $u - e = d - (n+1)e = 0$, which implies $d = (n+1)e$ as we wished to show.

Now we complete the proof of the first statement of the lemma. If $a \in D$ and $a \notin D^+$, then either $a = 0$ or $-a \in D^+$. If $a = 0$ then $a = 0 \cdot e$, so a is a multiple of e. If $-a \in D^+$ then, by what we have just proved, $-a = me$ for some positive integer m. It follows that $a = (-m)e$, so a is again an integer multiple of e.

Now suppose $m, n \in \mathbb{Z}$ and $m \neq n$. We can assume the notation is chosen so that $m > n$. By the part of the lemma already proved, $(m - n)e \in D^+$; therefore, in particular, $(m - n)e \neq 0$ since 0 is not in D^+. Thus, $me \neq ne$. In other words, if $me = ne$ then we cannot have $m \neq n$ and so $m = n$, as required to complete the lemma.

Now we complete the proof of Theorem 5.1. Define a mapping $\theta : \mathbb{Z} \to D$ by the rule $\theta(m) = me$ for each $m \in \mathbb{Z}$. We claim θ is an isomorphism. First we verify it is a homomorphism. This step is really just a restatement of properties of multiples:

$$\begin{aligned} \theta(m+n) &= (m+n)e = me + ne = \theta(m) + \theta(n), \\ \theta(mn) &= (mn)e = m(ne) = m(e(ne)) = (me)(ne) = \theta(m)\theta(n). \end{aligned}$$

Therefore, θ is a homomorphism. It is onto D because we have shown in the lemma that every element of D as the form $me = \theta(m)$ for some $m \in \mathbb{Z}$. That θ is one-to-one follows from the last statement of the lemma, which asserts that

$$\theta(m) = me = ne = \theta(n)$$

can hold only if $m = n$. This completes the proof. ■

EXERCISES

1. Let $R = \mathbb{Z} \oplus \mathbb{Z}$. Define an ordering on R by the rule $(a, b) < (c, d)$ if $a < c$ or if $a = c$ and $b < d$. (This is called the dictionary ordering on R since it resembles the rule for placing words in alphabetic order.) Describe the positive elements of R and show that R^+ does not have the well-ordering property. Give an explicit set of positive elements that does not have a least element.

2. Let $\mathbb{R}$ denote the ring of all real numbers with the usual ordering. Describe a set of positive elements that does not have the well-ordering property. [Hint: One way is to give a sequence of positive terms with a zero limit.]

6 A FUNDAMENTAL ISOMORPHISM

We have seen in the previous section that a homomorphism defined on a ring R has an associated ideal, namely, the kernel of the homomorphism, which determines how far the map is from being one-to-one. In this section we consider the relation between two homomorphisms having the same kernel. The result may seem surprising at first. We show that two homomorphisms with the same kernel have isomorphic images. Let us prove this immediately.

Theorem 6.1 *Let θ and τ be homomorphisms defined on a ring R such that $\ker \theta = \ker \tau$. Then there is an isomorphism between the image $\theta(R)$ of θ and the image $\tau(R)$ of τ.*

Proof: It should be noted that nothing is assumed about the "target" ring; it may be that θ maps R into one ring S and τ maps R into another ring S'. The theorem asserts that the images of the maps are isomorphic rings. We exhibit an isomorphism.

We attempt to define a function $\phi : \theta(R) \to \tau(R)$ by the rule

$$\phi : \theta(x) \longrightarrow \tau(x).$$

Realizing that there may be different elements of R with the same image under θ, we must consider the possibility that there is $y \in R$ with $\theta(x) = \theta(y)$. Then our proposed definition would require that

$$\tau(x) = \phi\big(\theta(x)\big) = \phi\big(\theta(y)\big) = \tau(y).$$

Let us verify that this is indeed the case. If $\theta(x) = \theta(y)$, then $\theta(x - y) = 0$ and so $x - y \in ker\ \theta$. By our assumption, $ker\ \theta = ker\ \tau$, so it follows that $x - y \in ker\ \tau$. Thus, $\tau(x - y) = 0$ and therefore $\tau(x) = \tau(y)$. Hence, ϕ is a well-defined function from $\theta(R)$ to $\tau(R)$.

We must now show that ϕ is an isomorphism. First we show it is a homomorphism. For any $x, y \in R$ we have

$$\phi\big(\theta(x) + \theta(y)\big) = \phi(\theta(x + y)) = \tau(x + y) = \tau(x) + \tau(y),$$

where we have used the homomorphism properties of θ and τ. Thus, ϕ preserves addition. For products we have, similarly,

$$\phi\big(\theta(x)\theta(y)\big) = \phi(\theta(xy)) = \tau(xy) = \tau(x)\tau(y).$$

Thus, ϕ is a homomorphism.

Next we show that ϕ is one-to-one. Suppose

$$\phi\left(\theta(a)\right) = \phi\left(\theta(b)\right).$$

Then, by definition, $\tau(a) = \tau(b)$ and $\tau(a - b) = 0$. Thus, $a - b$ is in the kernel of τ and, hence, in the kernel of θ. Thus, $\theta(a - b) = 0$ and $\theta(a) = \theta(b)$. Thus, ϕ is one-to-one.

Next we show ϕ is onto. Let $\tau(w)$ be any element in the image of τ. Then $\phi(\theta(w)) = \tau(w)$ so ϕ is onto. This proves that $\theta(R)$ is isomorphic to $\tau(R)$. ■

Now we know that, up to isomorphism at least, a homomorphic image of a ring is determined by the kernel. The next task is to make this dependence explicit. At the same time we will show that every ideal is the kernel of some homomorphism.

6.1 Congruence Modulo an Ideal

Let A be an ideal of a ring R; we use A to define an equivalence relation on R.

Definition 6.1 If A is an ideal of a ring R, we define the relation u **is congruent to** v **modulo** A to mean $u - v \in A$. We write $u \equiv v \pmod{A}$ to indicate that this holds.

The relation of congruence modulo A is an equivalence relation; the three conditions of Definition II.1.1 are easily verified. In the case $R = \mathbb{Z}$ and $A = (n)$, the relation just defined is exactly the same as the relation $u \equiv v \pmod{n}$ defined in Section II.2.

The use of the symbol $\equiv$ is intended to suggest equality since many of the properties which are taken for granted when using $=$ are also valid for $\equiv$. We make this explicit with some examples.

Theorem 6.2

Let u, v, u_1, and v_1 be elements of the ring R and let A be an ideal of R. If

$$u \equiv u_1 \pmod{A} \quad \textit{and} \quad v \equiv v_1 \pmod{A},$$

then

(i) $u + v \equiv u_1 + v_1 \pmod{A}$,
(ii) $uv \equiv u_1 v_1 \pmod{A}$.

Proof: The condition $u \equiv u_1 \pmod{A}$ implies $a = u - u_1$ is an element of A; similarly, $b = v - v_1$ is in A. Then we obtain the equations

$$\begin{aligned} u + v &= u_1 + v_1 + (a + b), \\ uv &= u_1 v_1 + (a v_1 + u_1 b + ab). \end{aligned}$$

The definition of ideal implies that $(a + b)$ and $(a v_1 + u_1 b + ab)$ are both elements of A. Thus, (i) and (ii) are valid. ■

6.2 Cosets

The equivalence class of v under the relation of congruence modulo A consists of all elements u such that $u - v = a$ is an element of A; equivalently, $u = v + a$. By selecting one element v from the equivalence class, we obtain all other elements in the class by adding arbitrary elements of A. This suggests the following notation for the equivalence class:

$$v + A = \{v + a : a \in A\}.$$

The set $v + A$ is called a **coset of** A, or more precisely the coset of A containing v. In order to gain some familiarity with this notation, we repeat some properties of equivalence classes:

Properties 6.1

(i) $u + A = v + A$ *if and only if* $u - v \in A$;
(ii) *If* $u + A$ *and* $v + A$ *have one element in common, then* $u + A = v + A$.

In particular, $0 + A = A$ is a coset of A and $u + A = A$ if and only if $u \in A$.

Here is a concrete example illustrating the cosets of an ideal:

Example 6.1

Let R be the ring $\mathbb{Z}_{20}$ of integers modulo 20 and let A be the principal ideal generated by $[4]$. Thus,

$$A = \{[0], [4], [8], [12], [16]\}.$$

The cosets of A are the sets obtained by adding one element to each of the elements of A. In this case we see there are four distinct cosets:

$$[0] + A = \{[0], [4], [8], [12], [16]\} \tag{2.2}$$
$$[1] + A = \{[1], [5], [9], [13], [17]\} \tag{2.3}$$
$$[2] + A = \{[2], [6], [10], [14], [18]\} \tag{2.4}$$
$$[3] + A = \{[3], [7], [11], [15], [19]\}. \tag{2.5}$$

Of course, we have relations such as $[4] + A = A$ or $[7] + A = [3] + A$ which can be verified quickly by noting that $[4] + A$ is a coset that contains $[4] + [0] = [4]$; Also, since $[4] \in A$ it follows that $[4] + A$ and A must be equal since they have a common element. Similarly, $[7] + A$ contains $[7]$ and $[3] + A$ contains $[7]$, so these cosets are the same.

Now return to the general case. Let R/A denote the set of cosets of A; that is, R/A is the set of equivalence classes with respect to the equivalence relation of congruence modulo A. The symbol R/A may be read as "R modulo A" or more briefly as "R mod A." Our plan is to define operations on R/A to make it into a ring. Once that is done, we will show that R/A is a homomorphic image of R and the kernel of the homomorphism is precisely A.

Theorem 6.3 *Let A be an ideal of the ring R. Operations of addition and multiplication are defined on the set R/A of cosets of A by the rules*

$$(u + A) + (v + A) = (u + v) + A, \tag{2.6}$$
$$(u + A)(v + A) = (uv) + A. \tag{2.7}$$

With these operations, R/A is a ring. Moreover, the mapping $\theta : R \to R/A$ defined by $\theta(u) = u + A$ is a homomorphism of R onto R/A with kernel A.

Proof: As we have seen several times previously, the main point is to show that addition and multiplication are well defined. To this end, let $u + A = u_1 + A$ and $v + A = v_1 + A$. Then $u \equiv u_1 \pmod{A}$ and $v \equiv v_1 \pmod{A}$ and so $u + v \equiv u_1 + v_1 \pmod{A}$ and $uv \equiv u_1 v_1 \pmod{A}$ by Theorem 6.2. It follows that

$$u + v + A = u_1 + v_1 + A \quad \text{and} \quad uv + A = u_1 v_1 + A.$$

As a result, the sum of the cosets as given in Eq. (2.6) is the same no matter which elements are used to represent the cosets. Similarly, the operation of multiplication is independent of the particular elements used. Thus, the set R/A of cosets of A has two operations defined on it. In order to show that R/A is a ring, we consider the mapping $\theta : R \to R/A$ defined by $\theta(u) = u + A$. Clearly, this maps R onto R/A. Equations (2.6) and (2.7) in the theorem can be rewritten as

$$\theta(u + v) = \theta(u) + \theta(v),$$

$$\theta(uv) = \theta(u)\theta(v).$$

The validity of the ring axioms in R now implies the validity of the same axioms in R/A. For example, to show the associative law for multiplication in R/A we have

$$\begin{aligned}(u+A)[(v+A)(w+A)] &= \theta(u)[\theta(v)\theta(w)] = \theta(u)\theta(vw)\\ &= \theta\big(u(vw)\big) = \theta\big((uv)w\big)\\ &= \theta(uv)\theta(w) = [\theta(u)\theta(v)]\theta(w)\\ &= [(u+A)(v+A)](w+A).\end{aligned}$$

In a similar way all the other ring axioms are seen to hold for R/A. Thus, R/A is a ring and θ is a homomorphism of R onto R/A. Now suppose $x \in R$ is in the kernel of θ. Then $\theta(x) = x + A = 0 + A$ implies $x \in A$. Conversely, if $x \in A$ then $\theta(x) = x + A = 0 + A$ and x is in the kernel. Thus, the kernel of θ is precisely A. ■

The ring R/A is called a **factor ring** of R. The map $\theta : R \to R/A$ defined by $\theta(u) = u + A$ is called the **canonical map** from R to R/A.

It is worth noting that $0 + A = A$ is the zero of R/A. If R has an identity e, then $e + A$ is the identity of R/A. If the ring R is commutative, then R/A is commutative. The converse need not be true, however. It is possible that R/A is commutative even if R is noncommutative. (See Exercises 9 and 10 that follow.)

The construction of factor rings and the earlier theorem about two homomorphisms with the same kernel give us a description of all homomorphic images of a ring.

Theorem 6.4

THE FIRST ISOMORPHISM THEOREM. *Let τ be a homomorphism of the ring R onto the ring S with kernel A. Then S is isomorphic to the factor ring R/A and there is a well-defined isomorphism $\alpha : R/A \to S$ that satisfies*

$$\alpha(u + A) = \tau(u).$$

Proof: Let $\theta : R \to R/A$ be the canonical map. The θ and τ are homomorphisms defined on R having the same kernel, A. By Theorem 6.1 we know $\theta(R)$ is isomorphic to $\tau(R)$; that is, R/A is isomorphic to S. One may verify that the isomorphism given in Theorem 6.1 is the same as the map α. ■

This theorem is called the First Isomorphism Theorem for Rings because it plays a fundamental role in the study of homomorphisms. It may be interpreted as saying that if isomorphic rings are not considered as different, then the only homomorphic images of R are the factor rings R/A, where A is an ideal of R. A few illustrations of the use of this theorem follow.

Example 6.2

Let θ be a homomorphism from the ring $\mathbb{Z}$ of integers onto a ring S. In order to describe S, we first look at the kernel of θ. The kernel is an ideal and so is a principal ideal (n) for some integer n. If $n = 0$, then θ is an isomorphism and S is isomorphic to $\mathbb{Z}$. If $n \neq 0$, then either n or $-n$ is positive and, since n and its negative generate the same ideal, $(n) = (-n)$, we may assume at the start that $n > 0$; otherwise, we could just replace n by $-n$. Then S is isomorphic to the factor ring $\mathbb{Z}/(n)$. We argue that $\mathbb{Z}/(n)$ is the ring $\mathbb{Z}_n$ previously introduced. The elements of $\mathbb{Z}_n$ are equivalence classes $[u] = \{u + nk : k \in \mathbb{Z}\}$. This equivalence class is precisely the coset $u + (n)$ and the operations in $\mathbb{Z}_n$ are exactly the same as the operations defined for $\mathbb{Z}/(n)$. Thus, S is isomorphic to $\mathbb{Z}_n$.

We have found all homomorphic images of $\mathbb{Z}$; they are the ring $\mathbb{Z}$, the zero ring, and the rings $\mathbb{Z}_n$ for $n > 1$.

Example 6.3

Let R be the ring of all rational numbers, all real numbers, or, in fact, any commutative ring with identity in which every nonzero element has an inverse. Let $\theta : R \to S$ be a homomorphism of R onto a ring S. In order to describe S we first obtain information about the kernel of θ. The kernel is an ideal but the only ideals of R are the two trivial ones, (0) and R. If the kernel is (0), then θ is an isomorphism. If the kernel is R, then $S = \{0\}$.

EXERCISES

1. Let A be the principal ideal generated by $[4]$ in the ring $\mathbb{Z}_{12}$. List all the elements of A. List the distinct cosets of A.
2. Show that there is a well-defined homomorphism $\theta : \mathbb{Z}_{100} \to \mathbb{Z}_{20}$ that satisfies $\theta[x]_{100} = [x]_{20}$. What is the kernel of this homomorphism?
3. Let $\theta : \mathbb{Z} \to \mathbb{Z}_{20}$ be the homomorphism $\theta(x) = [x]_{20}$. Show that every ideal of the ring $\mathbb{Z}_{20}$ has the form

$$\theta(A) = \{\theta(a) : a \in A\},$$

where A is an ideal of $\mathbb{Z}$ that contains 20. Argue that every ideal of $\mathbb{Z}_{20}$ is principal and give a generator for each ideal.

4. Let θ be a homomorphism from a ring R onto a ring S. For each ideal B of S, let $p(B)$ be the set

$$p(B) = \{x : x \in R \text{ and } \theta(x) \in B\}.$$

Show that $p(B)$ is an ideal of R containing the kernel of θ. Show, moreover, that this is a one-to-one correspondence between the set of ideals of S and set of ideals of R that contain the kernel of θ.

5. Let $n = md$, with m and d being positive integers. Show that the principal ideal generated by $[m]$ in $\mathbb{Z}_n$ has exactly d elements.

6. Suppose R is a commutative ring and every ideal of R is principal. Prove that every ideal in a homomorphic image of R is principal. [Hint: Use Exercise 4.]

7. Let A and B be ideals of a ring R with $A \subseteq B$. Let B/A denote the collection of cosets of A having the form $b + A$ with $b \in B$. Show that B/A is an ideal of the factor ring R/A.

8. (Continuation of previous exercise.) Since B/A is an ideal of R/A, the factor ring $(R/A)/(B/A)$ is defined. Show that $(R/A)/(B/A)$ is isomorphic to R/B. Carry this out as follows: Define a map $\alpha : R/A \to R/B$ by the rule $\alpha(u + A) = u + B$. Show that α is a well-defined homomorphism with kernel B/A; then apply the First Isomorphism Theorem. (This result is called the *Second Isomorphism Theorem for Rings.*)

9. Let A be an ideal of a ring R. Prove that R/A is commutative if and only if A contains $rs - sr$ for every $r, s \in R$.

10. Let R be the ring of two-by-two matrices of the form

$$r(x, y, z) = \begin{bmatrix} x & 0 \\ y & z \end{bmatrix}, \qquad x, y, z \in \mathbb{Z}.$$

Let A be the subset of R consisting of all elements $r(0, y, 0)$, $y \in \mathbb{Z}$. Prove (i) A is an ideal of R and (ii) R is a noncommutative ring but R/A is commutative.

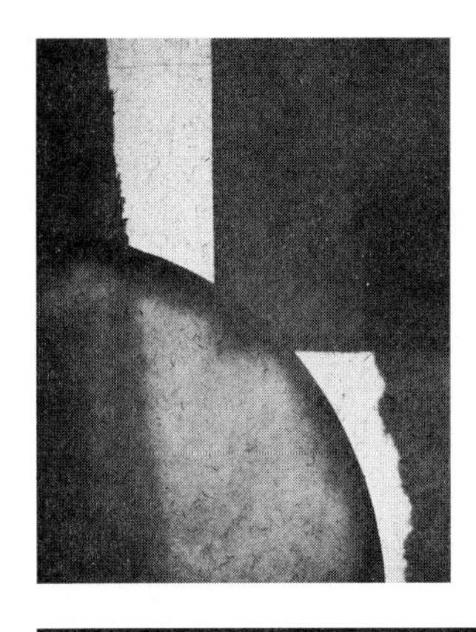

III INTEGRAL DOMAINS AND FIELDS

We study further properties of integral domains and introduce the notion of an abstract field. Several fields have already been used as examples, namely, the rational numbers and the real numbers, but the term is first defined here. We show that the class of integral domains coincides with the class of subrings of fields. The familiar fields encountered in elementary mathematics, the rational numbers, the real numbers, and the complex numbers, are introduced and some of their properties are studied.

1 FIELDS

We have already defined an integral domain as a commutative ring with identity in which the product of two nonzero elements is nonzero. A special case of this is a field which we now define.

Definition 1.1 A commutative ring F with more than one element and having an identity is a **field** if every nonzero element has a multiplicative inverse in F.

We have seen that if an element of a ring has an inverse, then the inverse is unique. If r is a nonzero element of a field with identity 1, then r^{-1} is the unique element of F that satisfies the equation $r^{-1}r = 1 = rr^{-1}$. Since the commutative law is always assumed to hold in a field, we may regard the inverse of r as defined by the single equation $rr^{-1} = 1$.

A field is necessarily an integral domain for if r and s are elements of a field and if $rs = 0$, then either $r = 0$ or $r \neq 0$ and so r^{-1} exists. Then

$$0 = r^{-1}(rs) = (r^{-1}r)s = s$$

and so $s = 0$. Hence, we have shown that the equation $rs = 0$ in F implies either $r = 0$ or $s = 0$. As an immediate consequence of this we obtain the following:

Theorem 1.1 *Every subring of a field is an integral domain.*

Proof: If R is a subring of a field F and if $r, s \in R$ with $rs = 0$, then either $r = 0$ or $s = 0$ because this conclusion holds for all pairs of elements in F. ∎

Now we prove a converse to this theorem, namely, every integral domain is a subring of a field (at least up to isomorphism).

1.1 Field of Quotients

Let D be any integral domain. We will show there is a field F that contains a subring isomorphic to D. The motivation comes, once again, from a study of the integers and the field of rational numbers. Every rational number is a fraction a/b with integers a and b and $b \neq 0$. Some of the maneuvers that we must use in case of a general integral domain D are due to the many representations of a given fraction; that is, $a/b = ac/bc$ for any integer $c \neq 0$. The rules for addition and multiplication of fractions are the familiar ones:

$$\frac{a}{b} + \frac{c}{d} = \frac{ad + bc}{bd} \quad \text{and} \quad \frac{a}{b}\frac{c}{d} = \frac{ac}{bd}.$$

We also notice the relation

$$\frac{a}{b} = \frac{c}{d} \quad \text{if and only if} \quad ad = bc.$$

These ideas are used in the general construction of a field containing the integral domain D. Let S denote the set of all ordered pairs (a, b) with $a, b \in D$ and $b \neq 0$; that is,

$$S = \{(a, b) : a, b \in D, b \neq 0\}.$$

Our next step is suggested by thinking of (a, b) as the familiar a/b, but we use an unfamiliar notation in order to clarify the logical procedure and to avoid using any property until we have actually proved it. At this stage, the symbol a/b has no meaning. Another point to be considered is that $a/b = au/bu$ for any nonzero element of D so that the pairs (a, b) and (au, bu) should be related. Accordingly, we define a relation on the set S. This will be an equivalence relation as defined in Chapter 2.

Property 1.1 *For elements (a, b) and (c, d) in S, define the relation $(a, b) \sim (c, d)$ to mean that $ad = bc$. Then this is an equivalence relation on S.*

Proof: There are three properties we must check to verify that this is an equivalence relation. The reflexive property $(a, b) \sim (a, b)$ holds because $ab = ba$ (D is commutative). The symmetric property also holds because D is commutative: $(a, b) \sim (c, d)$ if and only if $ad = bc$, whereas $(c, d) \sim (a, b)$

if and only if $cb = da$. These are equivalent statements and one holds if and only if the other does also. Finally, let us check the transitive property. Suppose that $(a, b) \sim (c, d)$ and $(c, d) \sim (e, f)$; it is necessary to show $(a, b) \sim (e, f)$. Since $(a, b) \sim (c, d)$, we have $ad = bc$; similarly, we have $cf = de$. Multiplication of these equations by f and b, respectively, yields $adf = bcf$ and $bcf = bde$. Thus, $adf = bde$ which can be written as $(af - be)d = 0$. Since D is an integral domain and since $d \neq 0$ (by definition of the pairs belonging to S) it follows that $af = be$; this means $(a, b) \sim (e, f)$. ■

Now that we have an equivalence relation defined on S, we may consider the equivalence classes. Our previous usage would dictate that the equivalence class containing (a, b) would be denoted by $[(a, b)]$. However, we will use the simpler notation $[a, b]$ to denote this class. Keep in mind that $[a, b] = [u, v]$ if and only if $(a, b) \sim (u, v)$, that is, if and only if $av = bu$. The equivalence class may be expressed as

$$[a, b] = \{(x, y) : (x, y) \in S, xb = ay\}.$$

Now let F denote the set of equivalence classes:

$$F = \{[a, b] : a, b \in D, b \neq 0\}.$$

We now define addition and multiplication on F and verify that we obtain a field. By keeping in mind that $[a, b]$ should eventually represent a/b, the definitions are suggested by the equations we used for the case $D = \mathbb{Z}$.

Theorem 1.2 *The following rules give well-defined operations of addition and multiplication on the set F of equivalence classes of S with respect to $\sim$:*

(i) $[a, b] + [c, d] = [ad + bc, bd]$,
(ii) $[a, b] \cdot [c, d] = [ac, bd]$.

Proof: First observe that $(a, b), (c, d) \in S$ imply $b \neq 0$ and $d \neq 0$. Hence, $bd \neq 0$ and so the elements on the right side of the Eqs. (i) and (ii) are indeed equivalence classes. There is an important point that must be considered before we may conclude that addition and multiplication have been defined. We must take into account the fact that there are many elements in an equivalence class and we have used one element from each class to write a formula for the sum and product. We must verify that the same class is obtained as the sum (or product) no matter which elements from the class are used. Let $X = [a, b]$ and $Y = [c, d]$ be equivalence classes and suppose that we also have $X = [a', b']$ and $Y = [c', d']$. Then our rules for addition would give us

$$X + Y = [ad + bc, bd]$$

and also

$$X + Y = [a'd' + b'c', b'd'].$$

If the rule for addition is a good one, then these elements must be the same. We set out to prove they are the same equivalence class, that is, we must show $(ad + bc)b'd' = (a'd' + b'c')bd$. The fact that $[a, b] = [a', b']$ implies $ab' = a'b$. Similarly, $[c, d] = [c', d']$ implies $cd' = c'd$. Multiply the first of these equations by dd' and the second by bb', and add the resulting two members to get

$$ab'dd' + cd'bb' = a'bdd' + c'dbb' \quad \text{or} \quad (ad + bc)b'd' = (a'd' + c'b')bd,$$

which is what we wanted to show. Thus, addition is well defined. The proof that multiplication is well defined is carried out in a similar manner and is left as an exercise. ∎

We may now state and prove the following theorem:

Theorem 1.3 *Let F denote the set of all equivalence classes of S relative to the equivalence relation $\sim$ defined previously. Then, with the operations of addition and multiplication defined in the previous theorem, F is a field. Moreover, the subset D' of all elements of the form $[a, 1]$, with $a \in D$, is a subring of F isomorphic to D. The isomorphism is given by $\theta : D \to D'$ with $\theta(a) = [a, 1]$. Every element of F has the form $\theta(a)\theta(b)^{-1}$ for some $a, b \in D$ and $b \neq 0$.*

Proof: It is first necessary to verify the defining properties of a ring; these are mostly routine exercises using the definitions of addition and multiplication of equivalence classes. We illustrate this by proving the associative law for addition. Let $[a, b], [c, d], [e, f]$ be elements of F. Then

$$\begin{aligned} ([a, b] + [c, d]) + [e, f] &= [ad + bc, bd] + [e, f] = [adf + bcf + bde, bdf] \\ [a, b] + ([c, d] + [e, f]) &= [a, b] + [cf + de, df] = [adf + bcf + bde, bdf], \end{aligned}$$

and therefore we have

$$([a, b] + [c, d]) + [e, f] = [a, b] + ([c, d] + [e, f]).$$

Since $[0, 1] + [a, b] = [a, b]$ and $[1, 1] \cdot [a, b] = [a, b]$, it follows that $[0, 1]$ is the zero of F and $[1, 1]$ is the identity element of F. However, if d is a nonzero element of D, we have $[d, d] = [1, 1]$ so $[d, d]$ is the identity. Similarly, $[0, 1] = [0, d]$ so the zero equals $[0, d]$ for any nonzero $d \in D$. We also observe that $[a, b] = [0, 1]$ implies $a \cdot 1 = b \cdot 0 = 0$ and so $a = 0$. In other words, the nonzero elements of F are the $[a, b]$ with $a \neq 0$ (and of course $b \neq 0$).

Since $[a, b] + [-a, b] = [0, b^2]$ and $[0, b^2]$ is the zero of F, it follows that the additive inverse of $[a, b]$ is $[-a, b]$. That is, we have $-[a, b] = [-a, b]$ so each element of F has an additive inverse.

One of the distributive laws is a consequence of the following calculations in which, at one point, we use the fact that $[b, b]$ is the identity of F:

$$\begin{aligned} [a, b]\,([c, d] + [e, f]) &= [a, b][cf + de, df] \\ &= [acf + ade, bdf], \end{aligned}$$

$$\begin{aligned} [a, b] \cdot [c, d] + [a, b] \cdot [e, f] &= [ac, bd] + [ae, bf] \\ &= [acbf + bdae, b^2 df] \\ &= [acf + ade, bdf][b, b] \\ &= [acf + ade, bdf]. \end{aligned}$$

The other distributive law is an immediate consequence of this one and the fact that multiplication is commutative.

Up to this point we have proved that F is a commutative ring with identity. To prove F is a field we must prove that every nonzero element has a multiplicative inverse in F. If $[a, b]$ is a nonzero element of F, then $a \neq 0$ and $b \neq 0$. Hence, $[b, a]$ is also an element of F. Moreover,

$$[a, b] \cdot [b, a] = [ab, ab] = [1, 1],$$

and the multiplicative inverse of $[a, b]$ is $[b, a]$. We may write $[a, b]^{-1} = [b, a]$. This completes the proof that F is a field.

Let D' be the subset of F consisting of all elements of the form $[a, 1]$, $a \in D$ and consider the function $\theta : D \to D'$ defined by $\theta(a) = [a, 1]$. This function maps D onto D' and is also one-to-one. If $\theta(a) = \theta(b)$, then $[a, 1] = [b, 1]$ and so $a \cdot 1 = b \cdot 1$ by definition of the equivalence relation. Thus, $a = b$. Now we show that θ is a homomorphism: For $a, b \in D$ we have

$$\begin{aligned} \theta(a + b) &=, [a + b, 1] = [a, 1] + [b, 1] = \theta(a) + \theta(b) \\ \theta(ab) &=, [ab, 1] = [a, 1][b, 1] = \theta(a)\theta(b). \end{aligned}$$

Thus, θ is an isomorphism of D with D'. The final statement of the theorem follows from the equation

$$[a, b] = [a, 1][1, b] = [a, 1] \cdot [b, 1]^{-1} = \theta(a)\theta(b)^{-1}.$$

This completes the proof of the theorem. ∎

We may simplify the notation by writing $[a, b] = a/b$. This amounts to using the simpler notation "a" in place of $[a, 1]$ and using a/b in place of $[a, 1][b, 1]^{-1}$. This justifies the familiar notation introduced at the beginning of this section.

The field F as constructed previously from the integral domain D is called **the field of quotients** of D.

In the special case in which $D = \mathbb{Z}$, the ring of integers, the field of quotients is the field of *rational numbers*, denoted by $\mathbb{Q}$. The elements of $\mathbb{Q}$ are written as a/b, with a and b integers and $b \neq 0$.

Let us emphasize the meaning of the notation we have introduced by considering, for example, the rational number $1/2$. We write $1/2$ for the equivalence class $[1, 2]$. We have $[1, 2] = [c, d]$ if and only if $d = 2c$; that is, $[1, 2] = [c, 2c]$ for any nonzero integer c. This agrees with the fraction notation that asserts $1/2 = c/2c$.

Since $(-a)/b = a/(-b)$, we see that every rational number can be written in the form c/d with $d > 0$. If c and d have a common factor so that $c = c'k$ and $d = d'k$, then $c/d = c'/d'$. It follows that every rational number can be expressed *in lowest terms* a/b with $b > 0$ and integers a and b having no common factor greater than 1.

EXERCISES

1. Examine the details of the construction of the field of quotients of the integral domain D and explain why it is necessary that D be an integral domain and not just a commutative ring with identity.

2. Suppose the integral domain D is already a field. Show that the construction of the field of quotients of D produces a field isomorphic to D.

3. Let n be a positive integer and let D be the subset of all rational numbers of the form a/n^k with $a \in \mathbb{Z}$ and k any positive integer. Show that D is an integral domain whose quotient field is isomorphic to the field of rational numbers.

4. Let R be an integral domain and let M be a nonempty subset of nonzero elements of R with the property $x, y \in M$ implying $xy \in M$. Thus, M is a multiplicatively closed set of nonzero elements. Let S be the set of all pairs (r, m) with $r \in R$ and $m \in M$. Define a relation on S by the rule $(r, m) \sim (r', m')$ if $rm' = r'm$. Imitate the construction of the field of quotients to obtain, in this case, a ring containing an isomorphic copy of R in which the image of every element of M has a multiplicative inverse. [The previous exercise is a special case of this in which R is the ring of integers and M is the set of all powers of n.]

2 PROPERTIES OF THE FIELD OF RATIONAL NUMBERS

We defined ordered integral domains in Chapter I, Section 5. Since a field is necessarily an integral domain, an *ordered field* naturally means a field which is an ordered integral domain. If a field happens to be the quotient field of an ordered integral domain, then the field is also ordered. The positive elements of the field are the quotients a/b, with a and b positive elements of the integral domain. We state these facts for the rational number field.

Theorem 2.1 *Let $\mathbb{Q}^+$ denote the set of all rational numbers a/b, where a and b are integers with $ab > 0$. Then $\mathbb{Q}^+$ has the properties given in Definition I.5.1 of an ordered integral domain, and therefore the field $\mathbb{Q}$ is an ordered field whose positive elements are the elements of $\mathbb{Q}^+$.*

Notice that when we write $ab > 0$, we mean that ab is a *positive* integer and we are only making use of the fact that $\mathbb{Z}$ is an ordered integral domain.

Proof: We first need to show that the definition of an element of $\mathbb{Q}^+$ does not depend upon the particular representation of a rational number. That is, we need to show that $a/b = c/d$ and $ab > 0$ implies $cd > 0$. The statement $ab > 0$ means a and b are either both positive or both negative. From $a/b = c/d$ we conclude $ad = bc$, and it follows that c and d are either both positive or both negative; hence, $cd > 0$.

Now we verify the three properties of $\mathbb{Q}^+$. Suppose a/b and c/d are rational numbers in $\mathbb{Q}^+$ and $a, b, c, d \in \mathbb{Z}$ have $ab > 0$ and $cd > 0$. If necessary, we may replace the representation of a/b with $(-a)/(-b)$ to achieve $a > 0$ and $b > 0$. Similarly, we may assume $c > 0$ and $d > 0$. Then the sum and product lie in $\mathbb{Q}^+$ for

$$\frac{a}{b} + \frac{c}{d} = \frac{ad + bc}{bd}, \qquad \left(\frac{a}{b}\right)\left(\frac{c}{d}\right) = \frac{ac}{bd}$$

from which we conclude both $(ad + bc)(bd) > 0$ and $acbd > 0$. Moreover, if p/q is any rational number with $p, q \in \mathbb{Z}$, then $pq > 0$, $pq = 0$, or $-pq > 0$. It follows that exactly one of the following holds:

$$\frac{p}{q} > 0, \quad \frac{p}{q} = 0, \quad -\frac{p}{q} > 0.$$

Hence, $\mathbb{Q}^+$ has the required properties and $\mathbb{Q}$ is an ordered field. ∎

Note that the same proof, with only the slightest change in the meaning of the symbols, shows that the quotient field of an ordered integral domain is an ordered field.

We have made use of the ordering on $\mathbb{Z}$ to describe the ordering on $\mathbb{Q}$. If we identify the integer a with the rational number $a/1$, it is clear that a is a positive integer if and only if a is a positive rational number. In other words, the ordering of $\mathbb{Q}$ is an *extension* of the ordering on $\mathbb{Z}$.

Given the ordering on $\mathbb{Q}$, we may now introduce inequalities involving rational numbers in the usual way. That is, if $r, s \in \mathbb{Q}$, then we write $r > s$ (or

$s < r$) to mean $r - s \in \mathbb{Q}^+$. We have available all the familiar properties of inequalities for rational numbers. In the future we make use of these without specific reference.

We have already observed that the set $\mathbb{Q}^+$ of positive elements is not well-ordered, because the set of all positive rational numbers does not have a least element. Another property somewhat more specific than this is given in the following.

Theorem 2.2

Between any two distinct rational numbers there is another rational number.

Proof: Let $r, s \in \mathbb{Q}$. We will show that the average, $(r + s)/2$, lies between r and s. Since $r \neq s$ let us assume the notation is selected so that $r < s$. Then $0 < s - r$ and also $0 < (s - r)/2$; adding r to both sides of this inequality yields

$$r < r + \frac{s - r}{2} = \frac{r + s}{2}.$$

In a similar manner, starting with $(r - s)/2 < 0$ and adding s to both sides shows

$$\frac{r + s}{2} < s,$$

which completes the proof. ■

The property expressed in this theorem is often expressed by saying that the rational numbers are *dense*. Here is another simple, but important, property of the rational numbers:

Theorem 2.3

ARCHIMEDEAN PROPERTY. *If r and s are any positive rational numbers, there is a positive integer n such that $nr > s$.*

Proof: Let $r = a/b$ and $s = c/d$, where a, b, c, d are positive integers. If n is a positive integer, then $n(a/b) > c/d$ is true if and only if $n(ad) > bc$. This last inequality is satisfied if we select $n = 2bc$. To see this, notice that $ad \geq 1$ because a and d are positive integers; it follows that $2ad > 1$. Now multiply by the positive integer bc to get $(2bc)ad > bc$. Hence, $n = 2bc$ satisfies our requirement. Of course, we do not suggest that this is necessarily the smallest possible choice of n. ■

3 THE FIELD OF REAL NUMBERS

The rational numbers are sufficient for use in simple applications of mathematics to physical problems. For example, measurements are usually given to a certain number of decimal places, and any finite decimal is a rational number. However, from a theoretical standpoint, the system of rational numbers is inadequate. The Pythagoreans made this discovery about 500 BC and were shocked by it. Consider, for example, an isosceles right triangle whose legs are 1 unit in length. Then, by the Pythagorean theorem, the hypotenuse has length $\sqrt{2}$; hence, these geometrical considerations suggest there must exist a "number" $\sqrt{2}$, although we will see later that it cannot be a rational number.

The inherent difficulty in extending the field of rational numbers to the field of real numbers is perhaps indicated by the fact that a satisfactory theory of the real numbers was not obtained until the latter half of the nineteenth century. Although other mathematicians also made contributions to the theory, it is usually attributed to the German mathematician R. Dedekind (1831–1916) and G. Cantor (1845–1918). We cannot present their work but will state without proof the fundamental theorem that was proved by each of them using quite different means. In order to state the theorem, we must first provide a few preliminary definitions.

So far, the only ordered field which we have studied is the rational number field. However, for the moment, suppose $(F, <)$ is an arbitrary ordered field.

Definition 3.1 Let S be a set of elements of F. If there is an element $b \in F$ such that $x \leq b$ for every element x in S, then b is called an **upper bound** of the set S in F.

For example, the set $S_1 = \{\frac{1}{2}, 2, 3, 17.33\}$ of elements of $\mathbb{Q}$ has an upper bound of 20. Also, 30 is an upper bound of S_1 as is 117, and so on. Thus, if a set has an upper bound, it has many upper bounds. The set $\mathbb{Z}$ of integers has no upper bound in $\mathbb{Q}$. As another example, consider the set

$$S_2 = \{a : a \in \mathbb{Q},\ a^2 < 2\}.$$

Then S_2 has upper bounds in $\mathbb{Q}$, one of them being 3. From the collection of all upper bounds of a given set, we single out one for particular attention.

Definition 3.2 Let S be a set of elements of an ordered field F. If there exists an upper bound c of S in F such that no smaller element of F is an upper bound of S, then c is called a **least upper bound** (LUB) of S in F.

Not every set has a LUB, but if one does exist, it is unique. To see this, suppose a set S has a LUB c. Let d also be a LUB of S. By the definition of an ordered integral domain, $d - c < 0$, $d = c$, or $d - c > 0$. If the first case is true, then $d < c$, and so d is an upperbound of S that is smaller than the LUB, c. This is impossible if c is also a LUB of S. Similarly, the third choice $c < d$ is impossible. The only choice is $d = c$ and so the LUB is unique.

For the set S_1 described previously, the element 17.33 is the LUB of S_1. However, for the set S_2, the situation is not quite so obvious. Although we do not give the details here, it is true that there is no rational number that is a LUB of S_2. That is, if $c \in \mathbb{Q}$ and c is an upper bound of S_2, then there exists $d \in \mathbb{Q}$ with $d < c$ and d is also an upper bound for S_2. Thus, S_2 has no *least* upper bound. (See the exercise at the end of this section.) Thus, we have an example of a set in $\mathbb{Q}$ that has upper bounds in $\mathbb{Q}$ but has no LUB in $\mathbb{Q}$. In the field of real numbers, whose existence is asserted in the next theorem, this situation cannot occur. In fact, the existence of an ordered field with this property is the principal contribution of Dedekind and Cantor to this subject. Here is a statement of the result:

Theorem 3.1 *There exists a field $\mathbb{R}$ called the field of real numbers, with the following properties:*

(i) *$\mathbb{R}$ contains the field $\mathbb{Q}$ of rational numbers as a subfield. Moreover, $\mathbb{R}$ is an ordered field and $\mathbb{Q}^+ \subset \mathbb{R}^+$.*

(ii) *If S is a nonempty subset of $\mathbb{R}$ which has an upper bound in $\mathbb{R}$, then S has a least upper bound in $\mathbb{R}$.*

The elements of $\mathbb{R}$ are called *real numbers*. An element of $\mathbb{R}$ which is not an element of $\mathbb{Q}$ is called an *irrational* number. The set S_2 considered previously has a LUB when it is considered a subset of $\mathbb{R}$; that LUB is traditionally denoted as $\sqrt{2}$. We may take this as a definition of the symbol $\sqrt{2}$. Then $\sqrt{2}$ is an irrational number and the fact that $\sqrt{2} \notin \mathbb{Q}$ prevents S_2 from having a LUB *in the field* $\mathbb{Q}$.

The fact that $\mathbb{Q}^+ \subset \mathbb{R}^+$ is sometimes expressed by saying that the ordering on $\mathbb{R}$ is an *extension* of the ordering on $\mathbb{Q}$. That is, a rational number is a positive rational number if an only if it is positive when considered as a real number. This is similar to the situation that arose when we passed from the integers to the rational numbers.

EXERCISES

1. Let $S_2 = \{a : a \in \mathbb{Q}, a > 0, a^2 < 2\}$. Show that a least upper bound for S (in $\mathbb{Q}$) would be a number c such that $c^2 = 2$. The following is an outline of how one proves there is no such rational number: We make use of the notion of factorization of integers to be discussed in Chapter 4. Let $c = a/b$ with positive integers a and b that have no common factor greater than 1. Then $2 = c^2 = a^2/b^2$ implies $2b^2 = a^2$. Since the left side is even, the right side is even. It follows that a is a multiple of 2 and hence a^2 is a multiple of 4. After canceling a factor 2, we discover that b^2 (and hence b) is even. This is in conflict with the assertion that a and b have no common factor.

4 PROPERTIES OF THE FIELD OF REAL NUMBERS

In this section we prove two fundamental properties of real numbers and state one additional property without proof. Of course, the proofs are based on the assumed properties (i) and (ii) stated in Theorem 3.1.

Throughout the rest of the book, $\mathbb{R}$ will denote the field of real numbers and $\mathbb{R}^+$ the positive elements of $\mathbb{R}$.

Theorem 4.1 ARCHIMEDEAN PROPERTY. *If $a, b \in \mathbb{R}^+$, there exists a positive integer n such that $na > b$.*

Proof: Let us assume that $ka \leq b$ for every positive integer k and seek a contradiction to show this assumption cannot stand. Another way to state this assumption is that b is an upper bound of the set $S = \{ka : k \in \mathbb{Z}^+\}$. Since S has an upper bound, it has a LUB, e.g., c. Then $c - a < c$ (because $a > 0$) and therefore $c - a$ is not an upper bound of S. Hence, some element of S is larger than $c - a$; let $c - a < ma$ for some $m \in \mathbb{Z}^+$. It then follows that $c < (m+1)a$. However, $(m+1)a \in S$ and so $(m+1)a \leq c$ by definition of c. We have reached conflicting conclusions so the original supposition cannot stand. ∎

It was shown in the previous section that between any two distinct rational numbers there is another rational number. Here is a generalization of that statement:

Theorem 4.2 *If $a, b \in \mathbb{R}$ with $a < b$, then there exists a rational number m/n ($m, n \in \mathbb{Z}$) such that*

$$a < \frac{m}{n} < b.$$

Proof: For simplicity we first consider the case $a > 0$.

Since $b - a > 0$ we may apply the Archimedean property to assert the existence of an integer $n \in \mathbb{Z}^+$ with $n(b - a) > 1$. Let n be a fixed integer with this property. Again applying the Archimedean property, this time to 1 and na, we get an integer $m \in \mathbb{Z}^+$ such that $m \cdot 1 > na$. Assume that m is the least positive integer with this property. Then $m > na$ implies $a < m/n$, which is half of what we want to prove. The other half is equivalent to $m < nb$. Suppose that this is not true and that $m \geq nb$. From the inequality $n(b - a) > 1$ we conclude

$$m \geq nb > 1 + na > 1.$$

Thus, $m > 1$ and $m - 1 \in \mathbb{Z}^+$ has the property $(m - 1) > na$. However, m was selected as the least positive integer greater than na, so we have reached a contradiction by making the assumption that $m \geq nb$. It follows that $m < nb$ and the proof is complete in the case $a > 0$.

Now suppose $a < 0$ and $b > 0$. Then 0 is a rational number between a and b. For the final case, suppose both $a < 0$ and $b < 0$. Then we have $(-b) < (-a)$ with both $(-b)$ and $(-a)$ positive. Hence, we can apply the case already proved to get a rational number between $-b$ and $-a$ and then the negative of that rational number lies between a and b. ∎

In particular, this theorem tells us that between any two irrational numbers there is a rational number. It is also true that between any two rational numbers there is an irrational number (see Exercise 2). Thus, the rational and irrational numbers are closely intertwined.

Although it is true that all the properties of the real numbers can be established using only the properties (i) and (ii) of Theorem 3.1, we will not give further proofs in this book. However, we conclude this brief discussion of the real numbers by stating without proof the following familiar and important result:

Theorem 4.3 *For each positive real number a and each positive integer n, there exists exactly one positive real number x such that $x^n = a$.*

The real number whose existence is asserted by this theorem is called the *principal* nth root of a and is denoted by $a^{1/n}$ or $\sqrt[n]{a}$.

EXERCISES

1. Define lower bound and greatest lower bound of a set of elements of an ordered field. Prove that if a nonempty set of elements of $\mathbb{R}$ has a lower bound, then it has a greatest lower bound in $\mathbb{R}$. [Hint: Consider the additive inverses of the elements of the set.]

2. Prove that if $a, b \in \mathbb{R}$ with $a < b$, then

$$a < a + \frac{b-a}{\sqrt{2}} < b.$$

Assuming that $\sqrt{2}$ is not rational, conclude that between any two distinct rational numbers there is an irrational number.

3. Let S_1 and S_2 be nonempty sets of real numbers having, respectively, b_1 and b_2 as least upper bounds. Let $S_3 = \{s_1 + s_2 : s_1 \in S_1,\ s_2 \in S_2\}$. Prove that $b_1 + b_2$ is the LUB of S_3.

4. For any two rational numbers a, b with $0 \leq a < b$ and any positive integer n, show that there is a positive rational number r such that $a < r^n < b$. [Hint: Use Theorem 4.3 about nth roots.]

5 THE FIELD OF COMPLEX NUMBERS

The field of complex numbers contains a subfield isomorphic to the field of real numbers. A general construction of fields that contain a given field will be given in the next chapter. Our purpose here is to give a construction of the field of complex numbers starting from the real number field.

We begin with the set of ordered pairs (a, b) with $a, b \in \mathbb{R}$. Our definitions of addition and multiplication will be motivated by the formal properties of expressions of the form $a + bi$, where $i^2 = -1$. In order not to make unwarranted assumptions about the existence of such a quantity, we must first show there is a field that contains a "number" whose square is -1. Accordingly, as in the case of the construction of the rational numbers, we begin with an unfamiliar notation in order to avoid using any property until we have established it. We remind the reader that the equality of ordered pairs, $(a, b) = (c, d)$, holds if and only if $a = c$ and $b = d$.

Theorem 5.1 *Let $\mathbb{C}$ be the set of all ordered pairs (a, b) with $a, b \in \mathbb{R}$, and let addition and multiplication be defined by the rules*

$$\begin{aligned}(a, b) + (c, d) &= (a + c, b + d)\\ (a, b) \cdot (c, d) &= (ac - bd, ad + bc).\end{aligned}$$

Then $\mathbb{C}$ is a field with respect to these definitions of addition and multiplication. Moreover, the set of all elements of $\mathbb{C}$ having the form $(a, 0)$, $a \in \mathbb{R}$ is a subfield of $\mathbb{C}$ isomorphic to the real number field $\mathbb{R}$.

Proof: The required properties of addition follow immediately by observing that the rules for addition of elements in $\mathbb{C}$ are exactly the same as the rules for addition in $\mathbb{R} \oplus \mathbb{R}$. It follows that addition is associative and commutative, $(0, 0)$ is the zero, and the additive inverse of (a, b) is $(-a, -b)$.

The associative law for multiplication is a consequence of the following computations:

$$\begin{aligned}((a, b)(c, d))\,(e, f) &= (ac - bd, ad + bc)(e, f)\\ &= (ace - bde - adf - bcf, acf - bdf + ade + bce),\\ (a, b)\,((c, d)(e, f)) &= (a, b)(ce - df, cf + de)\\ &= (ace - adf - bcf - bde, acf + ade + bce - bdf),\end{aligned}$$

and these are equal elements of $\mathbb{C}$.

Next, we verify one of the distributive laws as follows: On the one hand, we have

$$\begin{aligned}(a, b)\,((c, d) + (e, f)) &= (a, b)(c + e, d + f)\\ &= (ac + ae - bd - bf, ad + af + bc + be),\end{aligned}$$

and on the other hand, we have

$$\begin{aligned}(a, b)(c, d) + (a, b)(e, f) &= (ac - bd, ad + bc) + (ae - bf, af + be)\\ &= (ac - bd + ae - bf, ad + bc + af + be).\end{aligned}$$

These are equal elements of $\mathbb{C}$ giving us one of the distributive laws. The other distributive law follows from this one as soon as we observe that multiplication is commutative. To see this, apply the definition of multiplication to get

$$(c, d)(a, b) = (ca - db, cb + da) = (a, b)(c, d).$$

We have now proved that $\mathbb{C}$ is a commutative ring. The identity is easily verified to be $(1, 0)$. To show that $\mathbb{C}$ is a field, we have yet to prove that every nonzero element has a multiplicative inverse. Suppose that $(a, b) \neq (0, 0)$. Then a and b cannot both be zero; if $a \neq 0$, then $a^2 > 0$. If $b \neq 0$, then $b^2 > 0$ and so the element $a^2 + b^2$ is positive and hence nonzero. Thus,

$$x = \left(\frac{a}{a^2 + b^2}, \frac{-b}{a^2 + b^2} \right)$$

is an element of $\mathbb{C}$ and it may be verified by direct calculation that x is the inverse of (a, b); that is, $(a, b)x = (1, 0)$. We have shown that every nonzero element has an inverse and thus $\mathbb{C}$ is a field.

To complete the proof of the theorem, let $\mathbb{R}'$ be the set of all elements of $\mathbb{C}$ of the form $(a, 0)$, $a \in \mathbb{R}$. Then the mapping $\theta : \mathbb{R}' \to \mathbb{R}$ given by $\theta(a, 0) = a$, $a \in \mathbb{R}$, is a one-to-one mapping of $\mathbb{R}'$ onto $\mathbb{R}$. Moreover, θ preserves addition and multiplication:

$$\begin{aligned} \theta[(a,0) + (b,0)] &= \theta(a + b, 0) = a + b = \theta(a, 0) + \theta(b, 0) \\ \theta[(a,0)(b,0)] &= \theta(ab, 0) = ab = \theta(a, 0)\theta(b, 0). \end{aligned}$$

Hence, θ is an isomorphism of $\mathbb{R}'$ with $\mathbb{R}$. This completes the proof. ∎

An element of the field we have constructed is called a *complex number*, and $\mathbb{C}$ is the *field of complex numbers*.

The ordered pair notation is used for purposes of the definitions but a more familiar notation is commonly used. We begin by writing a in place of $(a, 0)$ for $a \in \mathbb{R}$. In particular, the identity $(1, 0)$ is written simply as 1. Thus, we have $\mathbb{R}$ actually contained in $\mathbb{C}$ rather than an isomorphic copy of $\mathbb{R}$. We now follow historical conventions and let $i = (0, 1)$ so that

$$i^2 = (0, 1)(0, 1) = (-1, 0) = -1.$$

Notice that using this notation gives

$$(a, b) = (a, 0)(1, 0) + (b, 0)(0, 1) = a(1, 0) + b(0, 1) = a + bi.$$

The rules for multiplication permit us to multiply complex numbers using the familiar properties

$$(a + bi)(c + di) = ac + (ad + bc)i + bdi^2 = (ac - bd) + (ad + bc)i,$$

where we have substituted -1 for i^2.

We have now extended the field of real numbers to the field of complex numbers.

5.1 The Complex Field Is Not Ordered

When we extended the rational field to the real number field, we saw that the order relation defined on the rational field extended to the real field. The order property of the rational field and the real field does not extend to the complex field, however. In fact, there is no order on the complex field; that is, $\mathbb{C}$ is not an ordered field. Let us verify this. In any ordered field the square of a nonzero element is positive (relative to the given ordering). If $\mathbb{C}$ did have an order defined on it, then the square $i^2 = -1$ would be positive. However, $1^2 = 1$ is also positive and so both 1 and -1 are positive and hence their sum $1 - 1 = 0$ is positive. This is an impossible situation and so $\mathbb{C}$ cannot be an ordered field.

The fact that $\mathbb{C}$ is not ordered means that inequalities cannot be used between complex numbers. In other words, it is meaningless to speak of one complex number as being greater than or less than another complex number.

5.2 The Conjugate of a Complex Number

We make the following definition:

Definition 5.1 If $u = a + bi$ is a complex number with $a, b \in \mathbb{R}$, then the **conjugate** of u is the number $\bar{u} = a - bi$.

For example, $\overline{(1 + 7i)} = 1 - 7i$, $\bar{3} = 3$, and $\bar{i} = -i$.

The mapping $u \to \bar{u}$ of $\mathbb{C} \to \mathbb{C}$ is a one-to-one map of $\mathbb{C}$ onto $\mathbb{C}$ as is easily seen by using the observation that taking the conjugate twice leaves the element unchanged. The conjugate mapping is an isomorphism of $\mathbb{C}$ with itself; that is, $\overline{u + v} = \bar{u} + \bar{v}$ and $\overline{uv} = \bar{u}\bar{v}$. We prove this directly: Let $u = a + bi$ and $v = c + di$ with $a, b, c, d \in \mathbb{R}$. Then

$$\begin{aligned} \overline{u+v} &= \overline{(a+c) + (b+d)i} = (a+c) - (b+d)i \\ &= (a - bi) + (c - di) = \bar{u} + \bar{v}, \\ \overline{uv} &= \overline{(ac - bd) + (ad + bc)i} = (ac - bd) - (ad + bc)i \\ &= (a - bi)(c - di) = \bar{u}\bar{v}. \end{aligned}$$

An isomorphism of a field (or any ring) with itself is called an *automorphism*. We have shown that conjugation is an automorphism of $\mathbb{C}$.

The conjugate is useful in a variety of ways. For example, if $u = a + bi$ with $a, b \in \mathbb{R}$, then

$$u\bar{u} = (a + bi)(a - bi) = a^2 + b^2.$$

In particular, $u\bar{u}$ is always a real number and if $u \neq 0$, then $u\bar{u} > 0$. Also, in view of the formula derived previously,

$$u^{-1} = \frac{\bar{u}}{u\bar{u}} = \frac{a}{a^2 + b^2} - \frac{b}{a^2 + b^2}i.$$

Many additional properties of the conjugate of a complex number are given in the exercises.

EXERCISES

1. Find the multiplicative inverse on the nonzero element (a, b) of $\mathbb{C}$ by assuming r and s are real numbers such that $(a, b)(r, s) = (1, 0)$, and solving for r and s.

2. Prove each of the following properties of conjugates of complex numbers:

(a) If $u \in \mathbb{C}$, then $u\bar{u} \in \mathbb{R}$ and $u + \bar{u} \in \mathbb{R}$; moreover, if $u \neq 0$, then $u\bar{u} > 0$.

(b) If $u \in \mathbb{C}$, then $\overline{(\bar{u})} = u$.

(c) If $u \in \mathbb{C}$ and $u \neq 0$, then $(\bar{u})^{-1} = \overline{(u^{-1})}$.

(d) If $u \in \mathbb{C}$, then $u = \bar{u}$ if and only if $u \in \mathbb{R}$.

(e) If $u \in \mathbb{C}$ and n is a positive integer, then $(\bar{u})^n = \overline{(u^n)}$.

3. Let $\alpha : \mathbb{C} \to \mathbb{C}$ be the conjugation map so that $\alpha(u) = \bar{u}$ for $u \in \mathbb{C}$. Then part d of the preceding exercise shows that $\alpha(u) = u$ if and only if $u \in \mathbb{R}$. Prove that if $\phi : \mathbb{C} \to \mathbb{C}$ is an isomorphism of $\mathbb{C}$ with itself such that $\phi(u) = u$ whenever $u \in \mathbb{R}$, then either $\phi = \alpha$ or $\phi(u) = u$ for every $u \in \mathbb{C}$. [Hint: Consider the possibilities for $\phi(i)$.]

4. Let S be an arbitrary commutative ring and T the set of ordered pairs (a, b) with $a, b \in S$. Define addition and multiplication of elements of T by the rules given in Eqs. (1) and (2) of Theorem 3.11.

(a) Verify that T is a commutative ring.

(b) Verify that T has an identity if and only if S has an identity.

(c) Assume that S is an integral domain. Show that T is an integral domain if there do not exist elements a and b in S such that $a^2 = -b^2 \neq 0$.

5. Use the construction of the preceding exercise with $S = \mathbb{Z}$, the ring of integers. Show that T is an integral domain. Is T an ordered integral domain?

6 GEOMETRIC REPRESENTATION AND TRIGONOMETRIC FORM

It is implicit in our construction of the complex numbers that the correspondence $a + bi \to (a, b)$ is a one-to-one mapping of $\mathbb{C}$ onto the set of all ordered pairs of real numbers. In ordinary plane analytic geometry, the ordered pairs of real numbers correspond to points in the xy-plane. Accordingly, we may represent each point in plane by a single complex number. In other words, we associate the complex number $a + bi$ with the point (a, b) in the xy-plane. A real number, $a + 0i$, corresponds to a point $(a, 0)$ on the x-axis. A number of the form $0 + bi$, called a *pure imaginary* complex number, corresponds to a point on the y-axis. Observe that the a complex number $a+bi$ and its conjugate $a - bi$ are symmetrically located on a line perpendicular to the x-axis.

Instead of specifying points in the plane by means of rectangular coordinates, we could use polar coordinates. If P is a point with nonzero coordinate $a + bi$, the distance from P to the origin O is the real number $r = \sqrt{a^2 + b^2}$. If θ is the angle through which the positive x-axis must be rotated to coincide with the line OP from the origin to the point P, then the definition of the trigonometric functions and polar coordinates gives the relations

$$a = r\cos\theta, \qquad b = r\sin\theta.$$

It follows that the complex number can be represented in the form

$$a + bi = r(\cos\theta + i\sin\theta).$$

We have assumed $a+bi \neq 0$. In the case $a+bi = 0$ (i.e., $P = O$), then $r = 0$ and θ is undetermined; that is, θ is not determined by the point and could be any angle.

We formalize this notation in the following definition.

Definition 6.1 For a complex number $a + bi$, the **absolute value** of $a + bi$ is the nonnegative real number $r = \sqrt{a^2 + b^2}$ and is denoted by $|a + bi|$. An angle θ such that $a = r\cos\theta$ and $b = r\sin\theta$ is called **an angle** of $a + bi$. The expression

$$a + bi = r(\cos\theta + i\sin\theta)$$

is called the **trigonometric form** of $a + bi$ (Fig. 3.1).

Clearly, the absolute value of $a+bi$ is uniquely determined by the complex number. However, the angle θ is not uniquely determined. In the first case with $r = 0$ the angle is arbitrary. If $r \neq 0$ and θ_1 and θ_2 are two angles of $a + bi$, then $\theta_1 = \theta_2 + n \cdot 2\pi$ if the angles are measured in radians (or $\theta_1 = \theta_2 + n \cdot 360°$ if the angles are measured in degrees) for some integer n.

As a consequence of these observations, we point out that if r and s are positive real numbers and if

$$r(\cos\theta + i\sin\theta) = s(\cos\phi + i\sin\phi),$$

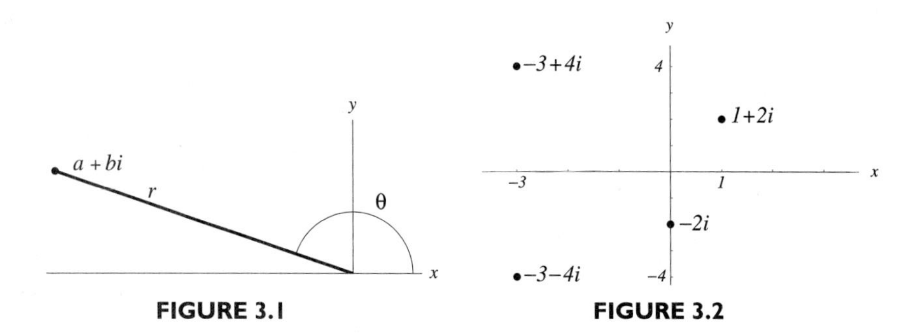

FIGURE 3.1

FIGURE 3.2

then $r = s$ and $\theta = \phi + 2\pi n$ for some integer n.

Our point of view is that the real number field $\mathbb{R}$ is contained in the complex number field $\mathbb{C}$. We have just defined the absolute value of a complex number and previously defined the absolute value of a real number. We point out here that the two definitions agree in the following sense: If a is a real number, then its absolute value (as a complex number) is $\sqrt{a^2}$. The definition of the square root symbol requires that $\sqrt{a^2} \geq 0$. Thus, if $a > 0$ we have $\sqrt{a^2} = a$. If $a < 0$, then, $\sqrt{a^2} = -a$. Thus, the definition of $|a|$ given in Definition 6.1 for a real number a agrees with this definition of $|a|$ when a is viewed as a complex number.

Now we illustrate the trigonometric form of a complex number by some examples. The complex number $-2+2i$ has absolute value $|-2+2i| = 2\sqrt{2}$ and angle $3\pi/4$ (or 135°). Thus,

$$-2 + 2i = 2\sqrt{2}(\cos 3\pi/4 + i \sin 3\pi/4).$$

Here are a few other examples which the reader may verify:

$$\begin{aligned} -4 &= 4(\cos \pi + i \sin \pi), \\ -3i &= 3(\cos \frac{3\pi}{2} + i \sin \frac{3\pi}{2}), \\ -2(\cos \theta + i \sin \theta) &= 2\big(\cos(\pi + \theta) + i \sin(\pi + \theta)\big). \end{aligned}$$

The angle of a given complex number can be found exactly in only special cases. Of course, an approximation may be found using a calculator with inverse trigonometric functions. For example, for the trigonometric form of the number $1 + 3i$, we find the absolute value is $\sqrt{10}$. However, the angle cannot be expressed exactly in radians or in degrees. If θ_1 is the positive solution of the equation $\tan \theta_1 = 3$ with $0 \leq \theta \leq \pi$, then θ is an angle for $1 + 3i$ (Fig. 3.2). We may then write

$$1 + 3i = \sqrt{10}(\cos \theta_1 + i \sin \theta_1)$$

as the trigonometric form of $1 + 3i$.

The trigonometric form of a complex number is significant largely because of the following theorem relating the trigonometric form of u, v, and uv:

Theorem 6.1

MULTIPLICATION THEOREM. *If u and v are complex numbers with trigonometric forms*

$$u = r(\cos\theta + i\sin\theta) \quad \textit{and} \quad v = s(\cos\phi + i\sin\phi),$$

then the trigonometric form of their product uv is given by

$$uv = rs\big(\cos(\theta+\phi) + i\sin(\theta+\phi)\big).$$

Otherwise expressed, $|uv| = |u| \cdot |v|$ and an angle of uv is the sum of an angle of u and an angle of v.

Proof: To establish this result, we multiply the trigonometric forms of the two numbers and then use the addition formulas of trigonometry. Thus,

$$\begin{aligned} uv &= rs(\cos\theta + i\sin\theta)(\cos\phi + i\sin\phi) \\ &= rs[(\cos\theta\cos\phi - \sin\theta\sin\phi) + i(\cos\theta\sin\phi + \sin\theta\cos\phi)] \\ &= rs[\cos(\theta+\phi) + i\sin(\theta+\phi)], \end{aligned}$$

from which the result follows immediately. ■

A special case of the preceding theorem in which $u = v$ shows at once that

$$u^2 = r^2(\cos 2\theta + i\sin 2\theta).$$

Here is an important generalization of this statement to more general powers of complex numbers:

Theorem 6.2

DEMOIVRE'S THEOREM. *If n is a positive integer and $u = r(\cos\theta + i\sin\theta)$ is the trigonometric form of the complex number u, then*

$$u^n = r^n(\cos n\theta + i\sin n\theta)$$

is the trigonometric form of u^n.

Proof: We checked the validity of the conclusion for the case $n = 2$ just before the statement of the theorem. We prove the general case by induction. Assume the conclusion is true for some positive integer k so that

$$u^k = r^k(\cos k\theta + i\sin k\theta).$$

We multiply both sides by u, using the trigonometric form to multiply on the right, and obtain

$$\begin{aligned} u^k \cdot u = u^{k+1} &= r^k \cdot r(\cos k\theta + i\sin k\theta)(\cos\theta + i\sin\theta) \\ &= r^{k+1}\big(\cos(k+1)\theta + i\sin(k+1)\theta\big) \end{aligned}$$

by Theorem 6.1. Hence, the theorem is true for all positive integers n. ■

6.1 Roots of Unity

For a given positive integer n, the complex number $\omega = \cos(2\pi/n) + i\sin(2\pi/n)$ satisfies the equation $\omega^n = 1$ as we see using DeMoivre's theorem. In fact, if k is an integer with $0 < k < n$, then

$$\omega^k = \cos\frac{2k\pi}{n} + i\sin\frac{2k\pi}{n}$$

also satisfies the equation $(\omega^k)^n = 1$. The n distinct powers

$$\omega, \omega^2, \ldots, \omega^{n-1}, \omega^n = 1$$

are called the complex nth *roots of unity* and give n distinct solutions of the equation $z^n = 1$. If we locate these n points in the plane, we see that they lie on the unit circle and form the vertices of a regular n-sided polygon. Moreover, these are the only complex numbers that satisfy the equation $z^n = 1$. We see this using the trigonometric form once again. If $z = r(\cos\theta + i\sin\theta)$ is a complex number whose nth power equals 1, then $r^n = 1$ and so $r = 1$. The angle of 1 is 0 and $n\theta$ is the angle of z^n. Thus, $n\theta$ must be an integer multiple of 2π; $n\theta = 2k\pi$ and $\theta = 2k\pi/n$ for some integer k. Thus, z is one of the powers of ω.

By using the roots of unity, we see that every nonzero complex number has exactly n nth roots; if $z^n = u$, then $(\omega^k z)^n = 1 \cdot z^n = u$ and so $\omega^k z$, $1 \leq k \leq n$ gives n solutions. Thus, we get n solutions if there is at least one solution. The complete story is given in the following theorem:

Theorem 6.3

Let $u = r(\cos\theta + i\sin\theta)$ be a nonzero complex number. Then the equation $z^n = u$ has exactly n distinct solutions given by

$$z = r^{1/n}\left(\cos\left(\frac{\theta + 2k\pi}{n}\right) + i\sin\left(\frac{\theta + 2k\pi}{n}\right)\right), \quad k = 1, 2, \ldots, n.$$

Proof: By DeMoivre's theorem, the numbers z given in the conclusion are solutions of the equation $z^n = u$. If z and y are two complex numbers having nth power equal to u, then

$$\left(\frac{z}{y}\right)^n = \frac{z^n}{y^n} = \frac{u}{u} = 1$$

and so z/y is an nth root of unity. Thus, $z/y = \omega^k$ for some k and $z = \omega^k y$. If

$$y = r^{1/n}\left(\cos\left(\frac{\theta}{n}\right) + i\sin\left(\frac{\theta}{n}\right)\right),$$

then

$$z = \omega^k y = r^{1/n}\left(\cos\left(\frac{\theta + 2k\pi}{n}\right) + i\sin\left(\frac{\theta + 2k\pi}{n}\right)\right)$$

by the multiplication theorem. ■

This theorem asserts that the equation $x^n - u = 0$ has a solution in $\mathbb{C}$ for every positive integer n and every complex number u and when $u \neq 0$ there are, in fact, n distinct solutions. A much more general statement is true about the complex numbers; namely every polynomial equation has a solution in $\mathbb{C}$. The proof of this is much deeper than the theorem about nth roots and will be given later. ■

Let us apply this last theorem to display the fifth roots of $2 + 2i$. We first write the number in trigonometric form as

$$2 + 2i = 2\sqrt{2}\left(\cos\frac{\pi}{4} + i\sin\frac{\pi}{4}\right).$$

Since $2\sqrt{2} = 2^{3/2}$, the fifth roots are the five complex numbers

$$z_k = 2^{3/10}\left(\cos\left(\frac{\pi}{4} + \frac{2k\pi}{5}\right) + i\sin\left(\frac{\pi}{4} + \frac{2k\pi}{5}\right)\right), \qquad k = 0, 1, 2, 3, 4.$$

EXERCISES

1. Express each of the following complex numbers in trigonometric form and indicate the points in a coordinate plane that have these numbers as coordinates:

   (a) $-1 - i$ (b) $-\sqrt{3} - i$ (c) $\sqrt{3} + i$
   (d) $-1 - \sqrt{3}i$ (e) -4 (f) $\cos 3\pi/8 - i\sin 3\pi/8$
   (g) $2 - 2i$ (h) $-2 - 2i$ (i) $\sin 3\pi/8 + i\cos 3\pi/8$

2. Use DeMoivre's theorem to compute each of the following and then express your answers in algebraic form by evaluating the necessary trigonometric functions:

   (a) $(1 + i)^5$ (b) $(\sqrt{3} - i)^8$ (c) $(-i)^{12}$
   (d) $(-1 + i)^{10}$ (e) $\left(-\frac{1}{2} - \frac{\sqrt{3}i}{2}\right)^6$
   (f) $(1 - \sqrt{3}i)^{11}$ (g) $\left(\frac{1}{\sqrt{2}} + \frac{i}{\sqrt{2}}\right)^{100}$
   (h) $(\cos\pi/6 + i\sin\pi/6)^{10}$ (i) $(-1 + \sqrt{3}i)^{11}$

3. Find the cube roots of 1 and express the answers in algebraic form. Draw a figure showing that these numbers are the coordinates of the vertices of an equilateral triangle,

4. Repeat the previous exercise for the fourth roots of 1 and for the eighth roots of 1.

5. Find the required roots and express them in algebraic form:
(a) The cube roots of $-2 + 2i$
(b) The cube roots of $-8i$
(c) The fourth roots of -4
(d) The sixth roots of $-i$
(e) The square roots of $-1 + \sqrt{3}i$

6. In each of the following, express the required roots in trigonometric form:
(a) The fifth roots of -1
(b) The fourth roots of $-1 + i$
(c) The square roots of $1 + i$
(d) The sixth roots of $1 - i$

7. Verify that the points with coordinates

$$(\cos \pi/6 + i \sin \pi/6)^n \qquad (n = 1, 2, 3, 4, 5, 6)$$

are the vertices of a regular hexagon inscribed in a circle of radius 1.

8. If $\bar{u}$ is the conjugate of the complex number u, verify the following:
(a) $|\bar{u}| = |u|$
(b) $u\bar{u} = |u|^2$
(c) $\bar{u} = u^{-1}$ if $|u| = 1$

9. Show that if $u \neq 0$, DeMoivre's theorem also holds for *negative* integers n.

10. Let $u, v \in \mathbb{C}$, and let P and Q be the points in the coordinate plane having respective coordinates u and v. Let R be the point with coordinate $u + v$. If O is the origin, show that OR is the diagonal of the parallelogram having OP and OQ as adjacent sides.

11. Show that if $u, v \in \mathbb{C}$, then the triangle inequality $|u + v| \leq |u| + |v|$ holds.

12. Expand $(\cos\theta + i \sin\theta)^3$ first by DeMoivre's theorem and then by the binomial expansion to get trigonometric identities for $\cos 3\theta$ and $\sin 3\theta$.

13. Expand $(\cos\theta + i \sin\theta)^4$ in two ways to get identities for $\cos 4\theta$ and $\sin 4\theta$.

14. Show that the multiplicative inverse of an nth root of 1 is also an nth root of 1.

15. If $t \in \mathbb{C}$ is an nth root of 1 but not an mth root of 1 for $0 < m < n$, then t is called a *primitive* nth root of 1. Show the following:
(a) $\omega = \cos(2\pi/n) + i \sin(2\pi/n)$ is a primitive nth root of 1.

(b) If t is any primitive nth root of 1, then $1, t, t^2, \ldots, t^{n-1}$ are all of the nth roots of 1.

(c) If t is a primitive nth root of 1, then t^k is also a primitive nth root of 1 if and only if k and n have no common factor greater than 1.

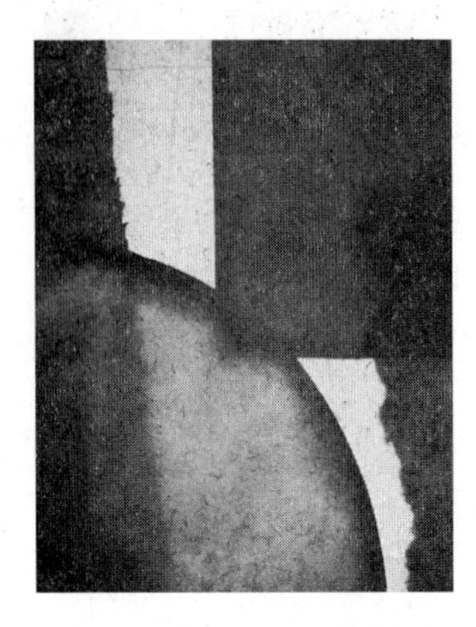

IV FACTORIZATION

The theme of the next few chapters is the factorization of elements as a product of other elements of the ring. We begin with familiar properties of factorization in the ring of integers. The most important theorem in this context is the uniqueness of the factorization of an intcgcr as a product of primes. One of the most important tools in the proof of this theorem is the division algorithm. The study leads us to consideration of finite fields as well as some other interesting number theoretic concepts.

In the next chapter we introduce the ring of polynomials and study factorization in that ring. There is a unique factorization theorem in this context and there is a division algorithm that plays an essential role in the proof. This study is important in the solution of polynomial equations. The ideas are based on those developed here for the ring of integers.

In addition to the theorems about factorization of integers and polynomials, many other important properties of these rings will be demonstrated.

1 DIVISORS IN $\mathbb{Z}$

We begin the study of factorization in the ring $\mathbb{Z}$ of integers with a definition.

Definition 1.1

If $a, d \in \mathbb{Z}$ with $d \neq 0$, we say d is a **divisor** of a (or is a **factor** of a) if there is an element $a_1 \in \mathbb{Z}$ such that $a = da_1$. If d is a divisor of a, we also say a is divisible by d or that a is a *multiple* of d. The notation $d|a$ is used to indicate that d divides a.

Many properties of divisibility follow directly from the definition:

Properties 1.1

(i) *If $d|a$ and $a|b$, then $d|b$.*

(ii) *If $d|a$, then $d|(-a)$.*

(iii) $1|a$ *for every* $a \in \mathbb{Z}$.
(iv) *Every integer divides 0.*
(v) *If* $d|1$, *then* $d = \pm 1$.
(vi) *If* $a|b$ *and* $b|a$, *then* $a = \pm b$.
(vii) *If* $d|a$ *and* $d|b$, *then* $d|ax + by$ *for all integers* x *and* y.

Proof: We prove only a few of these statements and leave the remaining ones as exercises for the reader.

(vi): If $a|b$, then $b = au$ for some integer u. If $b|a$, then $a = bv$ for some integer v. Substitute for b to conclude that $a = bv = auv$. After canceling a, we see that $uv = 1$ and so $v = \pm 1$ and $a = \pm b$.

(vii): If $d|a$ and $d|b$, then $a = du$ and $b = dv$ for some integers u and v. It follows that $ax + by = dux + dvy = d(ux + vy)$ and so d divides $ax + by$. ∎

Now we introduce the class of integers that play an important role in factorization.

Definition 1.2 A nonzero integer p that is not equal to 1 or -1 is called a **prime** if the only divisors of p are ± 1 and $\pm p$.

Note that if p is a prime, then $-p$ is also prime since p and $-p$ have the same divisors. The first few positive primes are

$$2, 3, 5, 7, 11, 13, 17, 19, 23, \ldots.$$

The list goes on indefinitely. Euclid showed that there are infinitely many primes. (See the exercises after the next section.)

If $n > 1$ is not a prime, then it must have divisors greater than 1 and smaller than n. Thus, a nonprime positive integer n has a factorization $n = uv$ with $1 < u, v < n$.

One of the principal reasons for introducing primes is that every integer other than 0, ± 1 is either a prime or a product of primes. This may seem obvious, and it is often taken for granted in arithmetic. However, the fact that every integer can be expressed in essentially only one way as a product of primes is not so obvious and is a very important fact about integers. The statement is often called the *fundamental theorem of arithmetic* and will be proved later in this chapter.

An important tool in this study is the division algorithm which was proved in Chapter II, Section 2.1. We restate it here for convenient reference:

Theorem 1.1 THE DIVISION ALGORITHM. *If* $a, b \in \mathbb{Z}$ *with* $b \geq 1$, *then there exists unique integers* q *and* r *such that*

$$a = bq + r, \qquad 0 \leq r < b.$$

1.1 Greatest Common Divisors

Given two integers a and b, an integer d that divides both a and b is called a *common divisor* of a and b. The set of common divisors of two nonzero integers a and b is finite and has a largest element. Keeping a view to more general situations in which the ring might not be ordered, we give a definition of the greatest common divisor that does not refer to the ordering of integers.

Definition 1.3 If a and b are nonzero integers, a **greatest common divisor** (GCD) of a and b is an integer d such that

(i) $d|a$ and $d|b$;

(ii) If c is any integer such that $c|a$ and $c|b$, then $c|d$.

Clearly, if d is a GCD of a and b, then so is $-d$. When dealing with the ring of integers, we will always assume that a GCD is positive. It is not obvious that a GCD exists, but we will prove that it does exist. We first remark that if a GCD exists, then it is unique. To see this suppose that d and d' are both GCDs of a and b. Then d' divides both a and b, so by (ii) we conclude $d'|d$. Reversing the roles of d and d' we may also conclude $d|d'$. It follows that $d' = \pm d$ and since we assume a GCD is positive, it follows that $d = d'$.

In the next theorem we prove the existence of the GCD of two nonzero integers. In addition the theorem shows that the GCD of a and b can be expressed in the form $d = ax + by$ for some integers x and y. The expression $ax + by$ is called a *linear combination* of a and b. The method of proof will show how to determine d and also the integers x and y. The method is called the Euclidean algorithm.

Theorem 1.2 EUCLIDEAN ALGORITHM. *Let a and b be nonzero integers. Then there exists a greatest common divisor d of a and b and d can be expressed as $d = ax + by$ for some integers x and y.*

Proof: We will use the division algorithm to define two finite sequences of integers $\{a_i\}$ and $\{q_i\}$ in the following way: Let $a_1 = a$ and $a_2 = b$. The integers q_1 and a_3 are defined by applying the division algorithm to divide a_1 by a_2 to obtain $a_1 = a_2q_1 + a_3$ with $0 \leq a_3 < |a_2|$. If $a_3 = 0$, then $a_2q_1 = a_1$. This implies $b|a$ and so $|b|$ is the GCD of a and b and the GCD exists. If $a_3 \neq 0$, then we define a series of equations:

$$\begin{array}{rcllr} a_1 &=& a_2q_1 + a_3, & 0 \leq a_3 < |a_2| & (1) \\ a_2 &=& a_3q_2 + a_4, & 0 \leq a_4 < a_3 & (2) \\ & \cdots & & \cdots\cdots\cdots & \\ a_{n-2} &=& a_{n-1}q_{n-2} + a_n, & 0 \leq a_n < a_{n-1} & (n-2) \\ a_{n-1} &=& a_nq_{n-1}. & & (n-1) \end{array} \qquad (4.1)$$

Each equation is obtained by applying the division algorithm to divide a_i by a_{i+1} provided both are nonzero. The integer n is selected so that $a_n \neq 0$, but the next remainder is equal to 0. This must happen eventually because

$$|a_2| > a_3 > \cdots > a_{n-1} > a_n \geq 0.$$

The series of positive integers cannot decrease forever, so eventually there is a 0 remainder.

We now show that a_n, the last nonzero remainder, is the GCD of a and b. The first step is to prove a_n divides both a and b. Remember that $a = a_1$ and $b = a_2$. From Eq. (4.1, $n-1$) we see that $a_n | a_{n-1}$. This implies a_n divides both $a_{n-1}q_{n-1}$ and a_n and so a_n divides their sum, which equals a_{n-2} by Eq. (4.1, $n-2$). Repeat this argument so that at some point we obtain a_n divides both $a_i q_i$ and a_{i+1}. Then a_n divides their sum which equals a_{i-1} by Eqs. (4.1, $i-1$). Eventually we reach $i = 1$ and obtain a_n divides $a_1 = a$ and $a_2 = b$. Thus, a_n is a common divisor of a and b.

Now we must show that a_n is a greatest common divisor; that is, if c is any common divisor of a and b, then $c | a_n$. So let c be an integer that divides both $a = a_1$ and $b = a_2$. From Eq. (4.1, 1) we conclude that $c | a_3$. Repeat this on successive equations. If at some point we know that c divides both a_{i-1} and a_i, then Eqs. (4.1, $i-1$) state that $a_{i-1} = a_i q_{i-1} + a_{i+1}$ and so $c | a_{i+1}$. Eventually we obtain $c | a_n$. Since $a_n > 0$ this proves a_n is the GCD of a and b.

Lastly we show that the GCD is a linear combination of a and b. We will show that each a_i, $i \geq 3$, is a linear combination of a and b. We have

$$a_3 = a_1 - a_2 q_1 = b - a q_1,$$

which shows that a_3 is a linear combination of a and b. Suppose we know that $a_{i-1} = ap + bs$ and $a_i = ap' + bs'$ are each linear combinations of a and b and we wish to show that a_{i+1} is also. From Eqs. $(i-1)$ we obtain

$$\begin{aligned} a_{i+1} &= a_{i-1} - a_i q_{i-1} = (ap + bs) - (ap' + bs')q_{i-1} \\ &= a(p - p'q_{i-1}) + b(s - s'q_{i-1}), \end{aligned}$$

which shows that a_{i+1} is also a linear combination of a and b. Eventually we reach $i + 1 = n$ to get the required property for a_n. ■

Here we give a numerical example. Find the GCD of 26 and 382 and express it as a linear combination of 26 and 382. By ordinary division we find that the Euclidean algorithm takes the form

$$\begin{aligned} 382 &= 26 \cdot 14 + 18 \\ 26 &= 18 \cdot 1 + 8 \end{aligned}$$

$$\begin{aligned} 18 &= 8 \cdot 2 + 2 \\ 8 &= 2 \cdot 4. \end{aligned}$$

The last nonzero remainder is 2 and so 2 is the GCD of 26 and 382. To express the GCD as a linear combination, it is somewhat clearer if we use symbols $a = 382$ and $b = 26$ during the computation because we want to carry out some, but not all, of the arithmetic operations. We take each of the equations just obtained and solve for the remainders and substitute from the higher equations into those below:

$$\begin{aligned} 18 &= a - 14b, \\ 8 &= b - 18 = b - (a - 14b) = 15b - a, \\ 2 &= 18 - 8 \cdot 2 = (a - 14b) - (15b - a)(2) = 3a - 44b. \end{aligned}$$

Hence, the GCD 2 is expressed as

$$2 = 3(382) - 44(26).$$

It will be convenient to have a compact notation for the GCD; we will write (a, b) to denote the GCD of a and b. Thus, for example, we have just shown that $(382, 26) = 2$. The context in which the notation is used should prevent any confusion with other possible uses of the ordered pair symbol.

There will be many times when we refer to a pair of integers whose GCD is 1. Accordingly, we make the following definition:

Definition 1.4 The nonzero integers a and b are said to be **relatively prime** if their GCD is 1; that is, if $(a, b) = 1$.

EXERCISES

1. Prove the statements in Properties 1.6 that were not proved in the text.
2. Show that if $x = y + z$ and if d is a divisor of two of the three integers x, y, z, then d is a divisor of the third.
3. Show that if p and q are positive prime integers and $p|q$, then $p = q$.
4. For each pair of integers a and b, find the quotient and remainder in the division of a by b:

   (a) $a = 1251, \quad b = 78$ (b) $a = 2357, \quad b = 99$

   (c) $a = 31, \quad b = 158$ (d) $a = -168, \quad b = 15$
5. If the GCD of a and b is 2, show that the GCD of $3a$ and $3b$ is 6.
6. Express the GCD as a linear combination of a and b:

   (a) $a = 52$ and $b = 28$ (b) $a = 320$ and $b = 112$

   (c) $a = 81$ and $b = 75$ (d) $a = 7469$ and $b = 2387$

7. Show that a and b are relatively prime if and only if 1 is a linear combination of a and b.

8. If $d = (a, b)$ and $a = da_1$ and $b = db_1$, show that a_1 and b_1 are relatively prime integers.

9. If m is a positive integer, show that $(ma, mb) = m(a, b)$.

10. If $x = yn + t$, prove that $(x, n) = (t, n)$. Express this property in words using the language of congruence modulo n and equivalence classes.

11. Prove that $(a, bc) = 1$ if and only if $(a, b) = 1$ and $(a, c) = 1$.

12. Let a, b, n be given integers. Prove that n is expressible as a linear combination of a and b if and only if $(a, b)|n$.

13. Define the GCD of three integers a, b, c and then prove it is equal to $((a, b), c)$.

14. Prove the following variant of the division algorithm: If $a, b \in \mathbb{Z}$ with $b \geq 1$, then there exists unique integers q and r such that

$$a = bq + r, \qquad -b/2 < r \leq b/2.$$

[When this variant of the division algorithm is used to find the GCD of two integers, it is called the **extended Euclidean algorithm.**]

15. Use the variant of the division algoithm given in the previous exercise to determine the gcd of the pairs in Exercise 6. This extended division algorithm may make the computation a bit easier compared to the conventional division algorithm because numerically smaller remainders arise.

2 THE FUNDAMENTAL THEOREM IN $\mathbb{Z}$

The principal theorem to be proved in this section has to do with the factorization of an integer into a product of primes. We begin with the following important property of primes called Euclid's lemma:

Lemma 2.1

EUCLID'S LEMMA. *If a and b are nonzero integers and p is a prime integer such that $p|ab$, then either $p|a$ or $p|b$.*

Proof: If $p|a$, then we are finished. So we suppose that p does not divide a and we must prove that $p|b$. Since p does not divide a, the definition of a prime implies that the only common divisors of a and p are ± 1. Thus, $(a, p) = 1$. Then Theorem 1.2 shows that there exist integers x and y such that $ax + py = 1$. Multiply by b to obtain

$$b = abx + pby.$$

The hypothesis implies that $p|ab$ and visibly $p|bpy$ and so p divides the right side of the equation and thus $p|b$ as we wished to prove. ∎

It is almost obvious that Euclid's lemma can be extended to products of more than two integers. We state this more general result for future reference:

Lemma 2.2 *Let p be a prime integer. If $a_1, \ldots, a_n$ are nonzero integers such that $p|(a_1 a_2 \cdots a_n)$, then $p|a_i$ for at least one i, $1 \leq i \leq n$.*

This lemma is easily established by induction and the proof is left as an exercise. The case $n = 2$ is covered by the preceding lemma.

It is easy to verify that each positive integer can be expressed as a product of positive prime integers. (In practice this might not be so easy if the integer is very large.) For example, $60 = 2 \cdot 2 \cdot 3 \cdot 5$, and so 60 is expressed as a product of positive prime integers. We could also write $60 = 2 \cdot 5 \cdot 3 \cdot 2$, but we do not consider the two factorizations as essentially different since they differ only in the order in which the primes are written. This factorization of 60 is a special case of the fundamental theorem of arithmetic which states that every positive integer can be expressed as a product of positive primes in essentially only one way.

Theorem 2.1 FUNDAMENTAL THEOREM OF ARITHMETIC. *Every integer $a > 1$ can be expressed as a product of positive primes in one and only one way, except for the order in which the factors are written.*

In the statement of the theorem it is to be understood that, as a special case, a "product" of primes may consist of a single prime. This takes care of the case in which a is itself a prime.

Proof: First we prove that every positive integer $z > 1$ can be expressed as a product of primes in at least one way. Let K be the set of all positive integers greater than 1 that *cannot* be so expressed. Our goal is to show K is the empty set. If K is not empty, then it contains a smallest element, call it c. Then c is the smallest positive integer that cannot be expressed as a product of primes. In particular, c is not itself a prime. Thus, $c = c_1 c_2$ with $1 < c_1 < c$ and $1 < c_2 < c$. Then neither c_1 nor c_2 lies in K since each is smaller than the smallest element of K. Thus, c_1 and c_2 *can* be expressed as a product of primes. However, then their product $c_1 c_2 = c$ can also be expressed as a product of primes. This is in conflict with the definition of c, however, and the supposition that K is not empty cannot stand. Thus, every integer $a > 1$ is a product of positive primes in at least one way.

Now we discuss the *uniqueness* of the factorization of an integer as a product of primes. Let L be the set of integers greater than 1 that can be expressed as a product of primes in two essentially different ways. Our goal is to show that L is empty. Suppose L is not empty. Then there is a smallest integer in L, call it m. By definition of L there exist two factorizations

$$m = p_1 p_2 \cdots p_r = q_1 q_2 \cdots q_s$$

into a product of positive primes that are essentially different. Since p_1 is a prime that divides m, it divides the right side $q_1 q_2 \cdots q_s$. By the extended form of Euclid's lemma, $p_1 | q_i$ for some i. However, q_i is also a positive prime and so $p_1 = q_i$. For simplicity of notation, let us assume that $i = 1$ (renumber the q_j to achieve this). Then we may cancel a common factor p_1 from each side to obtain

$$\frac{m}{p_1} = p_2 \cdots p_r = q_2 \cdots q_s.$$

However, $m/p_1 < m$ and m was the smallest integer with two essentially different factorizations. Thus, m/p_1 has essentially only one factorization. This means that we may renumber the q_j to achieve $p_j = q_j$ for $j = 2, 3, \ldots r$. At the same time we observe that $r = s$. Now, however, since we had already noted that $p_1 = q_1$, we discover that the two factorizations of m were essentially the same after all. Hence, L is the empty set. ∎

2.1 Standard Form

Of course, the primes occurring in a factorization of an integer into prime factors need not all be distinct. By collecting equal primes we see that every integer $a > 1$ can be uniquely expressed in the form

$$a = p_1^{n_1} p_2^{n_2} \cdots p_k^{n_k}, \tag{4.2}$$

where the p_j are distinct positive primes and each exponent n_j is a positive integer. We call this expression the *standard form* of the positive integer a. For example, the standard form of 60 is $2^2 \cdot 3 \cdot 5$.

Throughout this section we have considered factorization of positive integers only. However, this is not an essential restriction. If $a < -1$ is a negative integer, then $-a > 1$ and the fundamental theorem applies to ensure a factorization of $-a$ as a product of primes. It follows that a is uniquely expressible as (-1) times a product of positive primes. For example, $-60 = (-1) \cdot 2^2 \cdot 3 \cdot 5$.

EXERCISES

1. If we discard the assumption that the primes in the factorization of a must be positive (but keep the assumption that changing the order of the factors produces essentially the same factorization), give all the possible factorizations of (a) $a = 30$, (b) $a = 210$.
2. If p is a prime that divides a^2, for some integer a, prove that $p|a$.
3. If $p_1, p_2, \ldots, p_n$ are primes, show that the integer $1 + (p_1 p_2 \cdots p_n)$ is not divisible by any of the p_j.
4. Prove that there are infinitely many positive prime integers. [Hint: Use the previous exercise and consider the possibility that $p_1, p_2, \ldots, p_n$ is the set of all primes.]

5. Let n be any positive integer. Show that every prime divisor of $1 + (n!)$ is greater than n. [The factorial notation $n!$ was defined in Section 1.6.]

6. List the 25 positive primes less than 100. For the ambitious reader we mention that there are 95 positive primes less than 500. For the reader with a computer, we mention that there are 669 positive primes less than 5000.

3 APPLICATIONS OF THE FUNDAMENTAL THEOREM

If a and c are positive integers and c is a divisor of a, then $a = cd$ for some positive integer d. If c and d are expressed as products of prime factors, then clearly a is a product of all the prime factors of c times all prime factors of d. This gives the standard form for a and it follows that the only possible prime factors of c (or of d) are the primes that are factors of a. If the standard form of a is given in Eq. (4.2), then any divisor c of a must have the form

$$c = p_1^{m_1} p_2^{m_2} \cdots p_k^{m_k},$$

where the m_i are nonnegative integers satisfying $0 \leq m_i \leq n_i$, for $i = 1, 2, \ldots, k$. Conversely, any integer c of this form is a divisor of a.

If two integers a and b are expressed in standard form, it is easy to determine the GCD $d = (a, b)$. For each prime divisor p of a, let p^n be the exact power of p dividing a and let p^m be the exact power of p dividing b. Then the exact power of p dividing d is p^t, where t is the smaller of m and n. For example, if

$$a - 2^4 \cdot 3^2 \cdot 5^8 \cdot 7 \qquad b = 2^3 \cdot 3^6 \cdot 7^2 \cdot 13,$$

then the GCD of a and b is

$$d = 2^3 \cdot 3^2 \cdot 7.$$

Previously we showed how to find the GCD of two integers using the Euclidean algorithm; this method does not require finding prime factors of the number but only requires long division. From a computational standpoint, the method using the Euclidean algorithm may involve less work than the one just described. The problem is that it may be very difficult to find the prime factors of an integer if the integer is fairly large. The first exercise following this section is designed to convince you of this. In fact, the difficulty in finding the standard form of very large numbers is the basis for security in the electronic transmission of data. A very brief indication of this is given later in this chapter.

We conclude this section by giving a formal proof that certain equations do not have solutions in the field of rational numbers. The fundamental theorem of arithmetic plays an important part in the proof.

Theorem 3.1 *Let n be an integer with $n > 1$ and let m be any positive integer. If r is a rational number satisfying $r^n = m$, then r is an integer and so m is the nth power of an integer.*

Proof: Suppose a/b is a rational number which satisfies $(a/b)^n = m$. We may assume that a and b are positive integers having GCD equal to 1. If this were not the case, we could write a and b in standard form and then cancel any prime factors common to both, without changing the rational number a/b. Then we have

$$a^n = mb^n. \tag{4.3}$$

Our goal is to show that $b = 1$. If $b \neq 1$, then b is divisible by some prime p. Then Eq. (4.3) implies that p also divides a^n. By the fundamental theorem we reason that a and a^n have exactly the same prime divisors; only the powers of the primes are different in a and a^n. Let us make this argument explicit: If the standard form of a is

$$a = p_1^{e_1} p_2^{e_2} \cdots p_k^{e_k},$$

then the standard form of a^n is

$$a^n = p_1^{ne_1} p_2^{ne_2} \cdots p_k^{ne_k}.$$

Thus the divisor p of b also divides a; this conflicts with our choice that the GCD of a and b equals 1. It follows that $b = 1$ and m is the nth power of an integer. ■

Here is an application of this result: Let p be a prime, for example, $p = 2$. Then the real number $\sqrt{p}$ is not a rational number. The number $\sqrt{p}$ is a solution of the equation $x^2 - p = 0$. The theorem tells us that any rational solution must be an integer, but clearly p is not the square of any integer because its only positive divisors are 1 and p, whereas a square of an integer > 1 has, at least, the positive divisors $1, q, q^2$ for some prime q. Clearly p could be any integer that is not a square of an integer to conclude that $\sqrt{p}$ is not rational.

More general results along these lines are provided later.

EXERCISES

1. Express each integer a, b listed below in standard form and then use each to find the GCD of a and b. Compare the amount of calculation using this method with the amount required to find the GCD using the Euclidean algorithm.

a	120	970	53,599
b	4851	3201	133,331

2. (a) Use a method of proof similar to that used in the proof of Euclid's lemma to show that if a is a divisor of bc and if $(a, b) = 1$, then a divides c. (b) Prove the same result using the fundamental theorem.
3. Show that a positive integer $a > 1$ is the square of an integer if and only if all the exponents in the standard form of a are even integers.
4. Show that if b and c are positive integers such that bc is a square of an integer and if $(b, c) = 1$, then both b and c are squares of integers.
5. For a positive integer n, let $d(n)$ denote the number of positive integers that divide n.
   (a) If p is a prime number, show $d(p^k) = k + 1$.
   (b) If m and n are positive integers with a GCD equal 1, show $d(mn) = d(m)d(n)$. [Hint: Show that a divisor of mn can be uniquely expressed as a product ab with $a|m$ and $b|n$.]
   (c) If p, q are primes, show $d(p^k q^h) = (k + 1)(h + 1)$.
   (d) Generalize the previous part to find a general formula for $d(n)$ that can be determined from the standard form of n.

4 APPLICATIONS TO THE INTEGERS MODULO n

The fundamental theorem of arithmetic as well as information about greatest common divisors proved in the first few sections of this chapter can be interpreted to give useful information about the ring of integers modulo n. We begin by determining the n for which $\mathbb{Z}_n$ is a field.

Theorem 4.1 *Let $n > 1$ be an integer. Then $\mathbb{Z}_n$ is an integral domain if and only if n is prime. When $\mathbb{Z}_n$ is an integral domain, it is a field.*

Proof: Suppose that n is not a prime. Then there exist integers b and c with $n = bc$ and $1 < b, c < n$. In $\mathbb{Z}_n$ we have

$$[b][c] = [bc] = [n] = [0].$$

However, $[b] \neq [0]$ and $[c] \neq [0]$ because n does not divide either b or c. Thus, we have found two nonzero elements of $\mathbb{Z}_n$ with a product equal to zero. This shows that $\mathbb{Z}_n$ is not an integral domain when n is not prime.

Now suppose n is a prime integer; we show that $\mathbb{Z}_n$ is an integral domain. In fact, we will prove that $\mathbb{Z}_n$ is a field. To do this we must prove that every nonzero element has a multiplicative inverse. Suppose $[a] \neq [0]$. Then n does not divide the integer a and, since n is prime, the GCD of a and n is 1. By Theorem 1.2, there exist integers x and y such that $ax + ny = 1$; thus,

$$[1] = [ax + ny] = [ax] + [ny] = [a][x]$$

since $[ny] = [0]$. This proves that $[x]$ is the inverse of $[a]$. ■

This construction produces a new class of fields entirely different from the fields of rational, real, and complex numbers that appeared earlier in the text. We will study further properties of these finite fields in later chapters.

4.1 Units

The main step used to prove that $\mathbb{Z}_n$ is a field when n is prime can be applied to determine the elements of $\mathbb{Z}_n$ that have a multiplicative inverse when n is not a prime. An element of $\mathbb{Z}_n$ that has an inverse is called a *unit* and the set of all units in $\mathbb{Z}_n$ is denoted by $U(\mathbb{Z}_n)$.

Theorem 4.2 *The element $[u]$ of $\mathbb{Z}_n$ is a unit if and only if the GCD of u and n equals 1.*

Proof: Suppose that $[u]$ is a unit of $\mathbb{Z}_n$ and that $[v]$ is its inverse. Then $[u][v] = [1]$ and so there exists an integer x such that $uv = 1 + nx$. The GCD of u and n must divide $uv - nx = 1$ and so $(u, n) = 1$. Conversely, if $(u, n) = 1$, then by Theorem 1.2 there exist integers x and y with $ux + ny = 1$. Thus, $[u][x] = [1]$ and $[u]$ is a unit of $\mathbb{Z}_n$. ∎

4.2 Euler's Phi Function

For each positive integer n let $\varphi(n)$ denote the number of elements in the set $U(\mathbb{Z}_n)$ of units of $\mathbb{Z}_n$. In view of the last theorem, $\varphi(n)$ is the number of integers k which satisfy $0 < k < n$ and $(k, n) = 1$, provided $n > 1$. For $n = 1$, we have $\varphi(1) = 1$. This function occurs frequently in the study of elementary number theory; it is called the *Euler Phi Function*. We have just proved that for a prime p, $\varphi(p) = p - 1$. At the end of this section we give a convenient formula for $\varphi(n)$ in terms of the standard form of n. The following is an example of how this function can be used to prove a result about the divisibility of integers:

Theorem 4.3 EULER'S THEOREM. *If n is a positive integer and a is any nonzero integer with $(a, n) = 1$, then $a^{\varphi(n)} - 1$ is divisible by n.*

Proof: Let $s = \varphi(n)$. Notice that the conclusion of the theorem is equivalent to the statement that $[a]^s = [1]$ in $\mathbb{Z}_n$ and we will prove the theorem in this form. Since $(a, n) = 1$, it follows that $[a] \in U(\mathbb{Z}_n)$. Let $[a_1], \ldots, [a_s]$ be all the elements of $U(\mathbb{Z}_n)$. We know that the product of two units is again a unit, so in particular the elements $[aa_j] = [a][a_j]$, $(1 \le j \le s)$ are also units. Moreover, these products are distinct because

$$[a][a_i] = [a][a_j]$$

implies that $[a_i] = [a_j]$, since the $[a]$ can be canceled. It follows that the two sets

$$\{[a_1], \ldots, [a_s]\} \quad \text{and} \quad \{[aa_1], \ldots, [aa_s]\}$$

both equal $U(\mathbb{Z}_n)$ and contain exactly the same elements. Now let $[r]$ be the product of all the elements in $U(\mathbb{Z}_n)$. We compute $[r]$ in two ways:

$$[r] = [a_1][a_2]\cdots[a_s]$$

and

$$\begin{aligned}[r] &= [aa_1][aa_2]\cdots[aa_s] = [a]^s[a_1][a_2]\cdots[a_s] \\ &= [a]^s[r].\end{aligned}$$

Since $[r]$ is a product of units, $[r]$ is a unit and it may be canceled in the last equation to give $[a]^s = [1]$, which is equivalent to the conclusion of the theorem. ■

In the case in which n is a prime, the statement of Euler's theorem is somewhat simplified and is attributed to Pierre Fermat.

Theorem 4.4 FERMAT'S LITTLE THEOREM. *Let p be a positive prime integer and a any integer not divisible by p. Then $a^{p-1} - 1$ is divisible by p.*

Example 4.1 Let $n = 15$ so that $\varphi(15)$ counts the number of integers on the interval $1 \le k \le 14$ that are not divisible by 3 or 5. The units of $\mathbb{Z}_{15}$ are the classes containing $1, 2, 4, 7, 8, 11, 13$, or 14. Thus $\varphi(15) = 8$. It follows from Euler's theorem that $2^8 - 1$ is divisible by 15. This is easily verified by hand computation: $2^8 - 1 = 256 - 1 = 255 = 15 \cdot 17$.

Example 4.2 Let $p = 101$ and note that p is a prime. Then $2^{100} - 1$ is divisible by 101 according to Fermat's Little Theorem. It would take quite a while to verify this by hand computation!

In an earlier section we showed that $\mathbb{Z}_6$ is isomorphic to $\mathbb{Z}_2 \oplus \mathbb{Z}_3$. We can now prove a more general statement in this spirit.

Theorem 4.5 *Let m and n be positive integers with $(m,n) = 1$. There is a well-defined map*

$$\theta : [a]_{mn} \to ([a]_m, [a]_n) \tag{1}$$

that gives an isomorphism between $\mathbb{Z}_{mn}$ and $\mathbb{Z}_m \oplus \mathbb{Z}_n$.

Proof: The first step is to show that the function θ given by Eq. (1) is well defined. This follows from the equation

$$([a + mnt]_m, [a + mnt]_n) = ([a]_m, [a]_n).$$

It is then straightforward to verify that θ preserves addition and multiplication so that θ is a homomorphism. Next we show that it is one-to-one. Suppose $\theta[a]_{mn} = \theta[b]_{mn}$. Then

$$([a]_m, [a]_n) = ([b]_m, [b]_n),$$

which implies $a \equiv b \pmod{m}$ and $a \equiv b \pmod{n}$. Thus, $a - b$ is divisible by both m and n. Now we use the fact that $(m, n) = 1$ to conclude $a - b$ is divisible by mn. To see this, write the integers $a - b$, m, and n in standard form; say m is a product of certain prime powers $p_j^{e_j}$ and n is a product of prime powers $q_i^{c_i}$. Because $(m, n) = 1$, no p_j can equal any q_i. Since $a - b$ is divisible by m, every $p_j^{e_j}$ divides the standard form of $a - b$; similarly, every $q_i^{c_i}$ divides the standard form of $a - b$. The important point is that there is no duplication and so mn divides $a - b$. That is, $[a]_{mn} = [b]_{mn}$, and this shows θ is one-to-one. Since $\mathbb{Z}_{mn}$ and $\mathbb{Z}_m \oplus \mathbb{Z}_n$ have the same number of elements, the one-to-one map must be onto. Thus, θ is an isomorphism. ∎

There is a nice interpretation of this isomorphism for the Euler Phi Function.

Corollary 4.1 *Let φ denote the Euler Phi Function and let m and n be positive integers with $(m, n) = 1$. Then $\varphi(mn) = \varphi(m)\varphi(n)$.*

Proof: Under the isomorphism $\theta : \mathbb{Z}_{mn} \to \mathbb{Z}_m \oplus \mathbb{Z}_n$, units are mapped to units and so the number of units in $\mathbb{Z}_{mn}$, namely, $\varphi(mn)$, equals the number of units in $\mathbb{Z}_m \oplus \mathbb{Z}_n$. An element $(x, y) \in \mathbb{Z}_m \oplus \mathbb{Z}_n$ is a unit if and only if x is a unit of $\mathbb{Z}_m$ and y is a unit of $\mathbb{Z}_n$ [as is easily seen by writing the inverse of (x, y) as (u, v) and multiplying]. Thus, the number of units in $\mathbb{Z}_m \oplus \mathbb{Z}_n$ equals the product of the number of units in the two rings, namely, $\varphi(m)\varphi(n)$. Thus, $\varphi(mn) = \varphi(m)\varphi(n)$ as required. ∎

An extension of this corollary to the case of more than two factors can be proved by induction: If $m_1, \ldots, m_k$ are integers with $(m_i, m_j) = 1$ for $1 \leq i < j \leq k$, then

$$\varphi(m_1 m_2 \cdots m_k) = \varphi(m_1)\varphi(m_2) \cdots \varphi(m_k).$$

This property of the Euler Phi Function permits a simplification in its computation. If n has the standard form

$$n = p_1^{e_1} \cdots p_k^{e_k},$$

then

$$\varphi(n) = \varphi(p_1^{e_1} \cdots p_k^{e_k}) = \varphi(p_1^{e_1})\varphi(p_2^{e_2}) \cdots \varphi(p_k^{e_k}).$$

Next we evaluate $\varphi(p^e)$ for a prime p. This is quite easy because the numbers relatively prime to p^e are those not divisible by p. The only numbers k on the interval $1 \leq k \leq p^e$ that are divisible by p are the multiples $k = rp$ with $1 \leq r \leq p^{e-1}$. If we exclude these p^{e-1} numbers from the interval, all the remaining numbers are relatively prime to p^e. Hence,

$$\varphi(p^e) = p^e - p^{e-1} = p^{e-1}(p - 1).$$

We have proved the formula

$$\varphi(p_1^{e_1} \cdots p_k^{e_k}) = p_1^{e_1-1} \cdots p_k^{e_k-1}(p_1 - 1) \cdots (p_k - 1).$$

Example 4.3 The easy computation made earlier is even easier now:

$$\varphi(15) = \varphi(3)\varphi(5) = (3-1)(5-1) = 8.$$

Example 4.4 How many numbers between 1 and 100 are relatively prime to 100? Answer:

$$\varphi(100) = \varphi(2^2 \cdot 5^2) = 2 \cdot 5 \cdot 1 \cdot 4 = 40.$$

4.3 Primality Testing

Euler's theorem can sometimes be used to help determine if a given integer n is a prime. If n is prime, then $\varphi(n) = n - 1$ and so Euler's theorem implies

$$a^{n-1} \equiv 1 \pmod{n},$$

provided that a and n are relatively prime. If we happen to find some integer a that is relatively prime to n such that a^{n-1} is *not* congruent to 1 modulo n, then we may conclude that n must not be prime.

For example, let $n = 713$ and $a = 2$. The a and n are relatively prime and

$$2^{712} \equiv 624 \pmod{713}.$$

Thus we conclude that 713 is not prime. For some comments on the computation of 2^{712} modulo 713 and some reasons why one might want to determine the primality of an integer, see the next section.

EXERCISES

1. Let p be a prime and a and b elements of $\mathbb{Z}_p$. Show that the equation $ax = b$ has a solution $x \in \mathbb{Z}_p$ if and only if $a \neq 0$. [This can be restated in the language of congruences as follows: Let p be a prime and a and b integers. Show that the congruence $ax \equiv b \pmod{p}$ has a solution x if p does not divide a.]

2. Find the integers x that satisfy the following congruences:
(a) $3x \equiv 2 \pmod{7}$
(b) $12x \equiv 38 \pmod{5}$
(c) $12x \equiv 1 \pmod{13}$
(d) $6x \equiv 10 \pmod{17}$

3. What is the remainder after division of 2^{1000} by 13? [Hint: By Fermat's Little Theorem, $2^{12 \cdot k} \equiv 1 \pmod{13}$ so divide 1000 by 12 and use the remainder to make a hand computation that is manageable.]

4. What is the remainder after division of 3^{1001} by 99?

5. What are the two right-most digits of the number 7^{82} when written in the usual way in base 10?

6. Let $a \in \mathbb{Z}_p$ with p a prime integer. If $a \neq 0$, show $a^{-1} = a^{p-2}$.

7. Show that if n is a positive integer and $a \in U(\mathbb{Z}_n)$, then $a^{-1} = a^{\varphi(n)-1}$.

8. Prove that every unit of $\mathbb{Z}_7$ is a power of $[5]$.

9. Compute the product of all the units of $\mathbb{Z}_n$ (as in the proof of Euler's theorem) for the cases $n = 3, 5, 7, 11$.

10. For a prime p, use the facts that (i) $\mathbb{Z}_p$ is a field, (ii) every nonzero element of $\mathbb{Z}_p$ has an inverse, and (iii) at most two elements of $\mathbb{Z}_p$ are equal to their inverses to prove *Wilson's theorem:*

$$(p-1)! \equiv -1 \pmod p.$$

[Here we use the factorial notation $n! = 1 \cdot 2 \cdot 3 \cdots (n-1) \cdot n$.]

11. Let m and n be relatively prime positive integers and let x and y be integers such that $mx + ny = 1$. Show that there is a well-defined map $\gamma : \mathbb{Z}_m \oplus \mathbb{Z}_n \to \mathbb{Z}_{mn}$ which satisfies $\gamma([a]_m, [b]_n) = [any + bmx]_{mn}$. Then show that γ is an isomorphism and it is the inverse of the map $\theta : \mathbb{Z}_{mn} \to \mathbb{Z}_m \oplus \mathbb{Z}_n$ defined in the text.

12. Let m and n be positive integers with $d = (m, n) \neq 1$. Show that the there is a homomorphism $\theta : \mathbb{Z}_{mn} \to \mathbb{Z}_m \oplus \mathbb{Z}_n$ as defined in the proof of Theorem 4.5 but θ is neither one-to-one nor onto.

13. An element a of a ring R is called *idempotent* if $a^2 = a$. If m and n are relatively prime positive integers each greater than 1, show that $\mathbb{Z}_{mn}$ has at least two idempotent elements other than 0 and 1. [Hint: If $mx + ny = 1$, then consider $[mx]$.]

14. If p is a prime integer and n a positive integer, show that the ring $\mathbb{Z}_{p^n}$ has no idempotent elements other than $[0]$ and $[1]$. [Hint: A proof might use the fact that integers a and $a - 1$ are relatively prime.]

15. Let $m_1, \ldots, m_k$ be positive integers with $(m_i, m_j) = 1$ if $i \neq j$. Show that there is an isomorphism from $\mathbb{Z}_{m_1 m_2 \cdots m_k}$ to the direct sum of the rings $\mathbb{Z}_{m_i}$, $1 \leq i \leq k$.

16. Prove the *Chinese Remainder Theorem*: If $m_1, \ldots, m_k$ are positive integers with $(m_i, m_j) = 1$ if $i \neq j$ and if $b_1, \ldots, b_k$ are any given integers, then there is an integer x that satisfies the system of congruences

$$x \equiv b_1 (\text{mod } m_1), \quad x \equiv b_2 (\text{mod } m_2), \quad \ldots, \quad x \equiv b_k (\text{mod } m_k).$$

Moreover, any two integers that satisfy the system are congruent modulo the product of the m_i. [Hint: Use the previous exercise.]

17. Find an integer x (guaranteed to exist by the previous exercise) that satisfies the system of congruences:

(a) $x \equiv 1 \pmod 4, \quad x \equiv 3 \pmod 9$

(b) $x \equiv 1 \pmod 4, \quad x \equiv 3 \pmod 9, \quad x \equiv 4 \pmod{11}$

(c) $x \equiv 2 \pmod 8, \quad x \equiv -1 \pmod{15}$

18. There are many integers p such that both p and $p+2$ are prime numbers. For example, $p = 5$ and $p + 2 = 7$ or $p = 29$ and $p + 2 = 31$. It is unknown whether the number of such p is finite or infinite. This is called the twin prime problem. A consideration of triples of primes is easier to resolve. Prove that if p, $p + 2$, and $p + 4$ are prime numbers, then $p = 3$. [Hint: You might want to read these numbers modulo 3 and then use the observation that for any positive integer n, one of the numbers n, $n + 1$, and $n + 2$ is divisible by 3. Prove this first.]

5 RSA CRYPTOGRAPHY

Elementary number theory enters into the world of electronic communications in many ways, one being the idea of public key cryptography. We give a very brief indication of this idea here.

There are many occasions in which information must be transmitted from a sender to a receiver and the data should be kept secret from other potential receivers. Consider, for example, the situation in which the reader has made a purchase from an on-line distributor during which a credit card number was sent from the purchaser (sender) to the distributor (receiver). The credit card information is intended to be kept secret from any hacker (unauthorized receiver) who might be observing the data transmission. Another example comes from the banking industry in which there is daily transmission of financial information regarding transactions on accounts between financial institutions; this information is intended to be kept private. In all such instances, data are encrypted (or encoded) so that an unauthorized reader will not be able to interpret the data. There are many ways of encoding data so that they are sent in an unreadable form. Of course, data must be sent in such a way that the receiver will be able to decipher the information, i.e., decode the transmission.

One of the methods used to encode and decode data that is mathematically interesting is called the RSA Public Key System. This is an idea developed by R. Rivest, A. Shamir, and L. Adleman in 1977. The term "public key" means that the method of encoding the message is public information and not kept secret, but the method of decoding is known only to authorized individuals. This is a practical idea; for example, a central bank may have hundreds of branch locations with thousands of accounts on which reports are sent daily to a central location. Each sender knows how to encode the required data for transmission but only certain individuals at the receiving end know how to decode the information. In order for this to be successful, it must not be possible to determine the decoding method from the encoding method. This may seem implausible at first, but we will explain. First, we do some number theory based on ideas in this chapter.

Select two prime numbers p and q and let $n = pq$. Then $k = \varphi(n) = (p-1)(q-1)$ and we select any integer d that is relatively prime to k. The congruence $dx \equiv 1 \pmod{k}$ has a solution $x = e$ and we let e be a number such that $de \equiv 1 \pmod{k}$. In practice we select both d and e between 1 and k. Then we know the following facts:

(i) $de = kt + 1$ for some integer t.

(ii) For any integer z that is relatively prime to n, there is a congruence

$$z^{\varphi(n)} = z^k \equiv 1 \pmod{n} \quad \text{(by Euler's theorem)}$$

and so

$$z^{de} \equiv z^{kt+1} \equiv (z^k)^t z \equiv 1^t z \equiv z \pmod{n}.$$

(iii) For any integer z and any positive integer a, there is a unique integer t on the interval $0 \leq t \leq n-1$ such that $z^a \equiv t \pmod{n}$. We call t the **least positive residue** of the number z^a.

5.1 Encoding Scheme

Now we discuss the encoding scheme. Letters are to be represented as numbers; here we assign a two-digit number to each letter. In practice, other two-digit numbers represent spaces, digits, periods, commas, and other punctuation, but for simplicity we will not deal with this here. There is flexibility in doing this but there is no harm, as we will see, in using the obvious scheme for letters

$$A = 01, \ B = 02, \ \ldots, Y = 25, \ Z = 26.$$

Thus, the message SEND ME MONEY is translated into the numbers 19051404 1305 1315140525.

There would be very little security in simply transmitting these numbers. This first encryption is a simple substitution code with the letter A replaced by 01, etc. Any amateur code breaker can decipher a substitution code, given enough samples. Therefore, the next step in the encoding process is to modify these numbers. We must scramble the message so that it cannot be deciphered by an unauthorized observer without the knowledge of the decoding algorithm. We use the integer e (e for encode) and the integer n as follows. Form the least positive residues r_1, r_2, r_3 modulo n of the numbers

$$(19051404)^e \equiv r_1, \quad (1305)^e \equiv r_2, \quad (1315140525)^e \equiv r_3 \pmod{n}.$$

The numbers r_1, r_2, and r_3 are transmitted to the authorized receiver. The receiver must decode the message. Here, the method of decoding is quite easy. Form the least positive residue of the three numbers r_1^d, r_2^d, and r_3^d. Note that

$$r_1^d \equiv ((19051404)^e)^d \equiv (19051404)^{ed} \equiv (19051404) \pmod{n}.$$

Similarly,

$$r_2^d \equiv 1305, \quad r_3^d \equiv 1315140525 \pmod{n}.$$

Thus, the original message is recovered. There is one minor requirement. The original message numbers must equal their own least positive residues modulo n. In other words, we must agree to use words in a message that are numbers between 0 and $n - 1$. By taking n large and keeping words "short," relative to n, this is not a problem. Therefore, in our example, n must be at least 131510526. There are more important reasons for using large n.

In the implementation of this method, the integers n and e are made public so that anyone may encode a message. The primes p and q and the integer d are not made public so the decoding algorithm is kept private. Why does this work?

If n is public, it would seem that the potential code breaker would only have to factor n as a product of primes; that will determine p and q and then $k = (p - 1)(q - 1)$ from which d can be found, given e. The RSA method works because the first step, that of factoring the number n, is not easy when n is very large. In practice, p and q are very large primes, e.g., with 100–200 digits, and so n is an integer with approximately 200–400 digits.

How long does it take to factor an integer using a computer? Suppose we make a very crude estimate of the time required to factor a 400-digit integer n. The crude method of trial and error is to test the quotient n/i for $i = 3, 5, 7, \ldots$ and check to see if there is a remainder after division. If the remainder is 0, then i is a factor of n. Of course, there is no need to do trial division by any even number because it is easy to determine if n has a factor 2. We have to test only the odd numbers up to $\sqrt{n}$ because if n is not a prime, then one of its factors is at most $\sqrt{n}$. This means we must do about $\sqrt{(10^{400})}/2$ divisions. This number is greater than 10^{199}. How fast can a computer do division? Let us be generous and say that a fast computer can do 1 trillion divisions per second ($= 10^{12}$/sec). (No computer that we know of can reach anywhere close to this speed.) This means that a test for a prime factor of n might potentially take about $10^{199}/10^{12} = 10^{187}$ sec. The number of seconds in a year is $31,536,000$, which is less than 10^9. Thus, factoring n could potentially take about $10^{187}/10^9 = 10^{178}$ years. Of course, with sophisticated programming one might cut this down to a few thousand centuries, but who has the time to wait? The point is that this procedure for transmitting messages has very high security because it is very difficult to factor large integers.

One might also observe that there is potentially a great deal of computation involved in computing a number such as $(19051404)^e$. Fortunately, this number need not be computed. It is only necessary to know the least residue of this number modulo n and this can be computed in a very reasonable amount of time.

We give an example of this part of the computation with some small numbers. Select $p = 31$ and $q = 127$ so that $n = 3937$ and $k = \varphi(n) = 30 \cdot 126 = 3780$. For the numbers d and e we just randomly select $d = 143$ (so that d is relatively prime to k) and then compute a solution to $dx \equiv 1 \pmod{k}$. Even this step may seem troublesome if the work is done by hand, but the following is a reasonable way to find a solution. Since $k = 2^2 \cdot 3^3 \cdot 5 \cdot 7$, the congruence is equivalent to the series of congruences

$$\begin{aligned} 143x &\equiv 1 \pmod{4}, \quad 143x \equiv 1 \pmod{27}, \\ 143x &\equiv 1 \pmod{5}, \quad 143x \equiv 1 \pmod{7}. \end{aligned}$$

By replacing 143 with the least residue in each congruence, these can be simplified to the form

$$\begin{aligned} 3x &\equiv 1 \pmod{4}, \quad 8x \equiv 1 \pmod{27}, \\ 3x &\equiv 1 \pmod{5}, \quad 3x \equiv 1 \pmod{7}. \end{aligned}$$

From the first of these we find $x = -1 + 4t$ for some integer t. Substituting this value for x into the second congruence, we find $-8 + 32t \equiv 1 \pmod{27}$, or $5t \equiv 9 \pmod{27}$. Since $11 \cdot 5 \equiv 1 \pmod{27}$, we obtain $t \equiv 99 \equiv -9 \pmod{27}$. So far we know $t = -9 + 27t_2$ and $x = -37 + 4 \cdot 27t_2$. After two more substitutions one finds $x = 2987 + kt_4$ for some integer t_4. Thus, we select $e = 2987$.

For the purpose of illustration, we encode just one of the three words mentioned previously. After substitution of the two-digit numbers for letters, the word ME becomes 1305 which, in turn, gets encoded as the least positive residue of $(1305)^{2987}$ modulo 3937. Many computer packages will make this computation in a fraction of a second. If we were to do this by hand, the following method is fairly efficient. Start with the least residue of $(1305)^2 \equiv 2241 \pmod{n}$. Then square this residue to get

$$(1305)^4 \equiv (2241)^2 \equiv 2406 \pmod{n}.$$

Continue squaring the residues 11 times until reaching

$$(1305)^{2048} \equiv 144 \pmod{n}.$$

We keep the least positive residue r_i of all the powers $(1305)^{2^i} \equiv r_i \pmod{n}$. Then using the binary expansion $2987 = 1 + 2 + 2^3 + 2^5 + 2^7 + 2^8 + 2^9 + 2^{11}$

and the relation

$$(a)^{2987} = a \cdot a^2 \cdot a^8 \cdot a^{32} \cdot a^{128} \cdot a^{256} \cdot a^{512} \cdot a^{2048}$$

to finish the computation, using $a = 1305$. The result is

$$(1305)^{2987} \equiv r_0 r_1 r_3 r_5 r_7 r_8 r_9 r_{11} \equiv 425 \pmod{n}.$$

Thus the "message" 425 would be sent. It would be decoded by the receiver as

$$(425)^d = (425)^{143} \equiv 1305 \pmod{n}.$$

As a practical matter, the work can be simplified slightly more. Instead of using the least positive residues, r_1, r_2, r_3, it is sometimes more efficient to use the *numerically least residues* to deal with smaller numbers. The numerically least residue of an integer t is the unique integer t^* on the interval $-n/2 < t^* \leq n/2$ such that $t \equiv t^* \pmod{n}$. This may help in doing hand calculations of the powers but the least positive residue must be used at the end to decode the message.

Of course, in practice, no human does all this computation by hand. A computer program does the translation automatically and the computation is invisible to the users at each end of the communication channel.

The example given here only illustrates some of the computational ideas. The numbers selected for p and q are much too small for a secure method of data transmission since n can easily be factored.

EXERCISES

1. Use the method described in this section to compute, by hand, the least positive residues as indicated:
(a) $111^{13} \pmod{1200}$
(b) $7^{24} \pmod{2303}$

2. Use an encoding method in which letters A–Z are represented by two-digit numbers 01–26. Use the RSA method with $n = 713 = 23 \cdot 31$ and $e = 3$ to encode the message HELP ME. [This can (probably) be done without the use of a computer.]

3. Use an encoding method in which letters A–Z are represented by two-digit numbers 01–26. Use the RSA method with $n = 8,163,403$ and $e = 3$ to encode the message HELP ME. [The numbers are sufficiently large to require a computer or a sophisticated calculator.]

4. Suppose you discover that someone sent the message 6911791 5824815 using the encoding scheme given in exercise 3. Can you decode this message? [The numbers are sufficiently large to make this very difficult to do "by hand".]

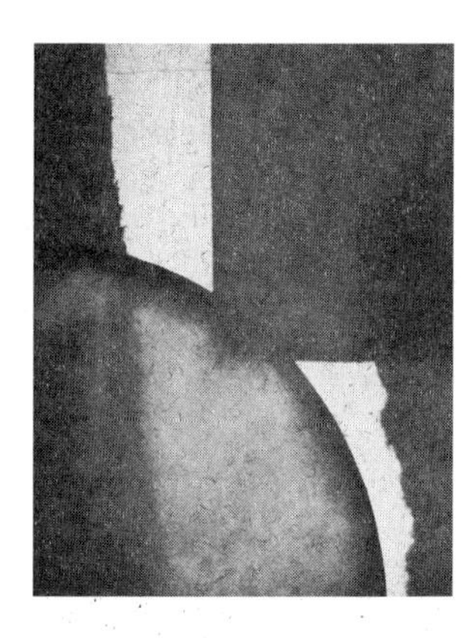

V POLYNOMIALS

We introduce the ring of polynomials with coefficients taken from a field. The study of polynomial equations is centuries old; we provide the basic construction that shows that every polynomial equation has a solution in some field containing the coefficients. Polynomials provide the main tool for the construction of field extensions. Starting from the field with p elements, we show how polynomials are used to construct fields with p^n elements. Although the ring of polynomials appears to be very different than the ring of integers, many of the factorization properties of the integers have analogs in the rings of polynomials. We examine this analogy very closely.

1 POLYNOMIAL RINGS

Let S be a commutative ring with identity. We intend to define the ring of polynomials with coefficients in S. The reader might expect a definition of objects such as $1 + x$ to be included in the definition, but we initially take a slightly different view. Rather than introduce the new symbol "x," we give a definition that uses only objects from S and then later introduce the more familiar notation.

Let P be the collection of all sequences $\mathbf{s} = (s_0, s_1, s_2, \ldots)$, where each s_i is an element of S and only a finite number of s_j are nonzero; in other words, there is an integer n (which may be different for different elements of P) such that $s_k = 0$ for all $k > n$. The element $s_i \in S$ is called the ith *coefficient* of the sequence. (Notice that the numbering begins at 0.) Two given sequences $\mathbf{s} = (s_0, s_1, s_2, \ldots)$ and $\mathbf{t} = (t_0, t_1, t_2, \ldots)$ are equal if and only if $s_i = t_i$ for all $i = 0, 1, 2, \ldots$.

The sequence $(0, 0, 0, \ldots)$ consisting of all zeros is called the *zero polynomial*. For a sequence $\mathbf{s}$ other than the zero polynomial, there is some largest index n such that $s_n \neq 0$. This n is called the *degree* of $\mathbf{s}$ and s_n is called the *leading coefficient* of $\mathbf{s}$. Note that the zero polynomial does not have a degree because it has no nonzero coefficients.

Now we make P into a ring by defining the operations. Addition of two elements is defined in the expected way:

$$\begin{aligned}(s_0, s_1, \ldots, s_m, \ldots) \; + \; & (t_0, t_1, \ldots, t_m, \ldots) \\ & = (s_0 + t_0, s_1 + t_1, \ldots, s_m + t_m, \ldots)\end{aligned}$$

and multiplication is given by the rule

$$(s_0, s_1, \ldots, s_m, \ldots) \cdot (t_0, t_1, \ldots, t_m, \ldots) = (u_0, u_1, \ldots, u_k, \ldots)$$

in which the terms u_k are defined by

$$u_k = s_0 t_k + s_1 t_{k-1} + \cdots + s_j t_{k-j} + \cdots + s_k t_0.$$

The verification of the axioms defining a ring can be done directly. Those involving only addition follow immediately from the corresponding axioms that hold in S. The associative law for multiplication can be verified with direct computation. For three polynomials $\mathbf{s}$, $\mathbf{t}$, $\mathbf{u}$, the rth coordinate of $(\mathbf{st})\mathbf{u}$ is

$$\sum_{i+j+k=r} (s_i t_j) u_k,$$

while the rth coordinate of $\mathbf{s}(\mathbf{tu})$ is

$$\sum_{i+j+k=r} s_i (t_j u_k).$$

The equality of these two expressions follows from the associative law in S. The other axioms are left to the reader. They will seem more natural when we use a more familiar notation, which we now introduce.

Let $x = (0, 1, 0, 0, \ldots)$. The multiplication rule is such that

$$\begin{aligned} x^2 &= (0, 0, 1, 0, 0, 0, \ldots) \\ x^3 &= (0, 0, 0, 1, 0, 0, \ldots) \end{aligned}$$

and more generally, $x^k = (0, 0, \ldots, 0, 1, 0, 0, \ldots)$, where the 1 occurs in the k–coordinate. More generally, multiplication by x produces a shift to the right as

$$x \cdot (s_0, s_1, \ldots, s_m, \ldots) = (0, s_0, s_1, \ldots, s_m, \ldots).$$

Next we identify a subring isomorphic to S. The map $s \to (s, 0, 0, \ldots)$ maps S to the collection of polynomials with zeros in all coordinates after the first. A polynomial with zero entries everywhere except possibly the first is called a *constant* polynomial. The definitions of addition and multiplication imply that this map is an isomorphism of S with the subring of constant polynomials. We identify the element of s in S with this polynomial and observe the multiplication rule:

$$s(t_0, t_1, t_2, \ldots) = (s, 0, 0, \ldots)(t_0, t_1, t_2, \ldots) = (st_0, st_1, st_2, \ldots).$$

With this observation, we may now produce the more familiar-looking polynomials.

Property 1.1 *If the elements of S are identified with the constant polynomials, then every polynomial $\mathbf{s} = (s_0, s_1, s_2, \ldots)$ can be expressed in the form*

$$\mathbf{s} = s_0 + s_1 x + s_2 x^2 + \cdots + s_n x^n,$$

where n is the degree of $\mathbf{s}$.

To see this, carry out the operations

$$\begin{aligned} s_0 + s_1 x + s_2 x^2 + \cdots + s_n x^n \quad = \quad & s_0(1, 0, 0, 0, \ldots) + \\ & s_1(0, 1, 0, 0, \ldots) + \\ & s_2(0, 0, 1, 0, \ldots) + \\ & \quad \vdots \qquad\qquad + \\ & s_n(0, 0, \ldots, 1, \ldots) = \\ & (s_0, s_1, s_2, \ldots, s_n, 0, \ldots) = \mathbf{s}. \end{aligned}$$

This more familiar notation will be used from now on. To emphasize the role played by x, we denote the ring of polynomials with coefficients from S by $S[x]$. We may summarize this development by listing some of the properties of $S[x]$:

Property 1.2 *Let S be a commutative ring with identity. A polynomial is an expression*

$$f(x) = s_0 + s_1 x + \cdots + s_n x^n, \qquad s_i \in S.$$

If $s_n \neq 0$, then the degree of $f(x)$ is n. If $g(x) = t_0 + t_1 x + \cdots + t_m x^m$ is a polynomial of degree m, then $f(x) = g(x)$ if and only if $n = m$ and $s_i = t_i$ for every i.

We denote the degree of a polynomial $f(x)$ by deg $(f(x))$. The following is an important property of the polynomial ring with coefficients in an integral domain:

Theorem 1.1 *Let S be an integral domain and $f(x)$ and $g(x)$ nonzero polynomials in $S[x]$. Then the product $f(x)g(x)$ is nonzero and*

$$\deg(f(x)g(x)) = \deg(f(x)) + \deg(g(x)).$$

In particular, $S[x]$ is an integral domain.

Proof: Let

$$\begin{aligned} f(x) &= s_0 + s_1 x + \cdots + s_n x^n, \qquad s_n \neq 0, \\ g(x) &= t_0 + t_1 x + \cdots + t_m x^m, \qquad t_m \neq 0. \end{aligned}$$

Then the highest power of x that could possibly appear in $f(x)g(x)$ with a nonzero coefficient is x^{n+m} and its coefficient is $s_n t_m$. This coefficient is nonzero because S is an integral domain so $f(x)g(x) \neq 0$, proving that $S[x]$ is an integral domain and that the degree of the product is $n+m$; the theorem is proved. The formula for the degree of the product may sometimes be referred to as the *degree equation.* ■

If F is a field, then the polynomial ring $F[x]$ is an integral domain but is not a field; clearly x is a nonzero element that has no inverse in $F[x]$ (by the degree equation). However, every integral domain has a field of quotients. The field of quotients of $F[x]$ is denoted by $F(x)$ and consists of all quotients $a(x)/b(x)$, with $a(x)$ and $b(x)$ polynomials in $F[x]$ and $b(x) \neq 0$. $F(x)$ is called the field *of rational functions over* F and its elements are called *rational functions*. In particular, if $F = \mathbb{Z}_p$ is a finite field with p elements, then $\mathbb{Z}_p(x)$ is an infinite field that contains $\mathbb{Z}_p$ as a subring.

In this section we introduced polynomials in *one* indeterminate x. However, this procedure can easily be extended as follows. If S is a commutative ring with identity, then $S[x]$ is also a commutative ring with identity. Now we use another symbol y to extend $S[x]$ to the ring $S[x][y]$ of polynomials in y with coefficients in $S[x]$. For brevity, we write $S[x,y]$ for this ring. If S is an integral domain, then so is $S[x]$, and thus so is $S[x,y]$. This construction may be repeated to obtain a polynomial ring $S[x_1, x_2, \ldots, x_m]$ in any number of variables, but we will study mainly polynomials in one variable.

The notation $S[x]$ is used for the ring of polynomials as just defined. The element x denotes a specific element in the collection of sequences and is usually called an *indeterminate* or a *variable*. The same notation $S[x]$ is sometimes used in a slightly different context. Suppose S is a subring of another commutative ring R and y is an element of R. Then $S[y]$ denotes the set of all polynomial expressions $s_0 + s_1 y + \cdots + s_n y^n$ with $s_i \in S$ and n any nonnegative integer. It might happen, in this context, that two polynomial expressions are equal even if the set of coefficients are not pairwise equal. For example, let S be the field of rational numbers, R the field of real numbers and $y = \sqrt{3}$. Then $\mathbb{Q}[\sqrt{3}]$ is the ring consisting of all expressions $a + b\sqrt{3}$ with $a, b \in \mathbb{Q}$. The expression y^2 is the same as the number 3, the expression y^3 is $3\sqrt{3}$, and so on. $S[y]$ can also be described as the smallest subring of R that contains both S and y. In the case of the polynomial ring, we say that x is an indeterminate to emphasize that we are not regarding x as some special element of a ring containing S. We might use a phrase such as "the ring generated by S and y" to describe $S[y]$ in case y is not an indeterminate.

1.1 Polynomial Functions

In calculus, a polynomial $f(x)$ with real number coefficients is usually regarded as a function defined for real numbers x and returning real number values. For a specific real number c, the value $f(c)$ is also a real number. For a general commutative ring, we may interpret polynomials in $S[x]$ as functions. Suppose R is a commutative ring that contains S as a subring. We allow the possibility that $R = S$. For any element $r \in R$ and any polynomial

$$f(x) = s_0 + s_1 x + s_2 x^2 + \cdots + s_n x^n \in S[x],$$

we define the value of f at r to be

$$f(r) = s_0 + s_1 r + s_2 r^2 + \cdots + s_n r^n.$$

With this definition, every polynomial $f(x)$ determines a function from R to R. Such a function is called *a polynomial function*. There is a difference between polynomials and polynomial functions. Consider the case $S = R = \mathbb{Z}_2$. There are only four functions from $\mathbb{Z}_2$ to itself, yet there are infinitely many polynomials with coefficients in $\mathbb{Z}_2$, each of which produces a function from $\mathbb{Z}_2$ to itself. Obviously, many different polynomials produce the same function. For example, the polynomial function determined by the polynomial x is the same function as the polynomial function determined by x^2. To see this, check the values at each point of $\mathbb{Z}_2$; both have the value 0 at $x = 0$ and the value 1 at $x = 1$ and hence they are equal functions. Of course, x and x^2 are not equal polynomials.

1.2 The Evaluation Homomorphism

If S is a subring of a commutative ring R, we may use any element of R to produce a homomorphism from $S[x]$ to R by evaluating a polynomial at the particular element. For any $r \in R$, define a mapping from the polynomial ring $S[x]$ to R by "evaluation" at r:

$$\theta_r : f(x) \longrightarrow f(r).$$

It is almost immediate that θ_r defines a homomorphism from $S[x]$ to R. Namely, if $f(x)$ and $g(x)$ are polynomials, then

$$\begin{aligned}\theta_r\left(f(x)+g(x)\right) &= f(r)+g(r) = \theta_r(f(x))+\theta_r(g(x)),\\ \theta_r\left(f(x)g(x)\right) &= f(r)g(r) = \theta_r(f(x))\cdot\theta_r(g(x)).\end{aligned}$$

It is worth noting that any element $r \in R$ can be used to produce a homomorphism from $S[x]$ to R as defined previously. The kernel of θ_r is the set of all polynomials in $S[x]$ that have the value 0 at r; that is, $f(x)$ is in the kernel if $f(r) = 0$. In particular, the set of all $f(x)$ in $S[x]$ such that $f(r) = 0$ is an ideal of $S[x]$.

EXERCISES

1. Give a complete proof that the ring S is isomorphic to the subring of $S[x]$ consisting of all constant polynomials.
2. If S is not an integral domain, show, by example, that the degree formula for products given in Theorem 1.1 can fail to hold.
3. For any polynomial $g(x)$, show that the polynomial $x(x-1)g(x)$ in $\mathbb{Z}_2[x]$ produces the polynomial function sending all elements of $\mathbb{Z}_2$ to 0.
4. List the four functions from $\mathbb{Z}_2$ to $\mathbb{Z}_2$ by providing a table of values. Show that the polynomial functions corresponding to the zero polynomial or polynomials of degree at most 1 give all the four functions from $\mathbb{Z}_2$ to itself.
5. Use the degree formula for products to show that for an integral domain S the units of the polynomial ring $S[x]$ are the units of S.
6. Use the fact that $\mathbb{Z}[x]$, the set of polynomials with integer coefficients, is a ring to prove that the binomial coefficients $C(n, j) = n!/(n-j)!j!$ are integers. [See Chapter 1, Section 6.]

2 DIVISORS IN $F[x]$

In this section and others in this chapter we study the ring of polynomials with coefficients in an arbitrary *field* F. We know then, by Theorem 1.1, that $F[x]$ is an integral domain. Our goal is to study in detail how elements of $F[x]$ are expressed as products. We begin with the following definition:

Definition 2.1 If $f(x)$ and $g(x)$ are polynomials in $F[x]$, we say that $g(x)$ **divides** $f(x)$ [or **is a factor of** $f(x)$] if there is $h(x) \in F[x]$ such that $f(x) = g(x)h(x)$. When $g(x)$ divides $f(x)$, we also say that $f(x)$ is **divisible** by $g(x)$ and we write $g(x)|f(x)$, which is read as "$g(x)$ divides $f(x)$."

It follows immediately from this definition that if c is a nonzero element of F (i.e., a polynomial of degree zero), then c is a divisor of every element in $F[x]$. Since c has a multiplicative inverse c^{-1} in F, we can write $f(x) = c[c^{-1}f(x)]$, which shows that c is a divisor of $f(x)$.

It is important to observe that if $f(x) = g(x)h(x)$, then $f(x) = cg(x) \cdot c^{-1}h(x)$ for any nonzero element $c \in F$. That is, if $g(x)|f(x)$, then $cg(x)|f(x)$ for every nonzero element c of F.

The next theorem states a property of $F[x]$ which will be used repeatedly to derive results about polynomial rings. The underlying idea of quotients and remainders was demonstrated in the division algorithm for the ring of integers. This is the first of several similarities between the ring of integers and the ring of polynomials with coefficients in a field.

Theorem 2.1 THE DIVISION ALGORITHM FOR POLYNOMIALS. *Let F be a field and $f(x), g(x) \in F[x]$ with $g(x) \neq 0$. Then there exists unique elements $q(x)$ and $r(x)$ of $F[x]$ such that*

$$f(x) = q(x)g(x) + r(x), \quad \text{and either } r(x) = 0 \text{ or } \deg r(x) < \deg g(x). \tag{5.1}$$

Proof: Let the polynomials be given by

$$\begin{aligned} f(x) &= a_0 + a_1x + \cdots + a_nx^n, \quad a_n \neq 0, \\ g(x) &= b_0 + b_1x + \cdots + b_mx^m, \quad b_m \neq 0. \end{aligned}$$

Consider the case in which $m > n$. Then the conclusion is true using $q(x) = 0$ and $r(x) = f(x)$. Therefore, from now on we may suppose $m \leq n$. There is one other trivial case, namely, that in which $\deg g(x) = 0$; in this case, $g(x)$ is a nonzero element c of F and so the conclusion holds with $q(x) = c^{-1}f(x)$ and $r(x) = 0$. Thus we may assume $m \geq 1$. We use mathematical induction to prove the theorem. Let S_n be the statement that the required $q(x)$ and $r(x)$ exist so that Eq. (5.1) is true if $\deg f(x) = n$. We know S_n is true for $n = 0, 1, \ldots, m-1$. Hence, assume $n \geq m$. Then $n - m \geq 0$ and we have a polynomial $f_1(x)$ defined by

$$f(x) = f_1(x) + b_m^{-1}a_nx^{n-m}g(x),$$

and $f_1(x)$ has degree at most $n-1$. We arranged the product $b_m^{-1}a_nx^{n-m}g(x)$ to have the same leading coefficient as $f(x)$ and so the highest terms canceled when we do the subtraction. Since $f_1(x)$ has degree less than n, the conclusion of the theorem is true for division of $f_1(x)$ by $g(x)$: There exist $q_1(x)$ and $r_1(x)$ such that

$$f_1(x) = q_1(x)g(x) + r_1(x), \quad \text{either } r_1(x) = 0 \text{ or } \deg r_1(x) < \deg g(x).$$

Now substitute into the defining equation for $f_1(x)$ to obtain

$$f(x) = \left(q_1(x) + b_m^{-1}a_nx^{n-m}\right) g(x) + r_1(x),$$

where $r_1(x)$ meets the conditions required by the conclusion. Setting

$$q(x) = q_1(x) + b_m^{-1} a_n x^{n-m} \quad \text{and} \quad r(x) = r_1(x)$$

gives us the conclusion stated in Eq. (5.1). We have proved that there exist $q(x)$ and $r(x)$ with the required properties.

Now we show the uniqueness of these two polynomials. Suppose

$$f(x) = q(x)g(x) + r(x) = q'(x)g(x) + r'(x),$$

where $r(x)$ and $r'(x)$ are either 0 or have degree less than the degree of $g(x)$. Then

$$(q(x) - q'(x))g(x) = r'(x) - r(x).$$

If the polynomials on each side of this equation are nonzero, then they have a degree; the right side has degree less than the degree of $g(x)$ because that is true of both $r(x)$ and $r'(x)$. However, the left-hand side has a degree at least as large as the degree of $g(x)$ and we have a conflict. The resolution is that both sides equal zero. Thus, $r(x) = r'(x)$ and $q(x) = q'(x)$, proving the uniqueness assertion. ∎

Here, we give an example to illustrate the style of the proof. Take the field to be $\mathbb{Q}$, the field of rational numbers, and let

$$f(x) = 2x^4 + x + 6, \quad g(x) = x^2 + 3.$$

We first multiply $g(x)$ by $2x^2$ to match the highest power in $f(x)$ and subtract, thereby obtaining a polynomial of degree less than 4:

$$f(x) - 2x^2 g(x) = 2x^4 + x + 6 - 2x^2(x^2 + 3) = -6x^2 + x + 6.$$

Next multiply $g(x)$ by -6 and subtract to get

$$f(x) - 2x^2 g(x) - (-6)g(x) = -6x^2 + x + 6 + 6x^2 + 18 = x + 24.$$

We have reached a polynomial of degree less than the degree of $g(x)$, so $q(x) = 2x^2 - 6$, $r(x) = x + 24$, and

$$f(x) = (2x^2 - 6)g(x) + x + 24 = q(x)g(x) + r(x).$$

As in the case with the division algorithm in the ring of integers, we call $q(x)$ the **quotient** and $r(x)$ the **remainder** after division of $f(x)$ by $g(x)$.

One of the applications of the division algorithm for the integers in Chapter II, theorem 4.4 was to prove that every ideal in $\mathbb{Z}$ is principal. The division algorithm for polynomials can be used in the same way to prove that every ideal in $F[x]$, for F a field, is principal (see exercises).

We have given this important result about division and mentioned divisors. Now we discuss the common divisors of two polynomials.

Definition 2.2 Let $f(x)$ and $g(x)$ be elements of $F[x]$ with at least one of them nonzero. A **common divisor** of $f(x)$ and $g(x)$ is a polynomial $h(x)$ that divides both $f(x)$ and $g(x)$. A polynomial $d(x)$ is a **greatest common divisor** (GCD) of $f(x)$ and $g(x)$

(i) If $d(x)$ is a common divisor of $f(x)$ and $g(x)$.

(ii) Whenever $h(x)$ is a common divisor of $f(x)$ and $g(x)$, then $h(x)|d(x)$.

It is not obvious from the definition that a GCD of $f(x)$ and $g(x)$ even exists. We will later prove that one does exists and will also show how to compute one. There will be many GCDs because of the observation made previously; namely, if $h(x)$ divides $f(x)$ and c is any nonzero element of F, then $ch(x)$ is also a divisor of $f(x)$. If $d(x)$ is a GCD of $f(x)$ and $g(x)$, then $cd(x)$ is one also. It is sometimes convenient to be able to refer to a unique member of this potentially large class of polynomials, so we introduce the following notion:

Definition 2.3 A nonzero element of $F[x]$ is called a **monic polynomial** if its leading coefficient equals 1, the identity of F.

If $h(x)$ is a nonzero polynomial with leading coefficient c, then $c^{-1}h(x)$ is a monic polynomial. If one insists that the GCD of two polynomials be a monic polynomial, then it is unique, as will be seen later.

Now we prove the existence of the GCD and give one of its important properties.

Theorem 2.2 EUCLIDEAN ALGORITHM FOR POLYNOMIALS. *Let $f(x)$ and $g(x)$ be elements of $F[x]$ with $g(x) \neq 0$. Then there is a greatest common divisor of $f(x)$ and $g(x)$. Moreover, one GCD is the nonzero monic polynomial of least degree which can be expressed as*

$$d(x) = f(x)a(x) + g(x)b(x)$$

for some elements $a(x)$ and $b(x)$ in $F[x]$.

Proof: We define two finite sequences $\{f_i\}$ and $\{q_i\}$ in the following way. Let $f_1 = f(x)$ and $f_2 = g(x)$. The other terms are defined by the series of equations

$$\begin{aligned}
f_1 &= f_2 q_1 + f_3, && \deg f_3 < \deg f_2 && (1)\\
f_2 &= f_3 q_2 + f_4, && \deg f_4 < \deg f_3 && (2)\\
&\vdots && \vdots && \vdots\\
f_{n-2} &= f_{n-1} q_{n-2} + f_n, && \deg f_n < \deg f_{n-1} && (n-2)\\
f_{n-1} &= f_n q_{n-1}. && && (n-1)
\end{aligned}$$

These equations are obtained by repeated application of the division algorithm for polynomials; it is assumed that $f_2, f_3, \ldots, f_n$ are nonzero and

$$\deg f_2 > \deg f_3 > \cdots > \deg f_n,$$

and that the remainder in Eq. $(n-1)$ is zero. Such an n surely exists because the degrees of the f_i form a strictly decreasing sequence of nonnegative integers; such a sequence cannot contain more than $\deg f_2$ terms. Now we prove that the last nonzero remainder, f_n, is a GCD of $f(x)$ and $g(x)$.

The proof uses the following idea: If u, v, w are elements of $F[x]$ with $u+v = w$ and if h is an element of $F[x]$ that divides two of the three elements, then h also divides the third element. For example, if $u = hu'$ and $w = hw'$, then $v = w - u = h(w' - u')$. We begin by showing that f_n is a common divisor of f_1 and f_2. From Eq. $(n-1)$ we conclude f_n divides f_{n-1}. Thus, f_n divides two of the three terms in Eq. $(n-2)$. Hence, we conclude f_n divides f_{n-2}. In the same way, this shows that f_n divides two of the three terms in Eq. $(n-3)$ and so f_n divides f_{n-3}. Continue to use equations until reaching Eq. (2) and then (1) to conclude that f_n divides f_1 and f_2. Thus, f_n is a common divisor of f_1 and f_2.

Next it is necessary to show that f_n is a greatest common divisor. Let h be any polynomial that divides both f_1 and f_2. Use the two-out-of-three idea again starting from Eq. (1) this time to conclude h divides f_3. Next use Eq. (2) and the fact that h divides f_2 and f_3 to conclude h divides f_4. Continue through the list of equations to conclude that h divides each f_i; in particular, h divides f_n and so f_n is a GCD of f_1 and f_2.

Lastly, we show that the GCD, f_n, is a linear combination of f_1 and f_2. We show that each f_i, $i \geq 3$, is a linear combination of f_1 and f_2. We have

$$f_3 = f_1 - f_2 q_1,$$

which shows that f_3 is a linear combination of f_1 and f_2. Suppose i is an index such that we know $f_{i-1} = f_1 p + f_2 s$ and $f_i = f_1 p' + f_2 s'$ are each linear combinations of f_1 and f_2. We wish to show that f_{i+1} is also a linear combination. From Eq. $(i-1)$ we obtain

$$\begin{aligned} f_{i+1} &= f_{i-1} - f_i q_{i-1} = (f_1 p + f_2 s) - (f_1 p' + f_2 s') q_{i-1} \\ &= f_1(p - p' q_{i-1}) + f_2(s - s' q_{i-1}), \end{aligned}$$

which shows that f_{i+1} is also a linear combination of f_1 and f_2. Eventually we reach $i+1 = n$ to get the required property for f_n.

The last statement to be proved is the assertion about the "least degree." Suppose $m(x)$ is a nonzero polynomial and that $m(x) = f_1 u + f_2 v$ for some $u, v \in F[x]$. Since f_n is a GCD of f_1 and f_2, f_n divides two of the three terms in this equation and so f_n also divides $m(x)$. If $m(x) = f_n(x)h(x)$, then

$$\deg m(x) = \deg f_n(x) + \deg h(x) \geq \deg f_n,$$

which shows that the degree of $m(x)$ cannot be less than the degree of a GCD, $f_n(x)$. Finally, all parts of the theorem have been proved. ■

The sequence of steps to obtain Eqs. (1)–$(n-1)$ is called **the Euclidean algorithm** for polynomials. It was also used for the ring of integers in Chapter 4.

Some concrete examples of these general results are now presented.

Example 2.1

Let $f(x) = x^3 + 2x^2 - 1$ and $g(x) = x^2 - 1$ be two polynomials in $\mathbb{Q}[x]$. Let us find their GCD and express it as a linear combination of $f(x)$ and $g(x)$. The sequence from the Euclidean algorithm is

$$\begin{aligned} f(x) &= g(x)(x+2) + (x+1), \\ g(x) &= (x+1)(x-1). \end{aligned}$$

The last nonzero remainder is $d(x) = x + 1$, and

$$d(x) = f(x) - g(x)(x+2)$$

expresses the GCD as a linear combination of $f(x)$ and $g(x)$.

Taylor's formula from the study of calculus is easily proved by induction for polynomials and can be of some computational value. The formula asserts that if $f(x)$ is a polynomial of degree n in $\mathbb{Q}[x]$ (or $\mathbb{R}[x]$) and if $f^{(k)}(x)$ denotes the kth derivative of $f(x)$, then

$$f(x) = \sum_{k=0}^{n} \frac{f^{(k)}(a)}{k!}(x-a)^k,$$

where $k! = 1 \cdot 2 \cdot 3 \cdots (k-1) \cdot k$ and a is any element of the coefficient field. We apply this in the next example.

Example 2.2

Let $f(x) = x^4 + 3x^2 + x + 1$ and $g(x) = x - 1$ be elements of $\mathbb{Q}[x]$. We compute the GCD by using the Taylor expansion of a polynomial. We expand in powers of $x - 1$ to obtain

$$\begin{aligned} f(x) &= \sum_{k=0}^{4} \frac{f^{(k)}(1)}{k!}(x-1)^k \\ &= 6 + 11(-1+x) + 9(-1+x)^2 + 4(-1+x)^3 + (-1+x)^4 \\ &= 6 + \left(11 + 9(-1+x) + 4(-1+x)^2 + (-1+x)^3\right) g(x). \end{aligned}$$

This shows that the polynomial 6 of degree 0 is a linear combination of $f(x)$ and $g(x)$ and so the GCD is a constant polynomial 1 (the monic version of 6).

We use the following terminology to keep the analogy with the terminology used with the ring of integers:

Definition 2.4

Two nonzero polynomials $f(x)$ and $g(x)$ in $F[x]$ are said to be **relatively prime** if their GCD is 1.

Thus, $f(x)$ and $g(x)$ are relatively prime if and only if it is possible to express 1 as a linear combination $1 = f(x)a(x) + g(x)b(x)$.

EXERCISES

1. Find the quotient and remainder after division of $f(x)$ by $g(x)$ for the following choices of polynomials in $\mathbb{Q}[x]$:
(a) $f(x) = x^3 + 2x + 1$, $g(x) = x^2 + 2$
(b) $f(x) = 4x^3 + 4x^2 + x - 1$, $g(x) = 2x + 5$
(c) $f(x) = (x - 4)^3$, $g(x) = (x - 2)$
(d) $f(x) = (x - 2)^5$, $g(x) = (x - 1)^2$

2. Let $g(x)$ be a nonzero element of $F[x]$ with $\deg g(x) \geq 1$. Show that for each $f(x) \in F[x]$ there exist polynomials $r_0(x), r_1(x), \ldots$ which are either equal to 0 or have degree less than $\deg g(x)$ and which satisfy

$$f(x) = r_0(x) + r_1(x)g(x) + r_2(x)g(x)^2 + \cdots + r_n(x)g(x)^n$$

for some nonnegative integer n. [This is the analog of the expansion of integers in base g.]

3. For each pair $f(x)$ and $g(x)$ listed in Exercise 1, find the polynomials $r_i(x)$ which satisfy the conditions in Exercise 2.

4. For the case $g(x) = x - a$ and any $f(x) \in \mathbb{Q}[x]$, use the theory of Taylor series from calculus to express the $r_i(x)$ in Exercise 2 in terms of the derivatives of $f(x)$.

5. Find the GCD of each of the following pairs of polynomials over the field $\mathbb{Q}$ of rational numbers and express it as a linear combination of the two polynomials:
(a) $2x^3 - 4x^2 + x - 2$ and $x^3 - x^2 - x - 2$
(b) $x^4 + x^3 + x^2 + x + 1$ and $x^3 - 1$
(c) $x^5 + x^4 + 2x^3 - x^2 - x - 2$ and $x^4 + 2x^3 + 5x^2 + 4x + 4$
(d) $(x - 3)^2(x^2 + x + 1)$ and $(x - 3)(x^3 - 1)$

6. Find the GCD of each of the pairs of polynomials over the field F.
(a) $x^3 + (2i + 1)x^2 + ix + i + 1$ and $x^2 + (i - 1)x - 2i - 2$; $F = \mathbb{C}$, the field of complex numbers

(b) $x^2 + (1 - \sqrt{2})x - \sqrt{2}$ and $x^2 - 2$; $F = \mathbb{R}$, the field of real numbers

7. Let n be an integer with $n \geq 2$ and let $g_1(x), \ldots, g_n(x)$ be n elements of $F[x]$, not all zero. Define the GCD of these n polynomials to be a polynomial that divides each $g_i(x)$ and is divisible by every polynomial in $F[x]$ that divides each $g_i(x)$. Use the results already proved for the case $n = 2$ and mathematical induction to show that a GCD of $g_1(x), \ldots, g_n(x)$ exists and can be expressed as a linear combination $u_1 g_1(x) + \cdots + u_n g_n(x)$ for some elements $u_i \in F[x]$.

8. Show that the GCD of n polynomials $g_1(x), \ldots, g_n(x)$ in $F[x]$ (as defined in the previous exercise) can be computed as follows: Let $d_2(x)$ be the GCD of $g_1(x)$ and $g_2(x)$. Let $d_3(x)$ be the GCD of $d_2(x)$ and $g_3(x)$. Continue this way until reaching $d_n(x) =$ GCD of $d_{n-1}(x)$ and $g_n(x)$. Then $d_n(x)$ is the GCD of the n polynomials.

9. Find the GCD of the three polynomials

$$\begin{aligned} g_1(x) &= x^3 - 7x - 6 \\ g_2(x) &= x^3 + 2x^2 - x - 2 \\ g_3(x) &= x^3 - 3x^2 - 4x + 12, \end{aligned}$$

which are elements of $\mathbb{Q}[x]$. Express the GCD as a linear combination of the three polynomials.

10. Let $f(x)$ and $g(x)$ be nonzero elements of $F[x]$. If F is a subfield of a larger field K, then $F[x] \subseteq K[x]$ and we may consider $f(x)$ and $g(x)$ as elements of $K[x]$. Show that the quotient and remainder in the division of $f(x)$ by $g(x)$ are the same whether these polynomials are regarded as elements of $F[x]$ or of $K[x]$. In particular, conclude that if there is a polynomial $h(x) \in K[x]$ such that $f(x) = g(x)h(x)$, then $h(x) \in F[x]$.

11. Verify that the statement of the division algorithm for polynomials (Theorem 2.2) remains true if the field F is replaced by any commutative ring S with identity, provided that the assumption $g(x) \neq 0$ is replaced by the assumption that the leading coefficient of $g(x)$ has a multiplicative inverse in S.

12. It was shown earlier that every ideal in the ring of integers is principal. Imitate that proof to verify that every ideal in $F[x]$, F a field is principal. [Hint: If A is a nonzero ideal of $F[x]$, take an element $f(x)$ in A having the smallest degree for elements of A and show $A = (f(x))$.]

3 UNIQUE FACTORIZATION IN $F[x]$

This section contains the main result about factorization of polynomials with coefficients in a field. In general, it says that a polynomial may be factored as a product of certain polynomials that cannot be factored further; in addition, there is essentially only one way this can be done. Throughout this section, F denotes an arbitrary field. We begin with a definition.

Definition 3.1 A nonzero polynomial $p(x)$ in $F[x]$ is **irreducible** if $\deg p(x)$ is positive and if $p(x)$ is not the product of two elements of $F[x]$ each having positive degree.

Thus, if $p(x)$ is irreducible and if $p(x) = f(x)g(x)$, then either $\deg f(x) = 0$ or $\deg g(x) = 0$. A polynomial of degree 0 is a nonzero element of F, and if c is such an element then $p(x) = c(c^{-1}p(x))$. Thus, every polynomial $cp(x)$ is a factor of $p(x)$ for trivial reasons. When $p(x)$ is irreducible, its only factors of positive degree are those of the form $cp(x)$, with c a nonzero element of F.

Every polynomial of degree 1 is irreducible by the degree formula. That is, if $p(x) = f(x)g(x)$ then

$$\deg p(x) = \deg f(x) + \deg g(x).$$

Hence, if both $f(x)$ and $g(x)$ have positive degree, then the degree of $p(x)$ is at least 2.

The notion of irreducibility for a polynomial depends on the field under consideration. For example, the polynomial $x^2 - 2$ is irreducible over the field $\mathbb{Q}$ of rational numbers. However, the same polynomial can be factored if we allow real number coefficients; that is, $x^2 - 2 = (x - \sqrt{2})(x + \sqrt{2})$ and these factors of degree 1 have coefficients in $\mathbb{R}$. More succinctly, we may say that $x^2 - 2$ is irreducible in $\mathbb{Q}[x]$ but is reducible in $\mathbb{R}[x]$. In later sections we will discuss irreducible polynomials over $\mathbb{Q}$, $\mathbb{R}$, and $\mathbb{C}$.

The following lemma is the analog for polynomials of Euclid's lemma for the integers. It is the key step of the proof of unique factorization for polynomials. We prove it now in order not to interrupt the argument later.

Lemma 3.1 EUCLID'S LEMMA FOR POLYNOMIALS. *Let $f(x)$, $g(x)$, and $p(x)$ be nonzero polynomials over the field F with $p(x)$ irreducible. If $p(x)|f(x)g(x)$, then either $p(x)|f(x)$ or $p(x)|g(x)$.*

Proof: If $p(x)|f(x)$, then the conclusion holds and we are done. Therefore, assume that $p(x)$ does not divide $f(x)$. Let $m(x)$ be the monic GCD of $p(x)$ and $f(x)$. We wish to show that $m(x) = 1$, the monic polynomial of degree 0. Suppose this is not the case. Then $m(x)$ is a divisor of $p(x)$ with degree greater than 0. By definition of irreducibility, we must have $m(x) = cp(x)$ for some nonzero element $c \in F$. However, $m(x)$ divides $f(x)$ and so $c^{-1}m(x) = p(x)$ also divides $f(x)$. We are assuming, however, that $p(x)$ does not divide $f(x)$, so the only option is that $m(x) = 1$. Since the GCD of $p(x)$ and $f(x)$ is 1, there exist elements $u(x)$ and $v(x)$ in $F[x]$ such that

$$1 = u(x)f(x) + v(x)p(x).$$

Multiply both sides of this equation by $g(x)$ to obtain

$$g(x) = u(x)f(x)g(x) + v(x)p(x)g(x).$$

By assumption, $p(x)$ divides $f(x)g(x)$ so $p(x)$ divides both terms on the right side of this equation and therefore $p(x)$ divides $g(x)$ as we wished to prove. ∎

If a polynomial $p(x)$ has the property that it divides a product $f(x)g(x)$ only when it divides one of the factors $f(x)$ or $g(x)$, it is called a **prime polynomial**. The lemma just proved shows that an irreducible polynomial is a prime polynomial. Conversely, if $p(x)$ is prime then it is irreducible. To see this, suppose $p(x) = u(x)v(x)$. By the definition of prime, $p(x)$ must divide either $u(x)$ or $v(x)$. By a degree count, we see that either $u(x)$ has degree 0 or $v(x)$ has degree 0. Thus, $p(x)$ is irreducible.

Since the terms *prime* and *irreducible* refer to the same polynomials, there is no need to use two different terms. However, when one considers commutative rings other than polynomial rings, the two concepts may not be the same. For now, we use the two terms interchangeably when considering polynomials.

The following is an easy extension of Euclid's lemma in which the irreducible polynomial divides a product of more than two factors:

Lemma 3.2 *If the product $f_1(x) \cdots f_k(x)$ of k polynomials is divisible by a prime polynomial $p(x)$, then $p(x)$ divides $f_i(x)$ for at least one index $i \leq k$.*

Proof: We use induction on the number of factors. The result is true for $k = 1$ in a trivial way; it is true for $k = 2$ by the previous form of Euclid's lemma. Assume the result is true when the number of factors is less than k and suppose $k > 2$. Let $g(x) = f_2(x) \cdots f_k(x)$ so that our assumption is that $p(x)$ divides

$$f_1(x)g(x) = f_1(x) \cdots f_k(x).$$

By Euclid's lemma (for the case $k = 2$) we see that either $p(x)|f_1(x)$ or $p(x)|g(x)$. In the first case we are finished. In the second case $p(x)$ divides the product of $k - 1$ factors and so, by induction, $p(x)$ divides one of the factors, $f_i(x)$. By the induction principle, the lemma is true for any number of factors. ∎

Now we may prove the main result.

Theorem 3.1 *If $f(x)$ is a polynomial of positive degree over the field F and a is its leading coefficient, then there exist distinct monic prime polynomials $p_1(x), \ldots, p_k(x)$, $k \geq 1$, such that*

$$f(x) = a[p_1(x)]^{n_1}[p_2(x)]^{n_2} \cdots [p_k(x)]^{n_k}, \tag{5.2}$$

where the exponents n_i are positive integers. Moreover, such a factorization is unique except for the order in which the factors are written.

Proof: The proof has two main parts; the first is to show that such a factorization exists, and the second is to show the uniqueness assertion. The first step is carried out by mathematical induction. Let M be the set of positive integers n for which the following statement is true:
S_n: If $f(x)$ is a polynomial of degree n, then $f(x)$ has a factorization in the form of Eq. 5.2 in which the $p_i(x)$ are monic prime polynomials.

We first show $1 \in M$. If $f(x)$ has degree 1, then $f(x) = ax + b$ with $a \neq 0$. Thus,

$$f(x) = a(x + a^{-1}b)$$

is a factorization in which $x + a^{-1}b$ is a monic prime polynomial. Thus, S_1 is true.

Now suppose that M contains $1, 2, \ldots, r-1$ for some integer $r \geq 2$. We must show that $r \in M$. Let $f(x)$ be any polynomial of degree r. If $f(x)$ is a prime polynomial with leading coefficient a, then $a^{-1}f(x) = p(x)$ is a monic prime polynomial and $f(x) = ap(x)$. This is a factorization of the required type so S_r is true for the case in which $f(x)$ is a prime polynomial of degree r. Now suppose $f(x)$ has degree r but is not a prime polynomial. Then $f(x) = u(x)v(x)$ for some elements $u(x), v(x) \in F[x]$ with $\deg u(x) = d < r$ and $\deg v(x) = d' < r$. Because u has degree d which is smaller than r, statement S_d is true and $u(x)$ has a factorization

$$u(x) = bg_1(x) \cdots g_s(x),$$

with each $g_i(x)$ a monic prime polynomial and b a nonzero element of F. Similarly,

$$v(x) = ch_1(x) \cdots h_t(x),$$

with each $h_i(x)$ a monic prime polynomial and $c \in F$. We take the product $f(x) = u(x)v(x)$ to obtain a factorization of $f(x)$ as a product of monic prime polynomials and a constant bc in F. After collecting equal primes, we see that $f(x)$ equals a product of the form of Eq. (5.2). Thus, S_n is true for every n.

Now we discuss the uniqueness assertion. Suppose we have two sets of monic prime polynomials $p_1(x), \ldots, p_k(x)$ and $q_1(x), \ldots, q_m(x)$ (repetitions allowed) such that

$$p_1(x) \cdots p_k(x) = q_1(x) \cdots q_m(x).$$

Since $p_1(x)$ visibly divides the left side of this equation, it must also divide the right side of the equation. By the extension of Euclid's lemma for polynomials, Lemma 3.2, $p_1(x)$ must divide one of the $q_j(x)$; let us say $p_1(x)$ divides $q_1(x)$ to be specific. By assumption, both $p_1(x)$ and $q_1(x)$ are monic prime polynomials. If one monic prime divides another, they are equal, so $p_1(x) = q_1(x)$. We may cancel this common term and obtain a shorter product:

$$p_2(x) \cdots p_k(x) = q_2(x) \cdots q_m(x).$$

Repeat this argument until all the $p_i(x)$ have been matched with some $q_j(x)$ and both are canceled; the left side equals 1 so all the $q_i(x)$ have been canceled. We have shown that $k = m$, and after renumbering the $q_i(x)$ we obtain $p_i(x) = q_i(x)$ for $i = 1, 2, \ldots, k$. This is sufficient to verify the uniqueness assertion of the theorem. ∎

We will see that this theorem has great theoretical significance, but in actual practice it may be quite difficult to find the factorization 5.2 for a given polynomial $f(x)$. It may be even quite difficult to determine if a given monic polynomial is prime. We will examine some special instances of this problem in later sections.

4 ROOTS OF POLYNOMIALS

In defining the polynomial ring $F[x]$, we emphasized that the symbol x is not an element of F. This differs from the point of view which is traditionally used in the study of function in calculus, for example. In that view, an expression $f(X) = X^2 - 2X + 5$ is interpreted as a function defined for all real numbers x by the rule $f(x) = x^2 - 2x + 5$. Thus, the value of the function f at the point x is the real number $f(x)$ and f is a function from the real numbers to the real numbers. Using this point of view, one may then discuss the problem of finding solutions to polynomial equations. That is, we pose the problem of finding all real numbers x such that $f(x) = 0$. For certain polynomials, such as the f just given, there is no real number x such that $f(x) = 0$. However, if we allow x to take on complex number values, then there are two solutions, namely, $x = 1 \pm 2i$. This is a situation frequently encountered and is one we will study carefully. The unique factorization of polynomials plays an important part in the discussion.

4.1 Evaluation Maps

We consider a fairly general point of view. In place of the real numbers and the complex numbers, we consider a field F and a larger field L containing F as a subfield. We also say that L is an *extension field* of F. Consider a polynomial $f(x) \in F[x]$; that is, f is a polynomial with coefficients in the smaller field F. For any element $c \in L$, we define $f(c)$ as follows:

If $f(x) = a_0 + a_1x + a_2x^2 + \cdots + a_nx^n$, then

$$f(c) = a_0 + a_1c + a_2c^2 + \cdots + a_nc^n$$

so that $f(c)$ is an element of L. We call the correspondence $f(x) \to f(c)$ the *evaluation map determined by* c. The evaluation map is a homomorphism from the polynomial ring $F[x]$ to the field L. The verification of this is not difficult once the notation is understood. We give the map a name, e.g., $\theta_c : f(x) \to f(c)$. To say that θ_c is a homomorphism means that it preserves addition and multiplication. We verify only one of these. Suppose

$$\begin{aligned} f(x) &= a_0 + a_1 x + a_2 x^2 + \cdots + a_n x^n \\ g(x) &= b_0 + b_1 x + b_2 x^2 + \cdots + b_n x^n \end{aligned}$$

(we allow some of the coefficients to be 0 in case the degrees are not equal). Then

$$\begin{aligned} \theta_c(f(x) + g(x)) &= \theta_c\left(\sum_{i=0}^{n}(a_i + b_i)x^i\right) = \sum_{i=0}^{n}(a_i + b_i)c^i \\ &= \sum_{i=0}^{n} a_i c^i + \sum_{i=0}^{n} b_i c^i \\ &= \theta_c(f(x)) + \theta_c(g(x)). \end{aligned}$$

Similarly, one shows that $\theta_c(f(x)g(x)) = f(c)g(c) = \theta_c(f(x))\,\theta_c(g(x))$. Thus, the evaluation map is a homomorphism.

For a given element $c \in L$, we write $F[c]$ for the set of all elements of L which equal $\theta_c(f(x))$ for some polynomial $f(x) \in F[x]$. Because θ_c is a homomorphism, $F[c]$ is a subring of L; that is, $F[c]$ is closed under addition, subtraction, and multiplication. $F[c]$ need not be a subfield of L, however. It will be seen later that $F[c]$ is a field if and only if c is the solution of some polynomial equation $g(x) = 0$ with $g(x)$ a nonzero polynomial in $F[x]$; that is, c is a root of $g(x)$ in the sense of the following definition:

Definition 4.1 If $f(x) \in F[x]$ and c is an element of an extension field of F such that $f(c) = 0$, then c is called a **root of $f(x)$**.

Consider the case $F = \mathbb{R}$, the field of real numbers, and $L = \mathbb{C}$, the field of complex numbers. Take $c = i$. Every complex number has the form $a + bi$ with $a, b \in \mathbb{R}$. Thus, any element $a + bi$ of $\mathbb{C}$ can be expressed as $f(i)$ by taking $f(x) = a + bx$; in other words, $\mathbb{R}[i] = \mathbb{C}$. Of course, a given element may be equal to $f(i)$ for many different choices of f. For example, let $h(x)$ be any polynomial in $\mathbb{R}[x]$ and let $f(x) = x + (1 + x^2)h(x)$. Then

$$\theta_i(f(x)) = f(i) = i + (1 + i^2)h(i) = i + 0 = i$$

since $1 + i^2 = 0$.

The crucial point of this illustration is that i is a root of $1 + x^2$ and hence also a root of $(1 + x^2)h(x)$ for any polynomial $h(x)$. Moreover, $1 + x^2$ is the monic polynomial of smallest degree having i as a root and $1 + x^2$ is an irreducible polynomial in $\mathbb{R}[x]$.

This example illustrates the statements of the next theorem.

Theorem 4.1 *Let L be an extension field of F and c an element of L which is a root of some nonzero polynomial in $F[x]$. Let $g(x)$ be a nonzero polynomial in $F[x]$ of least possible degree which has c as a root. Then*

(i) *$g(x)$ is an irreducible polynomial in $F[x]$;*
(ii) *if $h(x)$ is a polynomial in $F[x]$ such that $h(c) = 0$, then $h(x) = g(x)q(x)$ for some $q(x) \in F[x]$.*

Proof: In order to show that $g(x)$ is irreducible, we suppose that $g(x) = u(x)v(x)$ for two polynomials $u(x), v(x) \in F[x]$ and we must show that either $\deg u(x) = 0$ or $\deg v(x) = 0$. Evaluate at c to obtain $g(c) = 0 = u(c)v(c)$. Since $u(c)$ and $v(c)$ are two elements of a field L whose product equals zero, one of the elements is zero. To be specific, let us suppose that $u(c) = 0$. Then c is a root of $u(x)$. By assumption, $g(x)$ has the smallest degree of a polynomial in $F[x]$ that has c as a root. Hence, $\deg g(x) \leq \deg u(x)$. On the other hand, the equality $g(x) = u(x)v(x)$ implies $\deg g(x) = \deg u(x) + \deg v(x)$. Since the degree of the polynomial $v(x)$ is not negative, it follows that $\deg g(x) \geq \deg u(x)$. The combination of the two inequalities implies $\deg g(x) = \deg u(x)$ and also that $\deg v(x) = 0$, as we wished to prove.

Now suppose $h(x)$ is a polynomial in $F[x]$ with $h(c) = 0$. In order to show that $g(x)$ divides $h(x)$, we apply the division algorithm to obtain

$$h(x) = g(x)q(x) + r(x), \quad \text{with either } r(x) = 0 \text{ or } \deg r(x) < \deg g(x).$$

Now evaluate at c and use the fact that c is a root of both $g(x)$ and $h(x)$ to conclude

$$0 = h(c) = g(c)q(c) + r(c) = r(c).$$

Thus, $r(c) = 0$. If $r(x) \neq 0$, then $r(x)$ is a polynomial of degree smaller than the degree of $g(x)$ and also having c as a root. This is not possible by our choice of $g(x)$ as the polynomial of least degree having c as a root. The only alternative is $r(x) = 0$, from which we conclude that $g(x)q(x) = h(x)$. ■

This result gives a description of all the polynomials in $F[x]$ having c as a root; they are all the products formed with one particular irreducible polynomial and an arbitrary element of $F[x]$.

We can reformulate this statement using the language of ideals.

Corollary 4.1 *Let L be an extension field of F and c an element of L which is a root of some nonzero polynomial in $F[x]$. Let $g(x)$ be a nonzero polynomial in $F[x]$ of least possible degree which has c as a root. Then the principal ideal $(g(x))$ in the ring $F[x]$ is the kernel of the evaluation map $f(x) \to f(c)$.*

Proof: The statements previously made contain a proof of these assertions. ■

Let us apply these ideas to the special case in which c is an element of F rather than just an element of a larger field L.

Theorem 4.2

THE FACTOR THEOREM. *Let c be an element of F. The elements of $F[x]$ which have c as a root are those of the form $(x - c)h(x)$, where $h(x)$ is any polynomial in $F[x]$.*

Proof: Since $c \in F$, the polynomial $x - c$ belongs to $F[x]$ and clearly has c as a root. Moreover, it is a nonzero polynomial of least possible degree having c as a root. By Theorem 4.1, every polynomial having c as a root is divisible by $x - c$ and must have the form $(x - c)h(x)$ for some element $h(x) \in F[x]$. ■

As an application of this result, we obtain a restriction on the number of roots of a polynomial.

Theorem 4.3

A polynomial in $F[x]$ of degree n has at most n roots in an extension field of F.

Proof: Let L be an extension field of F. Since a polynomial with coefficients in F may be viewed as a polynomial with coefficients in L, we will prove the theorem in a slightly different form by showing that a polynomial with coefficients in L has no more roots than its degree.

We prove the theorem by using mathematical induction on the degree of polynomials. Let S_n be the statement

> S_n: If $f(x)$ is a polynomial in $L[x]$ with degree n, then $f(x)$ has at most n distinct roots in L.

First we show that S_1 is true. Let $f(x)$ be a polynomial of degree 1 so that $f(x) = ax + b$ with $a, b \in L$ and $a \neq 0$. If $f(c) = 0$, then $ac + b = 0$ and $c = -ba^{-1}$. Thus, there is exactly one root and so S_1 is true.

Now assume that the statements $S_1, S_2, \ldots, S_{n-1}$ are true. We will prove that S_n is true and so the induction principle allows us to conclude that S_m is true for all positive integers m.

Let $f(x) \in L[x]$ have degree n and let $c_1, \ldots, c_k$ be k distinct elements of L which are roots of $f(x)$. Since $f(c_k) = 0$, the factor theorem implies

$$f(x) = h(x)(x - c_k), \qquad h(x) \in L[x].$$

By assumption, $f(c_i) = h(c_i)(c_i - c_k) = 0$. Thus, c_i must be a root of $h(x)$ for $i = 1, 2, \ldots, k - 1$ since $c_i - c_k \neq 0$ (because the elements are distinct). Clearly, the degree of $h(x)$ is $n - 1$ and, since statement S_{n-1} is true, the number of distinct roots of $h(x)$ is at most $n - 1$. The $k - 1$ elements

$c_1, \ldots, c_{k-1}$ are roots of $h(x)$ so it follows that $k - 1 \leq n - 1$. Add 1 to conclude $k \leq n$ as we wished to prove. ■

The following are examples showing the relation between roots and degree -1 factors.

Example 4.1 Consider the rational number field $\mathbb{Q}$ and the polynomial $f(x) \in \mathbb{Q}[x]$ given by

$$f(x) = x^3 - x^2 - 3x - 1.$$

Suppose it is required to express $f(x)$ as a product of prime polynomials. It is necessary to search for prime factors of $f(x)$ and one way to do this is to search for roots in the rational number field. By inspection we discover that $f(-1) = 0$ so that -1 is a root. By the factor theorem, $x - (-1) = x + 1$ is a factor of $f(x)$ and so

$$f(x) = (x + 1)h(x)$$

for some polynomial $h(x)$. By long division we find $h(x)$ to obtain

$$f(x) = (x + 1)(x^2 - 2x - 1).$$

Next we must factor the quadratic term or determine that it is irreducible. By using the quadratic formula , the roots are found to be $1 \pm \sqrt{2}$. We have already seen that $\sqrt{2}$ is not a rational number and it follows that neither of the roots, $1 \pm \sqrt{2}$, are rational and so the polynomial $x^2 - 2x - 1$ has no degree 1 factors with rational coefficients.

Next we present an example of quite a different sort.

Example 4.2 Let $\mathbb{Z}_2$ be the ring of integers modulo 2. This is a field with two elements which we write as 0 and 1 rather than $[0]$ and $[1]$, which was written earlier. Let us show that the polynomial $f(x) = 1 + x + x^2$ is irreducible in $\mathbb{Z}_2[x]$. If it were reducible, it would have to be the product of two polynomials of degree 1. Any polynomial of degree 1 has a root in $\mathbb{Z}_2$. Let us test $f(x)$ to determine if it has any roots in $\mathbb{Z}_2$:

$$f(0) = 1, \quad \text{and} \quad f(1) = 1 + 1 + 1 = 1 \quad (\text{since } 1 + 1 = 0 \text{ in } \mathbb{Z}_2).$$

Thus, $f(x)$ has no roots in $\mathbb{Z}_2$ and it must be irreducible.

We conclude this section with a discussion of the ring $F[c]$ when c is taken from an extension field of F.

Theorem 4.4 *Let L be an extension field of F and c an element of L.*

(i) *If c is not the root of any nonzero polynomial in $F[x]$, then the ring $F[c]$ is isomorphic to $F[x]$. In particular, $F[c]$ is not a field.*

(ii) *If c is the root of a nonzero polynomial in $F[x]$, then $F[c]$ is a field consisting of all elements of the form $a_0 + a_1c + a_2c^2 + \cdots + a_{n-1}c^{n-1}$,* *where $a_i \in F$ and n is the smallest degree of a nonzero polynomial in $F[x]$ having c as a root.*

Proof: The evaluation map $\theta_c : f(x) \to f(c)$ is a homomorphism from $F[x]$ onto the subring $F[c]$. If c is not the root of any nonzero polynomial in $F[x]$, then the kernel of θ_c is (0). Since θ_c maps $F[x]$ onto $F[c]$ with kernel zero, it follows that θ_c is an isomorphism of rings.

Now suppose c is the root of some polynomial in $F[x]$ and let $g(x)$ be a polynomial of smallest degree having c as a root. Since $F[c]$ is a subring of the field L, it is an integral domain. To show that it is a field, we need to show that every nonzero element of $F[c]$ has a multiplicative inverse in $F[c]$. Let $z \in F[c]$ be a nonzero element and let $u(x) \in F[x]$ be any polynomial such that $z = u(c)$. Since $z \neq 0$, we know $g(x)$ does not divide $u(x)$ [otherwise, $u(c) = 0$]. By Theorem 4.1, we know that $g(x)$ is an irreducible polynomial. It follows that the GCD of $g(x)$ and $u(x)$ is 1. Now apply Theorem 2.2 to see that there exist polynomials $a(x)$ and $b(x)$ such that

$$u(x)a(x) + g(x)b(x) = 1.$$

Now evaluate at c to get

$$1 = u(c)a(c) + g(c)b(c) = za(c)$$

since $g(c) = 0$. Thus, z has an inverse and we have proved the theorem.

The last statement requires the division algorithm. Let $u(x)$ be any polynomial; by the division algorithm we may write $u(x) = g(x)q(x) + r(x)$, where either $r(x) = 0$ or $\deg r(x) \leq n - 1$, $n = \deg g(x)$. Evaluate at c to obtain $u(c) = r(c)$ since $g(c) = 0$. Thus, every nonzero element in $F[c]$ is equal to some $r(c)$ with $\deg r(x) \leq n - 1$. ∎

Let us illustrate the content of this theorem.

Example 4.3

Take $F = \mathbb{Q}$, the rational number field, and $L = \mathbb{C}$, the complex field. If $c = \pi$, the ratio of the circumference of a circle to its diameter, then $\mathbb{Q}[\pi]$ is isomorphic to the ring of polynomials $\mathbb{Q}[x]$ because π is not the root of any polynomial with rational number coefficients. This is not at all obvious and is considered quite a difficult theorem that requires fairly subtle theorems in analysis. We will not give a proof of this fact. A complex number that is not the root of any polynomial with rational coefficients is called a *transcendental* number. The base e of the natural logarithm is another example of a transcendental number.

On the other hand, $c = \sqrt[3]{2}$ is the root of the polynomial $g(x) = x^3 - 2 \in \mathbb{Q}[x]$ and so $\mathbb{Q}[c]$ is a field consisting of all numbers of the form

$$p + qc + rc^2 = p + q\sqrt[3]{2} + r(\sqrt[3]{2})^2, \qquad p, q, r \in \mathbb{Q}.$$

We leave it to the reader to verify (or take on faith) that c is not the root of any polynomial of degree 1 or 2 in $\mathbb{Q}[x]$ and so $g(x)$ is irreducible.

We illustrate another point of the proof by finding the inverse of the element $z = 1 + \sqrt[3]{2}$, which equals $h(c)$ when $h(x) = 1 + x$. Since $g(x)$ and $h(x)$ are relatively prime we find the linear combination

$$(x^3 - 2) \cdot 1 - (x^2 - x + 1)(x + 1) = (x^3 - 2) - (x^3 + 1) = -3.$$

Substitute c for x and divide by -3 to conclude

$$\frac{c^2 - c + 1}{3}(1 + c) = 1,$$

which gives the inverse of $1 + c$ as $(c^2 - c + 1)/3$. We will return to the investigation of fields constructed this way in the next section with a slightly different point of view.

EXERCISES

1. Prove that a polynomial of degree 2 or 3 over a field F is a prime polynomial if and only if the polynomial does not have a root in F.

2. Give an example of a polynomial of degree 4 over a field F that has no root in F but is not a prime polynomial.

3. Determine whether or not each of the following polynomials is prime over each of the given fields. If it is not prime, factor it as a product of prime polynomials over each field. As usual, $\mathbb{Q}$ is the field of rational numbers, $\mathbb{R}$ is the field of real numbers, and $\mathbb{C}$ is the field of complex numbers.

(a) $x^2 + x + 1$ over $\mathbb{Q}$, $\mathbb{R}$, and $\mathbb{C}$
(b) $x^2 + 2x - 1$ over $\mathbb{Q}$, $\mathbb{R}$, and $\mathbb{C}$
(c) $x^2 + 3x - 4$ over $\mathbb{Q}$, $\mathbb{R}$, and $\mathbb{C}$
(d) $x^2 + 2$ over $\mathbb{Q}$, $\mathbb{R}$, and $\mathbb{C}$
(e) $x^2 + \sqrt{15}$ over $\mathbb{R}$ and $\mathbb{C}$
(f) $x^2 + 4$ over $\mathbb{Q}$ and $\mathbb{Q}[i]$
(g) $x^2 + 2\sqrt{2}x + 2$ over $\mathbb{Q}[\sqrt{2}]$ and $\mathbb{R}$

4. Find an irreducible polynomial in $\mathbb{Q}[x]$ which has the given number as a root:

(a) $1 + \sqrt{2}$ (b) $3 - 4\sqrt{3}$ (c) $5 - 2i$ (d) $7 - \sqrt{5}$

5. Let $f(x) = a_0 + a_1x + \cdots + a_nx^n$ be a polynomial over the field F and let $b \in F$. Denote by $f(x + b)$ the polynomial obtained by substituting $x + b$ for x:

$$f(x + b) = a_0 + a_1(x + b) + \cdots + a_n(x + b)^n.$$

Prove that $f(x)$ is irreducible if and only if $f(x+b)$ is irreducible. Moreover, show that if c is a root of $f(x)$ in some extension field of F, then $c-b$ is a root of $f(x+b)$.

6. If c is a root of $f(x)$ and $f(x)=(x-c)^m p(x)$ for a polynomial $p(x)$ not having c as a root, then m is called the *multiplicity* of c. In each example, find the multiplicity of the given root.
(a) $x^4+x^3-3x^2-5x-2$ over $\mathbb{Q}$, root $c=-1$
(b) x^4+2x^2+1 over $\mathbb{C}$, root i

7. If $f(x)=a_0+a_1x+\cdots+a_nx^n$ is a polynomial over a commutative ring S, let us define the *derivative* $f'(x)$ to be the polynomial

$$f'(x)=a_1+2a_2x+\cdots+ja_jx^{j-1}+\cdots+na_nx^{n-1}.$$

Prove that the familiar rules for the derivative of a sum and product are valid in this general setting:

$$\begin{aligned}[f(x)+g(x)]' &= f'(x)+g'(x)\\ [f(x)g(x)]' &= f(x)g'(x)+f'(x)g(x).\end{aligned}$$

8. Suppose that $f(x)$ is a polynomial over a field F and let $f'(x)$ be its derivative (defined in the previous exercise). Prove each of the following:
(a) If the multiplicity of a root c of $f(x)$ is greater than one, then c is also a root of $f'(c)$.
(b) If b is a root of $f(x)$ and also a root of $f'(x)$, then the multiplicity of b as a root of $f(x)$ is greater than one.

9. Prove the remainder theorem: If $f(x)$ is a polynomial in $F[x]$ and c is any element of F, then $f(x)=(x-c)h(x)+f(c)$ for some element $h(x)\in F[x]$. [Hint: Apply the factor theorem to $f(x)-f(c)$.]

10. Each polynomial $f(x)$ in $F[x]$ determines a function from F to F by the rule $c\to f(c)$. Such a function is called a *polynomial function* from F to F. Prove that different polynomials determine different functions when F is an infinite field.

11. If $F=\mathbb{Z}_p$, the field with p elements, show that

$$g(x)=\prod_{j=0}^{p-1}(x-j)$$

is a nonzero polynomial that determines the same polynomial function as the zero polynomial. [Here we use the product symbol Π to indicate a product of the factors $(x-i)$ for $0\le i\le p-1$. It is the multiplicative analog of $\sum$ for sums.]

12. Let S be a commutative ring and $f(x)$ a polynomial of degree n in $S[x]$. Give an example to show that it is possible to have $f(c) = 0$ for more than n distinct elements $c \in S$. [Hint: Try $S = \mathbb{Z}_{25}$ and $f(x) = x^2$. Many other examples also illustrate this.]

5 FACTOR RINGS OF $F[x]$

In the previous section we considered a field F and an extension field L and proved facts about a polynomial in $F[x]$ that has an element $c \in L$ as a root. In that context, we selected an element c that was the root of some polynomial with coefficients in F. In this section we change the viewpoint and begin with a polynomial $f(x)$ having coefficients in F and ask if there is a field L containing F in which $f(x)$ has a root. For example, we know that the polynomial $x^2 + 1$ has no roots in the field $\mathbb{R}$ of real numbers but it does have a root in the field $\mathbb{C}$ of complex numbers which contains $\mathbb{R}$. What about the polynomial $x^8 + 3x^2 + 15$? Does it have a root in some extension field of $\mathbb{R}$? Does the polynomial $x^{25} + 17x^6 - 34x + 17 \in \mathbb{Q}[x]$ have a root in some extension field of $\mathbb{Q}$? We will prove that every polynomial in $F[x]$, for any field F, has a root in some extension field of F. This assertion will be proved using the polynomial $f(x)$ to construct a ring and then showing that this ring is actually a field that contains a root of $f(x)$. The notion of ideals of the polynomial ring and the factor ring modulo an ideal as described in Chapter II, Theorem 6.3 provides the necessary tools.

Let $f(x)$ be a nonzero polynomial in $F[x]$. The principal ideal $(f(x))$ is the set of all products $f(x)u(x)$, where $u(x)$ is an arbitrary element in $F[x]$. The factor ring $F[x]/(f(x))$ is the set of all cosets $v(x) + (f(x))$ with $v(x) \in F[x]$. We first state a fact about the representation of these cosets. The coset $0 + (f(x))$ is called the zero coset; all others are nonzero cosets.

Lemma 5.1 *Let $n = \deg f(x)$. Then every nonzero coset contains one and only one polynomial of degree less than n.*

Proof: Let $U = u(x) + (f(x))$ be any coset. Apply the division algorithm to divide $u(x)$ by $f(x)$ and get

$$u(x) = f(x)q(x) + r(x), \quad \text{with either} \quad r(x) = 0 \quad \text{or} \quad \deg r(x) < n.$$

Then $u(x) + (f(x)) = r(x) + (f(x))$. Since U is not the zero coset, we have $r(x) \neq 0$ and so the coset U contains at least one polynomial of degree less than n. If there were another polynomial $w(x)$ in U with $\deg w(x) < n$, then $r(x) - w(x) \in (f(x))$ and so $r(x) - w(x) = f(x)u(x)$ for some polynomial $u(x)$. However, the degrees of $r(x)$ and $w(x)$ are both smaller than n and therefore the degree of their difference is also smaller than n. If $u(x) \neq 0$, then $\deg f(x)u(x) = \deg f(x) + \deg u(x) \geq n$. This is an impossible situation and the only possibility is that $u(x) = 0$; therefore, $r(x) = w(x)$. Thus, the coset U contains only one polynomial with degree less than n. ∎

Next we introduce an alternative notation for the elements of the factor ring $F[x]/(f(x))$. We first denote the coset containing an element c of F as c'; thus, $c' = c + (f(x))$.

Lemma 5.2 *The mapping $c \rightarrow c'$ sends F onto an isomorphic subring of $F[x]/(f(x))$.*

Proof: We first argue that the correspondence is a homomorphism; for $c, d \in F$ we have

$$\begin{aligned} (c+d)' &= (c+d) + (f(x)) = c + (f(x)) + d + (f(x)) = c' + d' \\ (cd)' &= cd + (f(x)) = (c + (f(x)))\,(d + (f(x))) = c'd'. \end{aligned}$$

Next we argue that the homomorphism is one-to-one. It is sufficient to show that no nonzero element is mapped to zero. If $c \in F$ and $c \neq 0$, then $c + (f(x)) \neq 0 + (f(x))$ since $c - 0$ is not divisible by $f(x)$. Thus, the set of all cosets containing an element of F is a subring of $F[x]/(f(x))$ isomorphic to F. ∎

Next we introduce an abbreviated notation for the coset containing x: Let

$$j = x + (f(x)).$$

Then $j^2 = (x + (f(x)))(x + (f(x))) = x^2 + (f(x))$, and more generally for each positive integer m we have

$$j^m - x^m + (f(x)).$$

Every coset can be uniquely written in the form

$$b_0 + b_1 x + \cdots + b_{n-1} x^{n-1} + (f(x))$$

with $b_i \in F$. Each term $bx^i + (f(x))$ may be written more succinctly as

$$bx^i + (f(x)) = b' j^i,$$

and the general element of $F[x]/(f(x))$ may be expressed as

$$b'_0 + b'_1 j + \cdots + b'_{n-1} j^{n-1} = b_0 + b_1 x + \cdots + b_{n-1} x^{n-1} + (f(x)).$$

If $h(x) \in F[x]$, then the coset $h(x) + (f(x))$ may be regarded as the polynomial $h(x)$ evaluated at j; namely, $h(j)$. An example might help make this more clear. Suppose $h(x) = a + bx + cx^2$; then

$$\begin{aligned} h(x) + (f(x)) &= a + bx + cx^2 + (f(x)) \\ &= a'(1 + (f(x))) + b'(x + (f(x))) + c'(x + (f(x)))^2 \\ &= a' + b'j + c'j^2. \end{aligned}$$

The only point that requires some alteration is the replacement of the primed elements by elements of F; that is, we must agree that $b'j$ is the same as bj for $b \in F$. This amounts to identifying the field element b of F with its isomorphic image b'. With this understanding we see that

$$h(x) + (f(x)) = a' + b'j + c'j^2 = a + bj + cj^2 = h(j).$$

This is not just an exercise in manipulating symbols. We have constructed a ring that contains (an isomorphic copy of) F such that elements of the ring may be used to evaluate polynomials over F. Note what happens when we evaluate $f(x)$ at j:

$$f(j) = f(x) + (f(x)) = 0 + (f(x));$$

that is, $f(j) = 0$. In other words, we have a ring that contains F and a root of $f(x)$. What remains to do is to argue that we have constructed a field. However, this is not always true. The ring $F[x]/(f(x))$ is a field precisely when the polynomial $f(x)$ is irreducible, as we now prove.

Theorem 5.1 *Let F be any field and $f(x)$ a nonzero polynomial in $F[x]$. If $f(x)$ is irreducible, the ring $F[x]/(f(x))$ is a field containing F and a root of $f(x)$.*

Proof: Since $F[x]/(f(x))$ is a commutative ring, it is only necessary to show that every nonzero element has a multiplicative inverse. Let $u(x) + (f(x))$ be a nonzero coset so then $u(x)$ is a polynomial not divisible by $f(x)$. Since $f(x)$ is irreducible, the GCD of $f(x)$ and $u(x)$ is 1. By the Euclidean algorithm for polynomials (see Theorem 2.2), there exist $a(x)$ and $b(x)$ such that $a(x)u(x) + b(x)f(x) = 1$. Hence,

$$\begin{aligned} 1 + (f(x)) &= a(x)u(x) + b(x)f(x) + (f(x)) \\ &= a(x)u(x) + (f(x)) \\ &= \big(a(x) + (f(x))\big)\big((u(x) + (f(x))\big). \end{aligned}$$

This shows that the coset of $a(x)$ is the inverse of the coset of $u(x)$. Thus, every nonzero coset has an inverse and $F[x]/(f(x))$ is a field. The other assertions in the theorem have already been proved. ∎

Example 5.1 Let F be the field $\mathbb{R}$ of real numbers and let $f(x)$ be the polynomial $x^2 + 1$. Since $x^2 + 1$ is a prime polynomial, we know by Theorem 5.1 that the ring $\mathbb{R}[x]/(x^2 + 1)$ is a field. Moreover, every element in this field can be written as $a + bj$ with $a, b \in \mathbb{R}$ and where j is the coset $x + (x^2 + 1)$. Let us examine the multiplication rules for these elements. In particular,

$$j^2 = x^2 + (x^2 + 1) = -1 + x^2 + 1 + (x^2 + 1) = -1 + (x^2 + 1).$$

Since we agree to write the coset $-1 + (x^2 + 1)$ more simply as just -1, we find that $j^2 = -1$. In other words,

$$(a + bj)(c + dj) = ac + adj + bcj + bdj^2 = ac - bd + (ad + bc)j.$$

Along with the rule for addition,

$$(a + bj) + (c + dj) = (a + c) + (b + d)j,$$

we see that these rules are identical with the rules given earlier for multiplication and addition of complex numbers—with the sole change that we are now writing j where we wrote i previously. More formally, we may say that the map $\mathbb{R}[x]/(x^2 + 1) \longrightarrow \mathbb{C}$ given by $a + bj \to a + bi$ is an isomorphism of fields. Hence, we have another way of constructing the complex numbers from the field of real numbers.

Alternatively, one could reach this conclusion by considering the evaluation map $\theta_i : \mathbb{R}[x] \to \mathbb{C}$ that sends $f(x)$ to $f(i)$. Then $x^2 + 1$ is in the kernel of this homomorphism, and since $x^2 + 1$ is a prime polynomial it follows that the kernel of θ_i is the ideal $(x^2 + 1)$. The image of θ_i is all of $\mathbb{C}$ as can be seen after a short computation. Now the First Isomorphism Theorem asserts that $\mathbb{R}[x]/(x^2 + 1)$ is isomorphic to $\mathbb{C}$.

Example 5.2

Let F be the field $\mathbb{Z}_2$ of integers modulo 2, whose elements are written as 0 and 1. The polynomial $f(x) = 1 + x + x^2$ is irreducible in $\mathbb{Z}_2[x]$ because it has no roots in $\mathbb{Z}_2$ (as we see by testing the two elements of $\mathbb{Z}_2$). Hence, the ring $\mathbb{Z}_2[x]/(1+x+x^2)$ is a field, which we will denote as K. The polynomial $f(x)$ has degree 2 so the elements of K are (uniquely) expressible as

$$a + bj, \qquad a, b \in \mathbb{Z}_2.$$

There are two possibilities for both a and b so K is a field with four elements. As usual, addition of two of these elements is carried out as follows:

$$(a + bj) + (c + dj) = (a + c) + (b + d)j.$$

Regarding multiplication, we have

$$(a + bj)(c + dj) = ac + adj + bcj + bdj^2.$$

We know in general that $f(j) = 0$, so in this example that means $j^2+j+1 = 0$; equivalently, $j^2 = j + 1$. (Remember that in $\mathbb{Z}_2$, $+1 = -1$.) Therefore, the rule for multiplication is obtained by replacing j^2 with $j + 1$ to get

$$(a+bj)(c+dj) = ac+adj+bcj+bd(j+1) = ac+bd+(ad+bc+bd)j.$$

The four elements of K are $0, 1$, and $j, 1 + j$ and this field is small enough to express the addition and multiplication rules in explicit tables.

$(+)$	0	1	j	$1+j$
0	0	1	j	$1+j$
1	1	0	$1+j$	j
j	j	$1+j$	0	1
$1+j$	$1+j$	j	1	0

$(\cdot)$	0	1	j	$1+j$
0	0	0	0	0
1	0	1	j	$j+1$
j	0	j	$j+1$	1
$1+j$	0	$1+j$	1	j

It is not too difficult to show that any field containing exactly four elements is isomorphic to the field we have just constructed (see the exercises).

EXERCISES

1. Find the unique element in each coset z of $\mathbb{Q}[x]/(g(x))$ which has degree less than the degree of $g(x)$:
(a) $z = x^3 - 3x + 2 + (g(x))$, $g(x) = x^2 - 2$
(b) $z = x^4 + x^2 + 1 + (g(x))$, $g(x) = x^2 + x + 1$
(c) $z = x^4 + (g(x))$, $g(x) = x^2 - x + 5$
(d) $z = \big(x^2 + 4 + (g(x))\big)\big(x^3 - 3x + (g(x))\big)$, $g(x) = x^2 + 1$

2. In the field $\mathbb{Q}[x]/(g(x))$, find the multiplicative inverse of z:
(a) $z = x + 1 + (g(x))$, $g(x) = x^2 - 2$
(b) $z = x + 1 + (g(x))$, $g(x) = x^2 + 2$
(c) $z = 3x^2 - 5 + (g(x))$, $g(x) = x^3 - 5$

3. The polynomial $x^2 + x + 1$ is irreducible over the real number field $\mathbb{R}$ but it has a complex root. Use this root to establish an explicit isomorphism from $\mathbb{R}[x]/(x^2 + x + 1)$ to $\mathbb{C}$.

4. Let $g(x) = x^2 + ax + b$ be an irreducible polynomial over $\mathbb{R}$. Establish an explicit isomorphism of $\mathbb{R}[x]/(g(x))$ with $\mathbb{C}$.

5. Show that the polynomial $x^2 - 2$ is irreducible in $\mathbb{Q}[x]$. Verify that $\mathbb{Q}[x]/(x^2 - 2)$ is a field isomorphic to the set of all real numbers of the form $a + b\sqrt{2}$, $a, b \in \mathbb{Q}$.

6. Let L be a field with exactly four elements with zero element 0 and multiplicative identity 1.
(a) The map from $\mathbb{Z}$ to L defined by $m \to m \cdot 1$ is a homomorphism from the integers $\mathbb{Z}$ onto an integral domain. Argue that the kernel must be the ideal (2) and that $\{0, 1\}$ is a subfield of L isomorphic to $\mathbb{Z}_2$.
(b) Regarding $\mathbb{Z}_2$ as a subfield of L, pick an element $c \in L$ but $c \notin \mathbb{Z}_2$ and consider the evaluation map $\mathbb{Z}_2[x] \to L$ defined by $f(x) \to f(c)$. Show that this map is a homomorphism of $\mathbb{Z}_2[x]$ *onto* L and the

kernel must be an ideal $(f(x))$ with $f(x)$ an irreducible polynomial of degree 2.

(c) Verify the assertion that there is only one irreducible polynomial of degree 2 in $\mathbb{Z}_2[x]$ and then use this to verify the claim that there is only one field (up to isomorphism) with four elements as asserted in the body of this section.

7. The polynomial x^2 is not irreducible in $\mathbb{Q}[x]$ and so the ring $\mathbb{Q}[x]/(x^2)$ is not a field. Find all the nonzero elements of $\mathbb{Q}[x]/(x^2)$ that have multiplicative inverses.

8. Show that $g(x) = x^3 - x - 1$ is an irreducible polynomial in $\mathbb{Q}[x]$ and that the field $\mathbb{Q}[x]/(g(x))$ is isomorphic to the set of all real numbers of the form $a + b\sqrt[3]{2} + c\sqrt[3]{4}$, with $a, b, c \in \mathbb{Q}$.

9. Verify that $\mathbb{Q}[x]/(x^3 - p)$ is a field for any prime integer p and describe a field of real numbers to which it is isomorphic.

10. For the given field F and the given polynomial $f(x)$ over F, construct a multiplication table for the ring $F[x]/(f(x))$. Which of these rings are fields?

(a) $F = \mathbb{Z}_2$, $f(x) = x^2 + 1$ (b) $F = \mathbb{Z}_3$, $f(x) = x^2 + 1$

(c) $F = \mathbb{Z}_3$, $f(x) = x^2 + x + 1$ (d) $F = \mathbb{Z}_2$, $f(x) = x^3 + x + 1$

(e) $F = \mathbb{Z}_5$, $f(x) = x + 1$

11. An ideal P in a commutative ring R is called a *prime ideal* if $a, b \in R$ and $ab \in P$, then either $a \in P$ or $b \in P$. Show that a prime ideal in $F[x]$, F a field, is a principal ideal generated by an irreducible polynomial.

12. Prove that if R is a commutative ring with identity, an ideal $P \neq R$ is a prime ideal if and only if the factor ring R/P is an integral domain.

6 SPLITTING FIELDS

In the previous section, we determined that an irreducible polynomial $f(x) \in F[x]$ has a root in some field L containing F. In this section we complete this circle of ideas by showing that there is a field containing F and elements $c_1, \ldots, c_n$ such that $f(x)$ equals a product of the factors $(x - c_i)$ and some element of F. In other words, there is a field extension of F over which $f(x)$ factors as a product of factors of degree 1.

Theorem 6.1 *Let F be any field and $f(x)$ a polynomial of positive degree n in $F[x]$. Then there is a field extension L of F such that*

$$f(x) = a(x - c_1)(x - c_2)\cdots(x - c_n) \tag{5.3}$$

with a = leading coefficient of $f(x)$ and $c_1, c_2, \ldots, c_n$ being elements of L.

Proof: We do not insist that $f(x)$ is an irreducible polynomial, so let $f_1(x)$ be an irreducible factor of $f(x)$. Any root of $f_1(x)$ is also a root of $f(x)$ (but not the other way around, of course). By Theorem 5.1, there is a field extension L_1 of F that contains a root c_1 of $f_1(x)$, and hence also of $f(x)$. By the factor theorem, $f(x) = (x - c_1)h(x)$ for some polynomial $h(x)$ in $L_1[x]$. If $h(x)$ has positive degree, repeat this argument using $h(x)$ in place of $f(x)$ and L_1 in place of F. We conclude that there is a field L_2 containing L_1 and a root c_2 of $h(x)$. At this point we find

$$f(x) = (x - c_1)h(x) = (x - c_1)(x - c_2)h_2(x)$$

for some polynomial $h_2(x) \in L_2[x]$. After repeating this sufficiently often, we obtain the factorization 5.3 with the roots lying in some field containing F. ■

A field for which a given polynomial $f(x)$ has the factorization (5.3) is called a *splitting field* of $f(x)$. For example, the field $\mathbb{C}$ is a splitting field over $\mathbb{R}$ of the polynomial $x^2 + 1$ because

$$x^2 + 1 = (x + i)(x - i), \qquad \pm i \in \mathbb{C}.$$

As another example, consider the polynomial $x^3 - 1$ in $\mathbb{Q}[x]$. The polynomial is reducible, having 1 as a root which factors as $(x - 1)(x^2 + x + 1)$. The other two roots are found by the quadratic formula to be $w_1, w_2 = (-1 \pm i\sqrt{3})/2$, and so

$$x^3 - 1 = (x - 1)(x - w_1)(x - w_2).$$

Since the field $\mathbb{Q}[i\sqrt{3}]$ contains all three roots of the polynomial, $\mathbb{Q}[i\sqrt{3}]$ is a splitting field of $x^3 - 1$ over $\mathbb{Q}$.

We generalize these examples by considering the polynomial $x^n - 1 \in \mathbb{Q}[x]$ for any positive integer n. The roots of this polynomial are the nth roots of unity and are given by DeMoivre's formula,

$$\omega_k = \cos\left(\frac{2\pi k}{n}\right) + i\sin\left(\frac{2\pi k}{n}\right).$$

There are n distinct numbers $\omega_1, \omega_2, \ldots, \omega_{n-1}, \omega_n = 1$ and all are solutions of the equation $x^n - 1 = 0$. Since one easily verifies that

$$\omega_j = (\omega_1)^j, \qquad j = 1, 2, \ldots, n$$

it follows that $\mathbb{Q}[\omega_1]$ contains all the roots of $x^n - 1$. The factorization is

$$x^n - 1 = \prod(x - \omega_j) = \prod(x - \omega_1^j),$$

and it follows that $\mathbb{Q}[\omega_1]$ is the smallest field containing all the roots of $x^n - 1$ over $\mathbb{Q}$.

The field $\mathbb{Q}[\omega_1]$ is usually called the *cyclotomic* field of nth roots of unity over $\mathbb{Q}$.

In general, if $f(x)$ has the factorization 5.3 over F, then the smallest field containing F and the roots c_i is denoted by $F[c_1, \ldots, c_n]$. It can be shown that this field is uniquely determined up to isomorphism by $f(x)$ and F. We do not give the proof of this here. It may be considered the beginning of the subject known as Galois theory.

6.1 Finite Fields

Now we discuss finite fields. We have already seen that for each prime integer p, the ring $\mathbb{Z}_p$ of integers modulo p is a field with p elements. We will now produce other finite fields containing p^n elements for any positive integer n and prime p.

We are interested in fields L that contain an isomorphic copy of $\mathbb{Z}_p$. For such a field we know that $py = 0$ for every element $y \in L$. This follows because the identity 1 of $\mathbb{Z}_p$ must also be the identity of L (there is only one nonzero element in any field that satisfies $y^2 = y$). Thus,

$$py = p(1y) = (p1)y = 0y = 0$$

because $p1 = 0$ in $\mathbb{Z}_p$. The following lemma is a very useful consequence of this; it will be used frequently when dealing with fields that contain $\mathbb{Z}_p$. It is convenient to prove a slightly more general version that holds in any commutative ring S with the property $ps = 0$ for every $s \in S$.

Lemma 6.1 *Let a and b be any elements of a commutative ring in which $ps = 0$ for every element s in the ring. Then*

$$(a + b)^p = a^p + b^p.$$

Proof: By the binomial theorem (see Chapter I, Section 6) we have

$$(a + b)^p = a^p + pa^{p-1}b + \cdots + C(p, j)a^{p-j}b^j + \cdots + b^p,$$

where

$$C(p, j) = \frac{p!}{j!(p - j)!}.$$

We know that $C(p, j)$ is an integer expressed as a quotient of its numerator

$$p! = 2 \cdot 3 \cdots (p - 1) \cdot p$$

and its denominator

$$j!(p - j)! = 2 \cdot 3 \cdots (j - 1) \cdot j \cdot 2 \cdot 3 \cdots (p - 1 - j) \cdot (p - j).$$

Now assume that $j \neq 0$ or p. Then visibly the denominator is a product of integers each less than p, and so the denominator is not divisible by p. However, the numerator is clearly divisible by p. It follows that the factor p in the numerator is not canceled by any factor in the denominator and so the integer $C(p, j)$ is divisible by p if $j \neq 0, p$. From the previous discussion, it follows that $C(p, j) = p \cdot s$ for an integer s and

$$C(p, j)a^{p-j}b^j = p(sa^{p-j}b^j) = 0.$$

The only nonzero terms remaining in the binomial expansion of $(a + b)^p$ are the first and last corresponding to $j = 0$ and p; thus,

$$(a + b)^p = a^p + b^p$$

as required. ∎

As an easy application of mathematical induction, the reader may prove the following generalization:

Corollary 6.1 *Let a and b be any elements of a commutative ring in which $ps = 0$ for every element s. Then, for any positive integer k,*

$$(a + b)^{p^k} = a^{p^k} + b^{p^k}.$$

This corollary applies to elements of $\mathbb{Z}_p[x]$. Thus, we have expressions such as $(x + 1)^p = x^p + 1$. In fact, the elements $a \in \mathbb{Z}_p$ all satisfy $a^p = a$ (by Fermat's Little Theorem) so we get

$$(a_0+a_1x+\cdots+a_mx^m)^p = a_0^p+a_1^px^p+\cdots+a_m^px^{pm} = a_0+a_1x^p+\cdots+a_mx^{pm}.$$

Thus, it is very easy to compute pth powers of elements in $\mathbb{Z}_p[x]$.

We now use this computation in the case that the ring is the polynomial ring $\mathbb{Z}_p[x]$ to prove the existence of fields with p^n elements.

Theorem 6.2 *For each prime p and each positive integer n there is a field with exactly p^n elements.*

Proof: Let $h(x)$ be the polynomial in $\mathbb{Z}_p[x]$ defined by

$$h(x) = x^{p^n} - x.$$

By Theorem 6.1, there is a field L containing $\mathbb{Z}_p$ and containing elements c_i such that

$$h(x) = (x - c_1)(x - c_2)\cdots(x - c_{p^n}).$$

We will first show that the p^n elements $c_1, \ldots, c_{p^n}$ are distinct (no two are equal). One of the roots equals 0 since visibly $h(0) = 0$. The factor x corresponds to the root 0 and

$$h(x) = x(x^{p^n-1} - 1).$$

The second factor does not have 0 as a root so there is only one c_i equal to 0. Now let c be any of the nonzero c_i and suppose

$$h(x) = (x - c)^m k(x) \tag{5.4}$$

for some positive integer m and some polynomial $k(x)$. Thus, we are supposing that m of the c_i are equal to c and we must show $m = 1$.

Let us consider the polynomial $h(x + c)$ obtained by replacing x in $h(x)$ with $x + c$; we show that $h(x + c) = h(x)$,

$$h(x+c) = (x+c)^{p^n} - (x+c) = x^{p^n} + c^{p^n} - x - c = h(x) - h(c) = h(x)$$

using the previous corollary and the fact that $h(c) = 0$. Now we use the second form of $h(x)$ in Eq. (5.4) to obtain

$$h(x) = h(x + c) = x^m k(x + c).$$

This proves that x^m is a factor of $h(x)$. We have already seen that only the first power of x divides $h(x)$, so $m = 1$ as we wished to prove.

Thus, we have proved that $h(x)$ has exactly p^n distinct roots in some extension field of $\mathbb{Z}_p$. Now we prove that this set of p^n roots is a field. To do this we need to show that if a and b are roots of $h(x)$ then also $a + b$ and ab are roots of $h(x)$, and if $a \neq 0$ then a^{-1} is a root. The key feature is that an element a is a root of $h(x)$ if and only if

$$a^{p^n} = a.$$

Suppose a and b are roots; then

$$(a + b)^{p^n} = a^{p^n} + b^{p^n} = a + b,$$

which proves $a + b$ is a root of $h(x)$. For the product we have

$$(ab)^{p^n} = a^{p^n} b^{p^n} = ab,$$

showing that ab is a root of $h(x)$. Finally, a similar argument can be given to show that if $a \neq 0$ is a root of $h(x)$ then so is a^{-1}. These steps prove the set of roots of $h(x)$ is a field with exactly p^n elements. ■

There is a drawback to the previous proof in that it does not give a direct construction of a field with p^n elements—it merely proves that one exists. The following is a constructive method, although it requires finding a suitable polynomial. Suppose $g(x)$ is a polynomial in $\mathbb{Z}_p[x]$ and $g(x)$ is irreducible of degree n. Then the factor ring $\mathbb{Z}_p[x]/(g(x))$ is described in Theorem 5.1

and the discussion just before it; it is a field and every element in it can be uniquely written as

$$a_0 + a_1 j + a_2 j^2 + \cdots + a_{n-1} j^{n-1},$$

where j is the coset $j = x + (g(x))$ and the a_i are arbitrary elements of $\mathbb{Z}_p$. Since there are p choices for each a_i, there are p^n choices in all. Thus, $\mathbb{Z}_p[x]/(g(x))$ is a field with p^n elements.

Although it has not been proved that there is an irreducible polynomial of degree n in $\mathbb{Z}_p[x]$, this is a true statement. Thus (at least in principle) there is a way to construct a field with p^n elements. We gave the construction of a field with $2^2 = 4$ elements in the previous section. Let us give another example that will illustrate some additional points.

Example 6.1 A field with eight elements can be constructed as $\mathbb{Z}_2[x]/(g(x))$ if $g(x)$ is an irreducible polynomial of degree 3. The following is one way to find such a polynomial: First, find all the *reducible* polynomials of degree 3 and those left over must be irreducible. There are two irreducible polynomials of degree 1, namely, x and $x + 1$. There is only one irreducible polynomial of degree 2, namely, $x^2 + x + 1$. A reducible polynomial is either the product of three factors of degree 1 or a product of a degree 1 polynomial and the irreducible degree 2 polynomial. The products of three degree 1 factors are

$$x^3, \quad x^2(x+1) = x^3+x^2, \quad x(x+1)^2 = x^3+x, \quad (x+1)^3 = x^3+x^2+x+1$$

and the reducible degree 3 polynomials divisible by the irreducible degree 2 polynomial are

$$x(x^2 + x + 1) = x^3 + x^2 + x, \quad (x + 1)(x^2 + x + 1) = x^3 + 1.$$

The total number of polynomials of degree 3 is $2^3 = 8$. We have listed six reducible polynomials so there are exactly two irreducible polynomials of degree 3: $g_1(x) = x^3 + x + 1$ and $g_2(x) = x^3 + x^2 + 1$. It might appear at first that we can use $g_1(x)$ and $g_2(x)$ to construct two fields with eight elements; namely, $F_1 = \mathbb{Z}_2[x]/(g_1(x))$ and $F_2 = \mathbb{Z}_2[x]/(g_2(x))$. The field F_1 contains a root of $g_1(x)$ and F_2 contains a root of $g_2(x)$. However, these two fields are isomorphic as we now show.

We begin by showing that F_1 contains a root of $g_2(x)$. Let j be a root of $g_1(x)$ in F_1; then $j^3 = j + 1$. Now let $c = j^2 + 1$ so that c is an element of F_1. We will show that c is a root of $g_2(x)$. First observe that

$$j^3 = j + 1, \quad j^4 = j^2 + j, \quad j^6 = (j + 1)^2 = j^2 + 1.$$

Now we compute $g_2(c)$:

$$\begin{aligned} g_2(j^2+1) &= (j^2+1)^3 + (j^2+1)^2 + 1 \\ &= j^6 + 3j^4 + 3j^2 + 1 + j^4 + 1 + 1 \\ &= j^2 + 1 + j^2 + j + j^2 + 1 + j^2 + j \\ &= 0. \end{aligned}$$

Since $g_2(x)$ is irreducible, it follows that any polynomial in $\mathbb{Z}_2[x]$ having $1+j^2$ as a root must be divisible by $g_2(x)$.

Thus, we may define a homomorphism $\tau : \mathbb{Z}_2[x] \longrightarrow F_1$ by sending a polynomial $f(x)$ to $\tau(f(x)) = f(1+j^2) \in F_1$. We have just seen that $(g_2(x))$ is the kernel. The image of τ is isomorphic to $\mathbb{Z}_2[x]/(g_2(x))$ (by the First Isomorphism Theorem). Moreover, the image of τ must contain eight elements and therefore τ maps onto F_1. It follows that F_1 is isomorphic to F_2.

This example is a special case of a more general fact which asserts that any two fields with p^n elements are isomorphic. We do not prove this here, but see Chapter VII, Section 4. Now that we know there exist fields with p^n elements, it is reasonable to ask if there are any other finite fields with different numbers of elements. We show that no other numbers are possible.

Theorem 6.3 *Let K be a field with a finite number of elements. Then K has p^n elements for some prime p and positive integer n.*

Proof: Let e denote the multiplicative identity of K. Consider the map $\mathbb{Z} \to K$ defined by $m \to me$, where me means e added to itself m times (if m is positive) or the additive inverse of $(-m)e$ when m is negative. We set $0e = 0$. Then the correspondence $m \to me$ is a homomorphism of $\mathbb{Z}$ onto a subring of K. The kernel of this homomorphism cannot be (0) because then K would contain infinitely many elements. If the kernel is (p), then the image is isomorphic to $\mathbb{Z}/(p)$. Thus, $\mathbb{Z}/(p)$ is an integral domain because it is a subring of a field. Thus, p is a prime and we see that K contains an isomorphic copy of the finite field $\mathbb{Z}_p$. The proof is most easily completed if we use some facts from linear algebra. An alternative proof not using linear algebra is skecthed in the exercises. Since $\mathbb{Z}_p \subseteq K$, we may regard K as a vector space over $\mathbb{Z}_p$, and since K is finite it must have a finite basis $v_1,\ v_2, \ldots,\ v_n$. Then every element of K can be uniquely expressed in the form

$$a_1v_1 + a_2v_2 + \cdots + a_nv_n, \qquad a_i \in \mathbb{Z}_p.$$

The elements a_i may be freely selected from $\mathbb{Z}_p$—there are p choices for a_1, p choices for a_2, and so on. The total number of elements in K is p^n. ■

EXERCISES

1. Find the factorization of the given polynomials as a product of irreducible polynomials in the indicated $\mathbb{Z}_p[x]$:

(a) $x^9 + 1$ in $\mathbb{Z}_3[x]$ (b) $x^3 + x^2 + x + 1$ in $\mathbb{Z}_2[x]$

(c) $x^6 - x^3 + 1$ in $\mathbb{Z}_3[x]$ (d) $x^{2p} - (2x)^p + 1$ in $\mathbb{Z}_p[x]$

2. Let $g(x)$ be an irreducible polynomial of degree 2 in $F[x]$ and let L be a field that contains one root c of $g(x)$. Show that L is a splitting field for $g(x)$ over F. If $g(x) = x^2 + ax + b$, use long division to show that the other root of $g(x)$ is $-a - c$.

3. Let $f(x)$ be an irreducible polynomial of degree n in $F[x]$ for some field F. Let L be the field $F[x]/(f(x))$ and c an element of L. Show that c is the root of a polynomial $g(x)$ in $F[x]$. [Hint: Consider the homomorphism from $F[x] \to L$ given by evaluation at c.]

4. The polynomial $x^2 + 1$ is irreducible in $\mathbb{Z}_3[x]$, so $L = \mathbb{Z}_3[x]/(x^2 + 1)$ is a field with nine elements. Let c be a root of $x^2 + 1$ in L. Find the irreducible polynomial in $\mathbb{Z}_3[x]$ having $c + 1$ as a root. Do the same for $1 - c$. [Hint: Express $(c + 1)^2$ as a linear combination of 1 and c and then as a linear combination of 1 and $c + 1$.]

5. Continuing with the notation of the preceding problem, express the powers of $c + 1$ in the form $a + bc$, with $a, b \in \mathbb{Z}_3$.

6. The following is a sketch of an alternative proof of the fact that a finite field must contain exactly p^n elements for some prime p and positive integer n. As in the text, we may assume $\mathbb{Z}_p \subseteq K$. If equality holds, then K has p elements and we are done. If not, then there is an element $c \in K$ with $c \notin \mathbb{Z}_p$. Use evaluation maps to argue that $F = \mathbb{Z}_p[c]$ is a subfield of K with p^m elements for some m. If $F \neq K$, there is some $c_2 \in K$ with $c_2 \notin F$. Use evaluation maps to show that $F[c_2]$ has p^t elements for some positive t. This may be continued until reaching K.

7. Prove that there are $(p^2 - p)/2$ monic, irreducible polynomials of degree 2 in $\mathbb{Z}_p[x]$.

8. Prove that there are $(p^3 - p)/3$ monic, irreducible polynomials of degree 3 in $\mathbb{Z}_p[x]$.

7 POLYNOMIALS OVER THE RATIONAL FIELD

In this section we obtain some results about polynomials over the rational field by using some facts about polynomials over $\mathbb{Z}_p$. The connection between the rational field and $\mathbb{Z}_p$ is made through the ring of integers; $\mathbb{Q}$ is the quotient field of $\mathbb{Z}$, while $\mathbb{Z}_p$ is a homomorphic image of $\mathbb{Z}$. This proves to be a useful connection. We will prove theorems of two types: The first gives information about the rational roots of polynomials over $\mathbb{Q}$ and the second gives a test to prove that many polynomials are irreducible.

7.1 Rational Roots of Rational Polynomials

Let $f(x)$ be a polynomial with rational coefficients. There is an integer m such that $mf(x)$ has coefficients in $\mathbb{Z}$. For example, if

$$f(x) = \frac{1}{2}x^4 - \frac{3}{7}x^3 - x + \frac{1}{14}$$

then use $c = 14$ to determine that

$$cf(x) = 7x^4 - 6x^3 - 14x + 1.$$

Since the roots of $f(x)$ and $cf(x)$ are the same, the study of roots of polynomials with rational coefficients may be reduced to the study of roots of polynomials with integer coefficients. In this context, we prove the following about rational roots of such polynomials:

Theorem 7.1

RATIONAL ROOT TEST. *Let $g(x) = a_n x^n + a_{n-1}x^{n-1} + \cdots + a_1 x + a_0$ be a polynomial with integer coefficients. If p/q is a nonzero rational number such that $g(p/q) = 0$ and if $p, q \in \mathbb{Z}$ and p and q have no common factor larger than 1, then $p|a_0$ and $q|a_n$.*

Proof: The condition that $f(p/q) = 0$ implies the equation $q^n f(p/q) = 0$, which can be written as

$$a_n p^n + a_{n-1} q p^{n-1} + \cdots + a_1 q^{n-1} p + a_0 q^n = 0.$$

Move the left-most term to the right of the equal sign and factor out q to get

$$q(a_{n-1}p^{n-1} + \cdots + a_1 p q^{n-2} + a_0 q^{n-1}) = -a_n p^n.$$

Since the a_i are integers, q divides $a_n p^n$. By assumption, p and q have no common factor greater than 1 so q must divide a_n, which is one of the conclusions of the theorem. In a similar way we find that p divides a_0. ■

This theorem places restrictions on the possible rational numbers that need to be tested when searching for the roots of a polynomial. The following is a particularly simple example that reproves an earlier result:

Example 7.1

Let m be an integer that is not the nth power of any integer. Then $x^n - m$ has no rational roots. If p/q is a quotient of relatively prime integers p and q and if p/q is a root of $x^n - m$, then $q|1$ so we may assume that $q = 1$. Then $p^n - m = 0$ so m is the nth power of the integer p.

Example 7.2 Find the rational roots of $f(x) = x^3 - 7x^2 + 20x + 14$. A rational root of the form p/q must have $q|1$ and $p|14$. Thus, $q = \pm 1$ and $p|14$. We may assume $q = 1$ and p is limited to the possibilities ± 1, ± 2, ± 7, or ± 14. By testing these eight possibilities we discover that none is a root. Thus, the polynomial has no rational roots and, since it has degree 3, it is irreducible over $\mathbb{Q}$.

This example shows that testing for rational roots sometimes helps determine if the polynomial is irreducible. Of course, the knowledge that no rational root exists does not prove the irreducibility of the polynomial if the degree is 4 or more.

The hand work may be greatly reduced using ideas from calculus to examine real roots. In the previous example, the derivative $f'(x) = 3x^2 - 14x + 20 = 3(x - \frac{7}{3})^2 + \frac{11}{3}$ is positive when evaluated at every real number. Thus, $f(x)$ is an increasing function and can have at most one real root. Since $f(-1) < 0$ and $f(1) > 0$ there is a real root on the interval $(-1, 1)$. Hence, the only other rational numbers we need to test are 0 and ± 1.

7.2 Factorization of Polynomials over the Integers and the Rationals

A polynomial with integer coefficients may be regarded as a polynomial with rational coefficients. For any polynomial $f(x) \in \mathbb{Q}[x]$, there is an integer c such that $cf(x)$ has integer coefficients. Thus, the factorization of polynomials in $\mathbb{Q}[x]$ is closely related to factorization of polynomials in $\mathbb{Z}[x]$. We must be concerned about the possibility that a polynomial $f(x) \in \mathbb{Z}[x]$ can be factored as a product of polynomials in $\mathbb{Q}[x]$, each of positive degree, but no such factorization is possible in $\mathbb{Z}[x]$. This actually cannot happen, as we now begin to prove.

We use the following idea, which may be useful in several situations: Let p be a prime integer and $^{-}: \mathbb{Z} \longrightarrow \mathbb{Z}_p$ the natural map sending m to the coset $\bar{m} = m + (p)$. Then $^{-}$ induces a homomorphism from $\mathbb{Z}[x]$ to $\mathbb{Z}_p[x]$ by reducing the coefficients of a polynomial modulo p; that is, we define $\tau : \mathbb{Z}[x] \rightarrow \mathbb{Z}_p[x]$ by

$$\tau(a_n x^n + a_{n-1}x^{n-1} + \cdots + a_1 x + a_0) = \bar{a}_n x^n + \bar{a}_{n-1}x^{n-1} + \cdots + \bar{a}_1 x + \bar{a}_0.$$

We refer to the polynomial $\tau(f(x))$ as the *reduction mod* p of $f(x)$. It is left to the reader to verify that τ is indeed a homomorphism.

We give an application of this idea immediately:

Lemma 7.1 *Let $f(x), g(x), h(x)$ be elements of the ring $\mathbb{Z}[x]$ such that $f(x) = g(x)h(x)$. If p is a prime integer that divides every coefficient of $f(x)$, then either p divides every coefficient of $g(x)$ or p divides every coefficient of $h(x)$.*

Proof: Let τ be the homomorphism from $\mathbb{Z}[x]$ to $\mathbb{Z}_p[x]$ that reduces the coefficients modulo p. By assumption, $\tau(f(x)) = 0$ because every coefficient of $f(x)$ is divisible by p. Thus,

$$0 = \tau(f(x)) = \tau(g(x)h(x)) = \tau(g(x))\tau(h(x)).$$

Since $\mathbb{Z}_p[x]$ is an integral domain, either $\tau(g(x)) = 0$ or $\tau(h(x)) = 0$. If $\tau(g(x)) = 0$, then p divides every coefficient of $g(x)$. If $\tau(h(x)) = 0$, then p divides every coefficient of $h(x)$. ■

This lemma provides the main step in the next proof which shows that if a polynomial with integer coefficients can be factored as a product of polynomials with rational coefficients, then it can be factored as a polynomial with integer coefficients.

Lemma 7.2

GAUSS' LEMMA. *Let $f(x)$ be an element of $\mathbb{Z}[x]$ such that $f(x) = g(x)h(x)$ with $g(x)$ and $h(x)$ in $\mathbb{Q}[x]$. Then there exist polynomials $g_1(x), h_1(x) \in \mathbb{Z}[x]$ with $f(x) = g_1(x)h_1(x)$ and $\deg g(x) = \deg g_1(x)$, $\deg h(x) = \deg h_1(x)$.*

Proof: Let k be a positive integer divisible by the denominators of all the nonzero coefficients of $g(x)$ so that $G(x) = kg(x) \in \mathbb{Z}[x]$. Similarly, let l be a positive integer such that $H(x) = lh(x)$ has integer coefficients. Since $f(x) = g(x)h(x)$, it follows that

$$klf(x) = (kg(x)) \cdot (lh(x)) = G(x)H(x).$$

Now we may apply the preceding lemma as follows: If p is a prime divisor of kl, then it is a prime divisor of every coefficient of $klf(x)$, and hence p divides every coefficient of $G(x)$ or $H(x)$. Cancel p and repeat with each prime divisor of kl until all factors have been canceled. We finally reach $f(x) = g_1(x)h_1(x)$ with $g_1(x), h_1(x) \in \mathbb{Z}[x]$. Since $g_1(x)$ is a multiple of $g(x)$ by a nonzero rational number, it follows that $\deg g(x) = \deg g_1(x)$. Similarly, $\deg h_1(x) = \deg h(x)$. ■

We are now ready to prove Eisenstein's test that determines irreducibility of certain polynomials.

Theorem 7.2

EISENSTEIN'S CRITERION. *Let*

$$f(x) = a_n x^n + a_{n-1}x^{n-1} + \cdots + a_1 x + a_0$$

be a polynomial of positive degree n in $\mathbb{Z}[x]$, and let p be a prime integer such that $p \nmid a_n$, $p | a_i$, for $i = 0, 1, 2, \ldots, n-1$, and $p^2 \nmid a_0$. Then $f(x)$ is irreducible in $\mathbb{Q}[x]$.

Proof: If $f(x)$ can be factored over $\mathbb{Q}$, Gauss' lemma implies that it can be factored over $\mathbb{Z}$. Assume $f(x) = g(x)h(x)$ with $0 < \deg g(x) < n$ and $0 < \deg h(x) < n$ and both $g(x)$ and $h(x)$ in $\mathbb{Z}[x]$. We will reach an impossible situation based on this assumption. Let $\tau : \mathbb{Z}[x] \longrightarrow \mathbb{Z}_p[x]$ be the homomorphism that reduces coefficients modulo p. Then $\tau(f(x)) = \bar{a}_n x^n$, with $\bar{a}_n = a_n + (p)$, because all the other coefficients of $f(x)$ are divisible by p and so $\bar{a}_i = 0$, $i < n$. It follows that

$$\tau(f(x)) = \bar{a}_n x^n = \tau(g(x))\tau(h(x)).$$

We conclude that $\tau(g(x)) = \bar{b}x^r$ and $\tau(h(x)) = \bar{c}x^{n-r}$ for some nonnegative integer r since each of these is a divisor of $\bar{a}_n x^n$. From this we may conclude that all but one of the coefficients of $g(x)$ is divisible by p. Similarly, all but one of the coefficients of $h(x)$ is divisible by p. Neither $g(x)$ nor $h(x)$ has degree as large as n, so the particular coefficient not divisible by p cannot be the constant term, $g(0)$ or $h(0)$. Thus, p divides $g(0)$ and $h(0)$ and so p^2 divides their product $g(0)h(0)$; however, $g(0)h(0) = f(0) = a_0$, which is not divisible by p^2. We reach an impossible situation by assuming the polynomial is reducible. This contradiction implies that $f(x)$ cannot have the factorization that was assumed and $f(x)$ is irreducible. ∎

Corollary 7.1 *There exist irreducible polynomials in $\mathbb{Q}[x]$ of degree n for every positive integer n.*

Proof: We may give many examples for any n using Eisenstein's criterion to prove irreducibility. For example, $x^n - 2$ is irreducible over $\mathbb{Q}$ by using $p = 2$ in the previous theorem. For any prime p and any positive integer n, every polynomial of the form

$$x^n + pa_{n-1}x^{n-1} + \cdots + pa_1 x + p, \qquad a_1, \ldots, a_{n-1} \in \mathbb{Z}$$

is irreducible of degree n by Eisenstein's criterion. ∎

We give one last application to irreducibility testing of the homomorphism $\tau : \mathbb{Z}[x] \longrightarrow \mathbb{Z}_p[x]$ which reduces coefficients modulo p.

Theorem 7.3 *Let $f(x) = a_n x^n + a_{n-1}x^{n-1} + \cdots + a_1 x + a_0$ be a polynomial of positive degree n in $\mathbb{Z}[x]$, and let p be a prime not dividing the leading coefficient a_n of $f(x)$. If $\tau(f(x))$ is an irreducible polynomial in $\mathbb{Z}_p[x]$, then $f(x)$ is an irreducible polynomial in $\mathbb{Q}[x]$.*

Proof: If $f(x)$ is reducible in $\mathbb{Q}[x]$, then it has a factorization $f(x) = g(x)h(x)$ with $g(x)$ and $h(x)$ in $\mathbb{Z}[x]$ and $\deg g(x) = m > 0$ and $\deg h(x) = k > 0$. The leading coefficient a_n of $f(x)$ is not divisible by p and it equals the product of the leading coefficients of $g(x)$ and $h(x)$. Thus, neither of these leading coefficients is divisible by p. It follows that $\deg g(x) = \deg \tau(g(x))$ and $\deg h(x) = \deg \tau(h(x))$. However, now we have that

$$\tau(f(x)) = \tau(g(x))\tau(h(x))$$

is a factorization of the irreducible polynomial $\tau(f(x))$. This is an impossible situation caused by the assumption that $f(x)$ was reducible in $\mathbb{Q}[x]$ and so the theorem holds. ■

Example 7.3 The polynomial $x^3+4x^2+9x+225$ is irreducible in $\mathbb{Q}[x]$. To see this, reduce the coefficients modulo 2 to get the polynomial x^3+x+1 in $\mathbb{Z}_2[x]$. This is an irreducible polynomial in $\mathbb{Z}_2[x]$ so the original polynomial is irreducible in $\mathbb{Q}[x]$ by the previous theorem. More generally, the same reasoning shows that the polynomial $x^3+2ax^2+(2k+1)x+(2t+1)$ is irreducible in $\mathbb{Q}[x]$ for every choice of integers a, k, t.

The previous theorem is only useful for proving irreducibility of $f(x) \in \mathbb{Z}[x]$ over $\mathbb{Q}$ if a suitable prime p can be found such that $f(x)$ is irreducible molulo p. It can happen that $f(x)$ is irreducible but is reducible modulo p for many primes p. In fact, it is possible for an irreducible polynomial to be reducible modulo p for *every* prime. Thus, this theorem has only limited applicability to polynomials with fairly high degree.

EXERCISES

1. Express the polynomials as a product of degree 1 factors by finding the rational roots:

   (a) x^3-3x^2-x+3 (b) $x^3-3x^2-16x-12$

   (c) x^4-5x^2+6 (d) $2x^4+3x^3-12x^2-7x+6$

2. Express the polynomials as a product of irreducible polynomials in $\mathbb{Q}[x]$:

   (a) $2x^3+x^2+x-1$ (b) $2x^4-7x^3+8x^2-7x+6$

3. Let $f(x)$ be a monic polynomial in $\mathbb{Z}[x]$. If there is a rational root of $f(x)$, show that the root is an integer that divides $f(0)$.
4. Show by example that a polynomial $f(x) \in \mathbb{Z}[x]$ may simultaneously be irreducible in $\mathbb{Q}[x]$ and be reducible modulo p for some prime p.
5. Use Eisenstein's criterion to show each polynomial is irreducible in $\mathbb{Q}[x]$:

   (a) $x^4-21x^2+81x-6$ (b) $6x^5-10x^3+15x-30$

6. Show that $f(x) = x^4+4x+1$ is irreducible in $\mathbb{Q}[x]$ by applying Eisenstein's criterion to $f(x+1)$.
7. Apply Eisenstein's criterion to $f(x+1)$ to prove that $f(x) = 1+x+x^2+\cdots+x^{p-1} = (x^p-1)/(x-1)$ is irreducible in $\mathbb{Q}[x]$ for any prime number p.
8. The polynomial x^4+1 is irreducible in $\mathbb{Q}[x]$ but is reducible modulo for every prime. Verify that this polynomial is reducible modulo p for $p = 2, 3, 5, 7$, and 11.

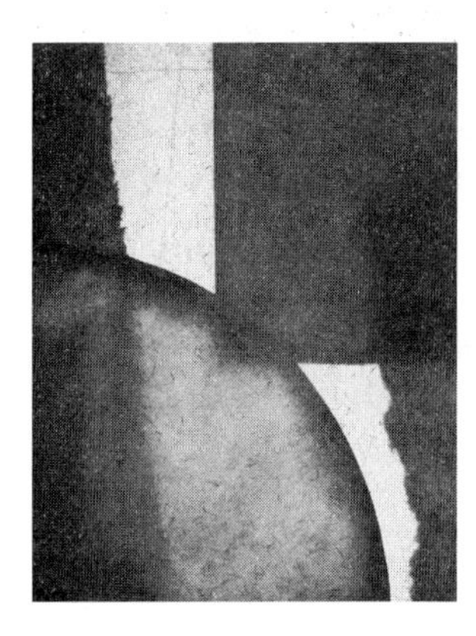

VI GROUPS

In the algebraic systems studied so far, we have had two operations, namely, addition and multiplication. The reader knew many examples of rings before they were defined: the integers and the real and complex numbers, for example. We now proceed to study an important class of systems in which there is only one operation. As soon as the definition of *group* is given in the first section, it will be apparent that we already have studied many examples of groups in previous chapters, although we have not used this terminology. The theory of groups is an important part of modern algebra, and many books have been written on the subject. In this chapter we will present only a few of the most fundamental properties of groups and give a number of examples that may serve to suggest the wide range of applications of the theory.

1 DEFINITION AND SIMPLE PROPERTIES

Let "$\circ$" be a binary operation defined on a nonempty set G. Recall that this statement means that if (a, b) is any ordered pair of elements of G, then $a \circ b$ is a uniquely determined element of G. It is most usual to call this operation either "addition" or "multiplication" and to use the familiar notation that is associated with these terms. However, we will give the definition in terms of "$\circ$" in order that we do not use any property of the operation that is not given as part of the definition that follows:

Definition 1.1 A nonempty set G on which there is defined a binary operation "$\circ$" is called a **group** (with respect to this operation) provided the following properties are satisfied:

(i) If $a, b, c \in G$, then $a \circ (b \circ c) = (a \circ b) \circ c$ *(associative law).*

(ii) There exists an element $e \in G$ such that $e \circ a = a \circ e = a$ for every element $a \in G$ *(existence of an identity).*

(iii) If $a \in G$, there exists an element $x \in G$ such that $a \circ x = x \circ a = e$ *(existence of inverses).*

The element e whose existence is asserted in (ii) is called an identity for the group. We will soon see that there is only one identity for a group so e can be called *the identity* of the group. The element x with the property asserted in (iii) is called *the inverse of* a. Notice that *every* element of the group has an inverse. It will be shown that the inverse of an element is unique.

Let us give some examples. In order to give an example of a group, it is necessary to specify the set and the operation which satisfy the conditions of the definition.

Example 1.1 The set $\mathbb{Z}$ of integers, with the operation $\circ$ taken as addition, is a group. First we must be certain that we have an "operation"; that is, if $a, b \in \mathbb{Z}$ it is necessary that $a + b$ is a uniquely defined element of $\mathbb{Z}$. This is certainly true. The first condition of the definition is the associative law which asserts

$$(a + b) + c = a + (b + c),$$

which we know to be true for addition of integers. The identity element is the integer 0 since $a + 0 = 0 + a = a$ for all $a \in \mathbb{Z}$. The inverse of an integer a is its negative, $-a$, because $a + (-a) = 0$. We have verified all three properties and we have a group. We refer to this as the *additive group of integers.*

This example can be put into a much more general context. Let S be the set of elements of any ring and let the operation $\circ$ be the addition defined for elements of S. The three properties in the definition of a group are stated as properties required of addition in any ring. We refer to this group as the *additive group of the ring* S.

Example 1.2 The set T of all nonzero rational numbers, with $\circ$ taken as multiplication, is a group. Clearly T is closed under multiplication since the product of nonzero rational numbers is again a nonzero rational number. The associative law for multiplication holds and the identity is the rational number 1. Every element of T has an inverse, namely, its reciprocal, which is also a nonzero rational. Thus, T is a group. We emphasize that every element of the group must have an inverse. This explains why the set of *all* rational numbers is not a group with multiplication as the operation. The problem is the number 0; it does not have an inverse.

This example can also be generalized. If F is an arbitrary field, the set of nonzero elements is a group with respect to the operation of multiplication in the field. This group is called *the multiplicative group* of the field.

Example 1.3 The set $\mathbb{R}^+$ of all *positive* real numbers with multiplication as the operation is a group. It is not the multiplicative group of the field $\mathbb{R}$ since that group includes the negative real numbers as well.

In like manner, $\mathbb{Q}^+$ is a group with respect to the multiplication of rational numbers. More generally, the positive elements of any ordered field are a group with respect to the multiplication defined in the field.

Example 1.4

For a positive integer n, the set of all units $U(\mathbb{Z}_n)$ of the ring of integers modulo n is a group. Its elements are the cosets $u + (n)$ with u any integer relatively prime to n. The product of two such cosets is again a coset of this type. The identity of the group is $1 + (n)$ and each element has an inverse by definition of units. This is an example of a group with a finite number of elements, that is, a *finite group*.

This example is a special instance of the group of units $U(R)$ of any ring with identity.

Example 1.5

Let $GL(2, \mathbb{Z})$ denote the set of all two-by-two matrices with integer entries

$$g = \begin{bmatrix} a & b \\ c & d \end{bmatrix},$$

where the integers a, b, c, d satisfy $ad - bc = \pm 1$.

It can be shown that this is the set of units of the ring $M_2(\mathbb{Z})$ of all two-by-two matrices over the integers. This example is a special case of the previous remark.

The notation may seem somewhat peculiar; GL stands for *general linear* and this is but one example of a group of linear transformations which are studied in the subject of linear algebra for which one is usually interested in the general linear group over a field. If F is any field, $GL(2, F)$ is the set of matrices of the form g given previously with $a, b, c, d \in F$ and $ad - bc \neq 0$.

Thus, for example, $GL(2, \mathbb{Z}_p)$ is the group of invertible two-by-two matrices over the field $\mathbb{Z}_p$. This gives examples of groups that have only a finite number of elements and the operation is not commutative. Most of the previous examples have the additional property that is not part of the definition of a group, namely, the operation is commutative and these are abelian groups according to the following definition:

Definition 1.2

Let G be a group with operation $\circ$. If $a \circ b = b \circ a$ for all $a, b \in G$, G is said to be an **abelian group** (or a commutative group).

The term *abelian group* is commonly used for this concept. The name pays tribute to Neils Henrik Abel (1802–1829), a famous Norwegian mathematician whose fundamental work provided inspiration for many later mathematicians. In the following sections we will study many examples of nonabelian groups but will discuss a detailed study of abelian groups in a later chapter.

As in the examples, we will usually call the operation in a group either addition or multiplication, and we will use the familiar notation associated with these names. Because of our familiarity with addition in rings, we will never use addition as the operation in a nonabelian group. If addition is used as the operation, we will always assume that the group is abelian. It follows that in an abelian group, all the properties of addition in a ring are satisfied. The identity will be denoted by 0 and called the "zero"; the inverse of an element a will be denoted as $-a$. We follow the conventions introduced for addition in rings, such as writing $b - a$ as a shorthand for $b + (-a)$.

When the operation in a group is called multiplication, the group may be either abelian or nonabelian. Accordingly, when we prove a property of an arbitrary group, we will usually think of the operation as multiplication and write ab in place of $a \circ b$. The identity of the group is usually written as e and the inverse of an element a is a^{-1}.

Now let G be an arbitrary group with the operation written as multiplication. The following properties are easily proved using the same ideas that appeared in the proofs of similar statements about rings. Accordingly, we give just an outline of the proofs leaving some details to be filled in by the reader.

Theorem 1.1 *The following hold in every group G:*

(i) *The identity of G is unique.*

(ii) *If $a \in G$, then a has a unique inverse.*

(iii) *If $a, b, c \in G$ and $ab = ac$, then $b = c$.*

(iv) *If $a, b, c \in G$ and $ba = ca$, then $b = c$.*

(v) *If $a, b \in G$, then there is a unique element $x \in G$ such that $ax = b$ and a unique element $y \in G$ such that $ya = b$. In fact, $x = a^{-1}b$ and $y = ba^{-1}$.*

(vi) *If $a, b \in G$, then $(ab)^{-1} = b^{-1}a^{-1}$.*

Proof: (i) If e and e' are identity elements, then $e = ee' = e'$. (ii) If p, q are both inverses of a, then $(qa)p = ep = p$ and $q(ap) = qe = q$ so $p = q$. (iii and iv) Multiply on the appropriate side by a^{-1}. The last two statements hold by just checking the products. Properties (iii) and (iv) are called the **cancelation laws.** ■

As in the case with multiplication of elements in a ring, there is a generalized associative law we assert—roughly, that a product of n elements of a group is the same no matter where the parentheses are inserted. Of course, the order in which the elements are written is important.

1.1 Powers of an Element

For an element $a \in G$ the powers of a are defined inductively: $a^1 = a$ and for $n \geq 2$ we set

$$a^n = a \cdot a^{n-1}.$$

For convenience we define $a^0 = e$, the identity of G, and $a^{-n} = (a^{-1})^n$. Thus, a^m is defined for all integers and the following laws of exponents hold:

$$a^m \cdot a^n = a^{m+n}, \qquad (a^m)^n = a^{mn}.$$

If G is an *abelian group*, we also have $(ab)^n = a^n \cdot b^n$, but this does not hold in general.

In cases in which the group is abelian and addition is the operation, we make use of multiples in place of powers; that is, na replaces a^n. We have the following analog of the previously mentioned laws:

$$\begin{aligned} ma + na &= (m+n)a, \\ n(ma) &= (nm)a, \\ n(a+b) &= na + nb. \end{aligned}$$

These properties are familiar for the case of addition of elements in a ring.

1.2 Subgroups

A set H of elements in a group G is called a *subgroup* if H is a group with respect to the operation already defined on G. In particular, this requires H be closed under the operation; if $x, y \in H$, then $xy \in H$. The other defining properties of a group must hold for H as well. A group G always has the two so-called *trivial* subgroups $H = \{e\}$ and $H = G$. Any other subgroup is called a *proper* subgroup.

Here is a concise way to determine if a subset of a group is a subgroup:

Theorem 1.2 *A nonempty subset K of a group G is a subgroup if and only if the following two conditions are satisfied:*

(i) *If $a, b \in K$, then $ab \in K$.*
(ii) *If $a \in K$, then $a^{-1} \in K$.*

If K has only a finite number of elements, condition (ii) is implied by condition (i).

Proof: Condition (i) implies that K is closed under multiplication. If $a \in K$, then $a^{-1} \in K$ by (ii) and so by (i) we have $aa^{-1} = e \in K$. Thus, K contains the identity element. The multiplication in G is associative and so, in particular, multiplication in K is associative. All conditions of the definition of a group are met, so K is a subgroup of G when (i) and (ii) hold.

Now suppose that K is a finite set and (i) holds. We must show that $a \in K$ implies $a^{-1} \in K$. Consider the powers $a, a^2, a^3, \ldots$ of a; by (i) these are all elements of K. Since K is finite, the powers cannot all be different and there must be two integers $r > s > 0$ such that $a^r = a^s$. From this we conclude $a^{-1} = a^{r-s-1}$. Since $r - s - 1 \geq 0$, a^{-1} is a positive power of a that lies in K. Thus, (ii) holds. ∎

Corollary 1.1 *If G is a finite group and $a \in G$, then the set of all powers $K = \{a, a^2, a^3, \ldots\}$ of a is a subgroup of G.*

Proof: This follows from the previous theorem because K is closed under multiplication. ■

If the group G is infinite, the set of powers of an element need not be a subgroup. That is, condition (ii) of the theorem need not follow from condition (i) in an infinite group. For example, the set of all positive integers is closed under addition but is not a subgroup of the additive group of integers.

1.3 Direct Products

We conclude this section by introducing for groups the analog of the direct sum of two rings. Suppose that G and H are groups. The Cartesian product $G \times H$ is made into a group in the obvious way using the operations defined for G and H:

$$(g_1, h_1) \circ (g_2, h_2) = (g_1 g_2, h_1 h_2), \qquad g_1, g_2 \in G;\ h_1, h_2 \in H.$$

The group defined in this way is called the *direct product* of G and H. If both G and H are abelian groups and addition is the operation in each, then we may write the operation in $G \times H$ as addition with the rule

$$(g_1, h_1) + (g_2, h_2) = (g_1 + g_2, h_1 + h_2), \qquad g_1, g_2 \in G;\ h_1, h_2 \in H.$$

We leave the verification to the reader that $G \times H$ is a group. Once this is verified, one may extend the notion to more than two groups. If $G_1, \ldots, G_n$ are groups, then the direct product of these n groups is denoted by $G_1 \times \cdots \times G_n$ and consists of all elements $(g_1, g_2, \ldots, g_n)$ with $g_i \in G_i$. The operation in this direct product is

$$(g_1, \ldots, g_i, \ldots, g_n)(g_1', \ldots, g_i', \ldots, g_n') = (g_1 g_1', \ldots, g_i g_i', \ldots, g_n g_n')$$

with $g_i, g_i' \in G_i$.

The terms *direct product* and *direct sum* are both used to denote this group. Clearly one could extend the idea to an infinite number of groups, but in that case the two terms traditionally refer to different objects. The direct product of an infinite number of groups $G_1, G_2, \cdots$ is the collection of all "rows" $(g_1, g_2, \cdots)$ with the operation defined in the obvious way. The direct sum, however, refers to the collection of all "rows" $(g_1, g_2, \cdots)$ in which $g_i = e_i$, the identity of G_i for all but a finite number of i. In the case of a finite collection of groups, these two constructions give the same groups but not otherwise.

EXERCISES

1. Which of the following sets are groups with respect to the indicated operation?

(a) The set $\{1, 3, 7, 9\}$ of elements of $\mathbb{Z}_{10}$ with operation multiplication

(b) The set $\{0, 2, 4, 6, 8\}$ of elements of $\mathbb{Z}_{10}$ with operation addition

(c) The set $\{1, 3, 9\}$ of elements of $\mathbb{Z}_{10}$ with operation multiplication

(d) The set of all rational numbers x with $0 < x \leq 1$, with operation multiplication

(e) The set of all positive irrational numbers with operation multiplication

(f) The set of all integers with an operation $\circ$ defined by $a \circ b = a + b + 1$

(g) The set of all integers with an operation $\circ$ defined by $a \circ b = a - b$

(h) The set of all rational numbers, other than 1, with the operation $\circ$ defined by $a \circ b = a + b - ab$

2. If H_1 and H_2 are subgroups of a group G, prove that $H_1 \cap H_2$ is a subgroup of G. Generalize this and show that the intersection of any number of subgroups of G is a subgroup of G.

3. Find all the subgroups of each of the following groups:

(a) The additive group of $\mathbb{Z}_7$

(b) The multiplicative group of $\mathbb{Z}_7$ (There are four subgroups.)

(c) The additive group of $\mathbb{Z}_{12}$ (There are six subgroups.)

(d) The multiplicative group of $\mathbb{Z}_{11}$ (There are four subgroups.)

4. Prove that $(ab)^2 = a^2b^2$ for all choices of a, b in a group G if and only if G is abelian.

5. Show that if G is a group having n elements, then the direct product $G \times G$ has n^2 elements. Show that the set $\{(a, a) : a \in G\}$ is a subgroup of $G \times G$. Is the set $\{(a, a^{-1}) : a \in G\}$ a subgroup of $G \times G$?

6. Give an example of a group G that has four elements and $x^2 = e$ holds for every $x \in G$.

7. Let $SL(2, F)$ denote the collection of all two-by-two matrices $g = \begin{bmatrix} a & b \\ c & d \end{bmatrix}$ with a, b, c, d elements of the field F such that $ad - bc = 1$. Show that $SL(2, F)$ is a subgroup of $GL(2, F)$, the group of two-by-two matrices for which $ad - bc \neq 0$. [$SL(2, F)$ is called the *special linear group* over F.]

8. Let a be a fixed element of a group G. Prove that the set

$$C_G(a) = \{x \in G : ax = xa\}$$

is a subgroup of G. $C_G(a)$ is called the *centralizer* of a in G.

9. For any group G the set $Z(G) = \{b \in G : bc = cb \text{ for all } c \in G\}$ is called the *center* of G. Prove $Z(G)$ is a subgroup of G. Show also that $Z(G) = G$ if and only if G is abelian.

2 MAPPINGS AND PERMUTATION GROUPS

One of the important ways that groups arise is in the consideration of mappings. Suppose that A, B, and C are sets and that we are given mappings $\alpha : A \to B$ and $\beta : B \to C$. It is then easy to define the composition of α and β as the mapping from A to C given by the rule

$$\beta\alpha : a \longrightarrow \beta(\alpha(a)), \qquad a \in A.$$

Note that $a \in A$ implies $\alpha(a) \in B$ and $\beta(\alpha(a)) \in C$. The composite mapping is denoted by $\beta\alpha$ and is called the *product* of the two maps. We give an example of the product of two maps. Let $A = \{1, 2, 3, 4\}$, $B = \{x, y, z, w\}$, and $C = \{r, s\}$. Define $\alpha : A \to B$ by

$$\alpha(1) = y, \quad \alpha(2) = z, \quad \alpha(3) = x \quad \alpha(4) = w.$$

Next define $\beta : B \to C$ by

$$\beta(x) = r, \quad \beta(y) = r, \quad \beta(z) = s, \quad \beta(w) = r.$$

Then the composite $\beta\alpha$ is given explicitly by the rules

$$\begin{aligned} \beta\alpha(1) &= \beta(\alpha(1)) = \beta(y) = r, \\ \beta\alpha(2) &= \beta(\alpha(2)) = \beta(z) = s, \\ \beta\alpha(3) &= \beta(\alpha(3)) = \beta(x) = r, \\ \beta\alpha(4) &= \beta(\alpha(4)) = \beta(w) = r. \end{aligned}$$

We take one more step now. Suppose we have mappings

$$\alpha : A \to B, \quad \beta : B \to C, \quad \gamma : C \to D.$$

Then $\beta\alpha$ is a mapping from A to C and so $\gamma(\beta\alpha)$ is a mapping from A to D. In like manner, $\gamma\beta$ is a mapping from B to D and $(\gamma\beta)\alpha$ is a mapping from A to D. An important fact is that these two mappings are equal, that is,

$$\gamma(\beta\alpha) = (\gamma\beta)\alpha.$$

Let us prove this statement. First we ask, How do we prove that two functions defined on A are equal? The answer is that we verify that the two functions have the same value at every point of A; that is, we must show

$$[\gamma(\beta\alpha)](a) = [(\gamma\beta)\alpha](a) \tag{6.1}$$

for each $a \in A$. First observe that by definition of the product of the maps γ and $\beta\alpha$ we have

$$[\gamma(\beta\alpha)](a) = \gamma[\beta\alpha(a)].$$

Then, by the definition of the product $\beta\alpha$ we have

$$\gamma[(\beta\alpha)(a)] = \gamma[\beta(\alpha(a))].$$

Thus, the left side of Eq. (6.1) equals $\gamma[\beta(\alpha(a))]$. In like manner we obtain

$$[(\gamma\beta)\alpha](a) = (\gamma\beta)(\alpha(a)) = \gamma[\beta(\alpha(a))].$$

Since both sides of Eq. (6.1) are equal to $\gamma[\beta(\alpha(a))]$, we have proved Eq. (6.1) and that $(\gamma\beta)\alpha = \gamma(\beta\alpha)$. This equality is merely the assertion that *multiplication of maps is always associative.*

In connection with the study of groups we are interested in the special case of mapping of a set A *onto the same set* A. Moreover, it is the one-to-one mappings of A *onto* A that we wish to study. The following is the terminology that is used:

Definition 2.1 A one-to-one mapping of a set A onto itself is called a **permutation** of the set A.

To show that a map γ is a permutation of A, we must verify that (i) γ is one-to-one—namely, $\gamma(a) = \gamma(b)$, for $a, b \in A$, implies $a = b$; and (ii) γ is onto, that is, for each $b \in A$ there is some $a \in A$ with $\gamma(a) = b$.

The next theorem shows why we have paused to study mappings in a discussion of groups.

Theorem 2.1 *The set S of all permutations of a set A is a group with respect to the multiplication of maps defined by $(\beta\alpha)(a) = \beta(\alpha(a))$, $\alpha, \beta \in S$, $a \in A$.*

Proof: First we must verify that S is closed under the operation of multiplication. In other words, it is necessary to show that when α and β are permutations of A then the product $\beta\alpha$ is also a permutation. We verify this for $\gamma = \beta\alpha$.

Suppose $\beta\alpha(a) = \beta\alpha(b)$. By definition of the product, this means

$$\beta(\alpha(a)) = \beta(\alpha(b)).$$

Since β is one-to-one, it follows that $\alpha(a) = \alpha(b)$. Now since α is one-to-one, it follows $a = b$ as required. Next let b be any element of A. There is some $c \in A$ such that $\beta(c) = b$ and there is some $a \in A$ such that $\alpha(a) = c$ because both α and β are onto mappings. Thus,

$$\beta\alpha(a) = \beta(\alpha(a)) = \beta(c) = b.$$

This shows that $\beta\alpha$ is onto and $\beta\alpha$ is a permutation, that is, $\beta\alpha \in S$. Therefore, S is closed under the operation.

We have already proved that multiplication of mappings is associative, and hence the first requirement of a group is satisfied.

The identity mapping is the function $\epsilon : A \to A$ by $\epsilon(a) = a$ for every $a \in A$. This is the identity in S. To see this, let $\alpha \in S$. Then

$$\epsilon\alpha(a) = \epsilon(\alpha(a)) = \alpha(a)$$

by definition of ϵ. Thus, $\epsilon\alpha = \alpha$. Similarly, $\alpha\epsilon = \alpha$. This shows that ϵ is the identity of S. There remains only the proof that every element of S has an inverse in S.

Let $\alpha \in S$. For each x in A there is one, and only one, element $y \in A$ such that $\alpha(y) = x$. Let β denote the function that assigns y to x; that is, $\beta(x)$ is the unique element of A that is mapped to x by α. We need to verify that β is the inverse of α. The defining conditions for β are

$$\alpha(x) = y \quad \text{if and only if} \quad \beta(y) = x, \quad x, y \in A.$$

Substitute y from the first equation into the second to get

$$\beta(\alpha(x)) = x. \tag{6.2}$$

This is equivalent to $\beta\alpha(x) = x$ and so $\beta\alpha = \epsilon$. Similarly, substitute x from the second equation into the first to get

$$\alpha(\beta(y)) = y, \quad y \in A.$$

This implies $\alpha\beta = \epsilon$ and so β is indeed the inverse of α. There is still the requirement that $\beta \in S$. This follows easily. For any $x \in A$, Eq. (6.2) shows that x is a value of β and so β is onto. If $\beta(a) = \beta(b)$, then we may apply α to both sides to conclude $a = b$; thus, β is one-to-one. This completes the proof that α has an inverse in S and that S is a group. ■

For a set A, the set of all permutations of A is called *the group of permutations of A*. Note that if A is an infinite set, the group of permutations of A is an infinite group. If A is a finite set, the group of permutations of A is a finite group. We will study this finite group in detail.

Definition 2.2 Let n be a positive integer. The group of all permutations of a set with n elements is called the **symmetric group** on n symbols and will be denoted as S_n.

Let us consider an example in which $A = \{1, 2, 3\}$, a set with three elements. The symmetric group S_3 consists of all permutations of A and contains six elements; here are two of them:

$$\begin{array}{lll} \alpha(1) = 2, & \alpha(2) = 1, & \alpha(3) = 3, \\ \beta(1) = 2, & \beta(2) = 3, & \beta(3) = 1. \end{array} \tag{6.3}$$

The product of these two mappings may be computed using the definition of the product of mappings. We have

$$\begin{aligned} \alpha\beta(1) &= \alpha(\beta(1)) = \alpha(2) = 1, \\ \alpha\beta(2) &= \alpha(\beta(2)) = \alpha(3) = 3, \\ \alpha\beta(3) &= \alpha(\beta(3)) = \alpha(1) = 2. \end{aligned}$$

The use of the definition to compute products becomes very tedious when the number of elements in the set A is fairly large. Even writing down the elements in the format of Eq. (6.3) is time-consuming. We will introduce some efficiency with notation that makes the description of elements of S_n somewhat more workable. Maps will be written using two rows, with the elements of the set written in the first row (in any order) and under each element of A we place its image under the map. We illustrate this notation with the elements of S_3. The mappings α and β defined previously are written as

$$\alpha = \begin{pmatrix} 1 & 2 & 3 \\ 2 & 1 & 3 \end{pmatrix}, \qquad \beta = \begin{pmatrix} 1 & 2 & 3 \\ 2 & 3 & 1 \end{pmatrix}.$$

Using this notation, one may compute products efficiently. In order to compute $\alpha\beta$, first write the permutations in the proper order using the double-row notation and a partially blank permutation which will be the product:

$$\alpha\beta = \begin{pmatrix} 1 & 2 & 3 \\ 2 & 1 & 3 \end{pmatrix} \begin{pmatrix} 1 & 2 & 3 \\ 2 & 3 & 1 \end{pmatrix} = \begin{pmatrix} 1 & 2 & 3 \\ a & b & c \end{pmatrix}.$$

The spaces now filled with a, b, c are left blank; our task is to find the correct elements of A to be placed there. To find the image of 1 (i.e., a), first look at the image of 1 under β and find it to be 2; then find the image of 2 under α and find it to be 1. The eye may be aided by following $\beta : 1^* \rightarrow 2^{**}$, $\alpha : 2^{**} \rightarrow 1^{***}$ in the product

$$\alpha\beta = \begin{pmatrix} 1 & 2^{**} & 3 \\ 2 & 1^{***} & 3 \end{pmatrix} \begin{pmatrix} 1^* & 2 & 3 \\ 2^{**} & 3 & 1 \end{pmatrix} = \begin{pmatrix} 1 & 2 & 3 \\ 1 & b & c \end{pmatrix}.$$

In the analogous way we determine the images of 2 and 3 in the product: β sends 2 to 3 and α sends 3 to 3 so that $b = 3$; β sends 3 to 1 and α sends 1 to 2 so that $c = 2$.

The double-row notation makes it easy to write the six elements of S_3. In doing so, we compute several other products of α and β so that every element of S_3 is expressed as a power of α times a power of β:

$$e = \begin{pmatrix} 1 & 2 & 3 \\ 1 & 2 & 3 \end{pmatrix}, \quad \beta = \begin{pmatrix} 1 & 2 & 3 \\ 2 & 3 & 1 \end{pmatrix}, \quad \beta^2 = \begin{pmatrix} 1 & 2 & 3 \\ 3 & 1 & 2 \end{pmatrix},$$
$$\alpha = \begin{pmatrix} 1 & 2 & 3 \\ 2 & 1 & 3 \end{pmatrix}, \quad \alpha\beta = \begin{pmatrix} 1 & 2 & 3 \\ 1 & 3 & 2 \end{pmatrix}, \quad \alpha\beta^2 = \begin{pmatrix} 1 & 2 & 3 \\ 3 & 2 & 1 \end{pmatrix}.$$

Notice that $\alpha^2 = \beta^3 = e$. Suppose we compute one more product:

$$(\alpha\beta^2)\alpha = \begin{pmatrix} 1 & 2 & 3 \\ 3 & 2 & 1 \end{pmatrix}\begin{pmatrix} 1 & 2 & 3 \\ 2 & 1 & 3 \end{pmatrix} = \begin{pmatrix} 1 & 2 & 3 \\ 2 & 3 & 1 \end{pmatrix} = \beta.$$

Thus, we have discovered the relation $(\alpha\beta^2)\alpha = \beta$. By similar computations, the entire multiplication table for this group may be determined and the products expressed as the products of α and β listed previously.

Multiplication Table for S_3

	e	β	β^2	α	$\alpha\beta$	$\alpha\beta^2$
e	e	β	β^2	α	$\alpha\beta$	$\alpha\beta^2$
β	β	β^2	e	$\alpha\beta^2$	α	$\alpha\beta$
β^2	β^2	e	β	$\alpha\beta$	$\alpha\beta^2$	α
α	α	$\alpha\beta$	$\alpha\beta^2$	e	β	β^2
$\alpha\beta$	$\alpha\beta$	$\alpha\beta^2$	α	β^2	e	β
$\alpha\beta^2$	$\alpha\beta^2$	α	$\alpha\beta$	β	β^2	e

As is usual for multiplication tables, rows and columns are indexed by the elements in the group. The entry at the intersection of row x and column y is the product xy. The product of $\alpha\beta^2$ and α should be written as $(\alpha\beta^2)\alpha$; the actual entry is β because the elements of S_3 are listed in the form $\alpha^i\beta^j$, with $0 \le i \le 1$ and $0 \le j \le 2$. Of course, some other selection of element expressions could be used to list the elements.

We can use the double-row notation to denote any permutation of a finite set. In general, if $\{i_1, i_2, \ldots, i_n\}$ is some arrangement of the integers $1, 2, \ldots, n$, then

$$\alpha = \begin{pmatrix} 1 & 2 & 3 & \cdots & n \\ i_1 & i_2 & i_3 & \cdots & i_n \end{pmatrix}$$

is the permutation of the set $A = \{1, 2, 3, \ldots, n\}$ such that

$$\alpha(1) = i_1, \quad \alpha(2) = i_2, \quad \ldots, \quad \alpha(n) = i_n.$$

We previously determined that S_3 has six elements. Let us determine the number of elements in the symmetric group S_n, that is, the number of permutations of the set $A = \{1, 2, 3, \ldots, n\}$ with n elements.

Theorem 2.2 *The number of elements in the symmetric group S_n is $n! = 1 \cdot 2 \cdots (n-1) \cdot n$.*

Proof: An element of S_n is completely determined by specifying the images of the elements $1, 2, \ldots, n$. We count the number of elements in S_n by counting the number of choices for the images. The image of 1 can be any of the n elements of A. Once the image of 1 is selected, any of the remaining $n-1$ elements can be the image of 2. So far there are $n(n-1)$ choices for the images of 1 and 2. Once these two images have been selected, there remain $n-2$ elements that can be the image of 3, and so on. It follows that there are $n(n-1)(n-2) \cdots 2 \cdot 1 = n!$ different permutations of A and so S_n has $n!$ elements. ∎

Any group whose elements are permutations is called a *permutation group* or a *group of permutations*. Any subgroup of the symmetric group S_n is a permutation group. For example, from the multiplication table for S_3 we conclude that $\{e, \beta, \beta^2\}$ is a subgroup of S_3. This is an example of a permutation group which is not a symmetric group since it is not the group of *all* permutations of some set.

2.1 Rigid Motions

We conclude this section with a brief indication of how one may obtain some groups by considering the symmetry of certain geometric figures. We consider a square which we imagine is a movable piece of rigid material. The square may be rotated any number of times; it may be flipped over and/or rotated any number of times so long as it is returned to occupy its original position but not necessarily in its original orientation. We call such a move a *rigid motion* of the square. For purpose of identification, we label the four corners so we may recognize the starting position and final position after a rigid motion. The starting position will always be taken as

1	4
2	3

. (6.4)

We distinguish two basic motions: the rotation, denoted by ρ, and a "flip," denoted by λ. The rotation ρ is made through $90°$ in the counterclockwise direction keeping the square in the same plane throughout. The flip λ turns the square over by rotating it about the horizontal line through its center. We symbolically represent these motions in the following figures:

1	4		4	3		1	4		2	3
		$\xrightarrow{\rho}$			,			$\xrightarrow{\lambda}$		
2	3		1	2		2	3		1	4

If α and β are any two rigid motions, we define their product $\alpha\beta$ as the outcome of first applying the motion β and then applying α to the square. Thus, for example, $\rho\rho = \rho^2$ is the effect of rotation first through $90°$ and then rotating through $90°$ again. Thus, ρ^2 is a rotation through $180°$. Similarly, ρ^3 is a rotation through $270°$ and ρ^4 is a rotation through $360°$, which is the same as not moving the square at all. The rigid motion that does not move the square is denoted by e, the identity motion; we may write $\rho^4 = e$. Here we use the idea that two rigid motions are equal if they produce the same final position of the square. We also have $\lambda^2 = e$ since two flips return the square to its original position. The product $\rho\lambda$ first flips the square over about the horizontal line and then rotates it through $90°$. There are exactly eight possible positions of the square after a rigid motion:

1	4	4	3	3	2	2	1
2	3	1	2	4	1	3	4
2	3	1	2	4	1	3	4
1	4	4	3	3	2	2	1

The eight rigid motions of the square are achieved by the products

$$e, \quad \rho, \quad \rho^2, \quad \rho^3, \quad \lambda, \quad \lambda\rho, \quad \lambda\rho^2, \quad \lambda\rho^3.$$

What is the outcome of $\rho\lambda$? This expression is not in the previous list but the effect on the square is exactly the same as the rigid motion $\lambda\rho^3$ as is seen by applying each product to the square in its starting position. Thus, $\lambda\rho^3 = \rho\lambda$. In fact, the product of any two of the eight rigid motions will be another one of the eight. In other words, the set of rigid motions of the square is closed under this product. In fact, it is not difficult to verify that the set of rigid motions with the product operation is a group. This group is one member of a class of groups called *dihedral groups*. The group of rigid motions of the square is the **dihedral group** with eight elements and is denoted by D_8. With some computation with these symbols, the entire multiplication table of the group may be determined. We leave the verification of the entries in the following table as an exercise for the reader:

Multiplication Table for D_8

	e	ρ	ρ^2	ρ^3	λ	$\lambda\rho$	$\lambda\rho^2$	$\lambda\rho^3$
e	e	ρ	ρ^2	ρ^3	λ	$\lambda\rho$	$\lambda\rho^2$	$\lambda\rho^3$
ρ	ρ	ρ^2	ρ^3	e	$\lambda\rho^3$	λ	$\lambda\rho$	$\lambda\rho^2$
ρ^2	ρ^2	ρ^3	e	ρ	$\lambda\rho^2$	$\lambda\rho^3$	λ	$\lambda\rho$
ρ^3	ρ^3	e	ρ	ρ^2	$\lambda\rho$	$\lambda\rho^2$	$\lambda\rho^3$	λ
λ	λ	$\lambda\rho$	$\lambda\rho^2$	$\lambda\rho^3$	e	ρ	ρ^2	ρ^3
$\lambda\rho$	$\lambda\rho$	$\lambda\rho^2$	$\lambda\rho^3$	λ	ρ^3	e	ρ	ρ^2
$\lambda\rho^2$	$\lambda\rho^2$	$\lambda\rho^3$	λ	$\lambda\rho$	ρ^2	ρ^3	e	ρ
$\lambda\rho^3$	$\lambda\rho^3$	λ	$\lambda\rho$	$\lambda\rho^2$	ρ	ρ^2	ρ^3	e

Next we show how the rigid motions may be represented by permutations. We use the symbols 1,2,3,4 to denote locations as well as corners of the square. When the square is in its original position, corner 1 is in location 1, corner 2 is in location 2, and so on. When the square is rotated by applying ρ, then corner from location 4 is moved to location 1, the corner in location 1 is moved to location 2, the corner in location 2 is moved to location 3, and the corner in location 3 is moved to location 4. We record these data as a permutation $\pi(\rho)$ given by

$$\rho \longrightarrow \pi(\rho) = \begin{pmatrix} 1 & 2 & 3 & 4 \\ 2 & 3 & 4 & 1 \end{pmatrix}.$$

More generally, we assign a permutation $\pi = \pi_\alpha$ to a rigid motion α as follows: For $i = 1, 2, 3$, or 4, let the rigid motion α move the corner in location i to location j. Then π_α is the permutation sending i to j. For example, the flip motion λ produces the permutation $1 \to 2$, $2 \to 1$, $3 \to 4$, and $4 \to 3$; that is,

$$\lambda \longrightarrow \pi_\lambda = \begin{pmatrix} 1 & 2 & 3 & 4 \\ 2 & 1 & 4 & 3 \end{pmatrix}.$$

Thus, each rigid motion is represented by a permutation of $\{1, 2, 3, 4\}$. Since the product of two rigid motions is also a rigid motion, one might suspect that the associated permutations multiply correctly. Here, "correctly" means that the permutation associated with the product $\alpha\beta$ of two rigid motions equals the product of the permutations associated with α and β; in symbols this is written as $\pi_{\alpha\beta} = \pi_\alpha \pi_\beta$ for any two rigid motions α and β. That they do satisfy this as multiplicative condition is equivalent to the statement that π is a homomorphism of groups which will be defined in the next section.

The set of eight permutations formed by taking all possible products of the two permutations $\pi(\rho)$ and $\pi(\lambda)$ is called the *octic group* and gives an example of a permutation group on the symbols $\{1, 2, 3, 4\}$ that is not the full symmetric group.

EXERCISES

1. In the following, α and β are permutations of the set $\{1, 2, 3, 4, 5\}$. In each case, compute $\alpha\beta$, $\beta\alpha$, α^2, and β^2.

(a) $\begin{array}{lllll} \alpha(1) = 2, & \alpha(2) = 1, & \alpha(3) = 3, & \alpha(4) = 5, & \alpha(5) = 4; \\ \beta(1) = 1, & \beta(2) = 4, & \beta(3) = 2, & \beta(4) = 3, & \beta(5) = 5. \end{array}$

(b) $\begin{array}{lllll} \alpha(1) = 4, & \alpha(2) = 3, & \alpha(3) = 5, & \alpha(4) = 1, & \alpha(5) = 2; \\ \beta(1) = 2, & \beta(2) = 3, & \beta(3) = 1, & \beta(4) = 4, & \beta(5) = 5. \end{array}$

(c) $\begin{array}{lllll} \alpha(1) = 2, & \alpha(2) = 1, & \alpha(3) = 4, & \alpha(4) = 5, & \alpha(5) = 3; \\ \beta(1) = 2, & \beta(2) = 3, & \beta(3) = 4, & \beta(4) = 5, & \beta(5) = 1. \end{array}$

(d) $\alpha = \begin{pmatrix} 1 & 2 & 3 & 4 & 5 \\ 5 & 4 & 3 & 1 & 2 \end{pmatrix}$, $\beta = \begin{pmatrix} 1 & 2 & 3 & 4 & 5 \\ 3 & 2 & 1 & 5 & 4 \end{pmatrix}$.

(e) $\alpha = \begin{pmatrix} 1 & 2 & 3 & 4 & 5 \\ 1 & 3 & 2 & 5 & 4 \end{pmatrix}$, $\beta = \begin{pmatrix} 1 & 2 & 3 & 4 & 5 \\ 2 & 3 & 1 & 4 & 5 \end{pmatrix}$.

(f) $\alpha = \begin{pmatrix} 1 & 2 & 3 & 4 & 5 \\ 5 & 4 & 3 & 2 & 1 \end{pmatrix}$, $\beta = \begin{pmatrix} 1 & 2 & 3 & 4 & 5 \\ 5 & 4 & 2 & 1 & 3 \end{pmatrix}$.

2. In the multiplication table of D_8, verify the entries giving the products $(\lambda\rho)(\lambda\rho^2)$ and $(\lambda\rho^3)(\lambda\rho)$.

3. The square in Eq. (6.4) is flipped about the diagonal through the corners marked 2 and 4 producing a rigid motion of the square. Express this motion as a product using the rotations ρ^i and the flip λ as defined in this section. Do a similar exercise for the motion that flips the square around the diagonal through corners 1 and 3.

4. Find all the elements α of the group D_8 of rigid motions of the square that satisfy $\alpha^2 = e$. [Hint: Look at the multiplication table for e on the diagonal.]

5. Find all elements θ of S_3 that satisfy $\theta^2 = e$. Do the same for all elements that satisfy $\theta^3 = e$.

6. Describe the group of rigid motions of a rectangle that is not a square and show that it has four elements.

7. How many elements are in the group of rigid motions of a regular pentagon? A regular hexagon?
8. Let $\alpha : A \to B$ be a given mapping. Prove each of the following:
   (a) If ϵ_A is the identity mapping on the set A, there exists a mapping $\beta : B \to A$ such that $\beta\alpha = \epsilon_A$ if and only if α is a one-to-one mapping.
   (b) There exists a mapping $\gamma : B \to A$ such that $\alpha\gamma = \epsilon_B$ if and only if α is an onto mapping.
   (c) If both the mappings β and γ exist, as defined in parts a and b, then $\beta = \gamma$.

3 HOMOMORPHISMS AND ISOMORPHISMS

One of the most important tools for studying groups is the mappings between two groups that preserve the structure imposed by the definition of group. We give the definition of these maps in a very general form that takes into account the possibility that the operations in the two groups may not be the same.

Definition 3.1

Let G be a group with operation $\circ$ and H a group with operation $*$. A mapping $\theta : G \to H$ is called a **homomorphism** if for every $a, b \in G$ we have

$$\theta(a \circ b) = \theta(a) * \theta(b).$$

If there is a homomorphism of G *onto* H we say H is a **homomorphic image** of G.

If θ is a homomorphism from G to H, the set of elements $\{\theta(g) : g \in G\}$ is called the *image of* θ. Thus, a homomorphism θ is onto H if and only if the image of θ equals H.

The special case in which the mapping is one-to-one and onto is of such importance that we use a special term to distinguish it.

Definition 3.2

A homomorphism θ from G to H is called an **isomorphism** if θ is one-to-one and onto. When there is an isomorphism from G to H, we say G and H are **isomorphic** groups.

If $\theta : G \to H$ is an isomorphism, then there is an isomorphism, denoted by θ^{-1}, from H to G such that $\theta\theta^{-1} = \epsilon_H$ and $\theta^{-1}\theta = \epsilon_G$, where ϵ_G denotes the identity map on G defined by $\epsilon_G(x) = x$ for all $x \in G$. Let us verify this statement. For $h \in H$, define $\theta^{-1}(h)$ to be the unique element $g \in G$ such that $\theta(g) = h$. Since θ is onto, there is such a g; since θ is one-to-one, there is only one such g. It is now evident that $\theta(\theta^{-1}(h)) = h$ [by substituting $\theta^{-1}(h)$ for g in $\theta(g) = h$] and that $\theta^{-1}(\theta(g)) = g$ [by substituting $\theta(g)$ for h in $\theta^{-1}(h) = g$]. It takes just a few more lines to verify that θ^{-1} is a homomorphism of groups and that it is indeed an isomorphism. These details are left to the reader to verify. Thus, if G and H are groups and if there is an isomorphism from G to H, then there is an isomorphism from H to G.

We give some examples of homomorphisms and isomorphisms of groups.

Example 3.1 Let $(\mathbb{Z}, +)$ be the additive group of integers and $(\mathbb{Z}_4, +)$ the additive group of integers modulo 4. The natural map $\theta : m \longrightarrow m + (4)$ is a homomorphism of $(\mathbb{Z}, +)$ onto $(\mathbb{Z}_4, +)$.

Example 3.2 Let G be the additive group of $\mathbb{Z}_4$ with elements written as $\{0, 1, 2, 3\}$ and let H be the additive group of $\mathbb{Z}_8$ with elements written as $\{0^*, 1^*, \ldots, 7^*\}$. Define a mapping $\theta : G \longrightarrow H$ by the rules

$$\theta(0) = 0^*, \quad \theta(1) = 2^*, \quad \theta(2) = 4^*, \quad \theta(3) = 6^*.$$

We show this is a homomorphism. One simply checks the sums:

$$\begin{aligned} \theta(1+1) &= \theta(2) = 4^* = 2^* + 2^* = \theta(1) + \theta(1) \\ \theta(1+2) &= \theta(3) = 6^* = 2^* + 4^* = \theta(1) + \theta(2) \\ \theta(1+3) &= \theta(0) = 0^* = 2^* + 6^* = \theta(1) + \theta(3). \end{aligned}$$

The three remaining sums are done the same way to verify that $\theta(a + b) = \theta(a) + \theta(b)$ for all $a, b \in G$. In this example, θ is one-to-one but is not onto H. There is no element $a \in G$ such that $\theta(a) = 1^*$, for example. Of course, we could predict that there is no homomorphism from G onto H because G has four elements and H has eight elements (see the exercises).

Example 3.3 Let $\mathbb{R}^+$ be the set of positive real numbers with multiplication as the operation and let L be the additive group of the field of all real numbers. (We determined in Section 6.1 that $\mathbb{R}^+$ is a group.) Define a function $\phi : \mathbb{R}^+ \longrightarrow L$ by the rule

$$\phi(x) = \log_{10}(x), \qquad x \in \mathbb{R}^+.$$

The familiar properties of the logarithm function assure us that

$$\phi(xy) = \log_{10}(xy) = \log_{10}(x) + \log_{10}(y) = \phi(x) + \phi(y), \qquad x, y \in \mathbb{R}^+.$$

This shows that the group operation is preserved under the mapping ϕ and therefore ϕ is a homomorphism. We argue that ϕ is an isomorphism by producing an inverse homomorphism. (See Exercise 4 following this section.) For any positive real numbers b, c we know from the definition of the logarithm function that

$$10^{\log_{10}(b)} = b, \qquad \log_{10}(10^c) = c.$$

Thus, the function $\gamma : c \longrightarrow 10^c$ is the inverse of ϕ . Clearly, $\gamma : L \longrightarrow \mathbb{R}^+$ is a homomorphism because

$$\gamma(b + c) = 10^{b+c} = 10^b \cdot 10^c = \gamma(b)\gamma(c).$$

It follows that both ϕ and γ are isomorphisms. Of course, one could use logarithms to other bases to produce other isomorphisms between these groups.

Next we discuss some very basic properties of homomorphisms. For ease of notation, we assume both groups have operation written as multiplication.

Theorem 3.1 *Let $\theta : G \to H$ be a homomorphism from the group G into the group H. Then*

(i) *If e is the identity of G, then $\theta(e)$ is the identity of H.*
(ii) *If $a \in G$, then $\theta(a^{-1}) = \theta(a)^{-1}$.*
(iii) *If G is abelian, then the image of θ is abelian. In particular, if θ is onto and G is abelian, then H is abelian.*

Proof: (i) Since $e^2 = e$ and θ is a homomorphism, we have

$$\theta(e) = \theta(e^2) = \theta(e)^2 = \theta(e)\theta(e).$$

By the cancelation law in H we obtain $\theta(e) = e_H$, the identity of H. (ii) From the equation

$$\theta(a^{-1})\theta(a) = \theta(a^{-1}a) = \theta(e) = e_H,$$

we conclude that $\theta(a^{-1})$ is the inverse in H of $\theta(a)$. (iii) Since $ab = ba$ for all $a, b \in G$ we must have

$$\theta(ab) = \theta(ba)$$

and so

$$\theta(a)\theta(b) = \theta(b)\theta(a).$$

Thus, multiplication of any pair of elements in the image of θ gives the same product in either order.

The reader may have to make a suitable modification in the statement of this theorem in case the operation in one of the groups is addition. For example, if H is an additive group and G is multiplicative, the conclusion of statement (ii) should read $\theta(a^{-1}) = -\theta(a)$ since $-\theta(a)$ is the inverse of $\theta(a)$ in H.

3.1 Kernels

Homomorphisms between groups are an important tool. The next idea indicates how homomorphisms are related to certain subgroups.

Definition 3.3 Let $\theta : G \to H$ be a homomorphism of the group G into the group H. The set of elements of G mapped to the identity of H by θ is called the **kernel of θ** and is denoted by $\ker\theta$.

Example 3.4 If A and B are two groups and $A \times B$ is their direct product, the map $\theta : A \times B \to A$ defined by $\theta(a, b) = a$ is a homomorphism. Its kernel is the set of all elements (e_A, b), $b \in B$.

Example 3.5

If $A = (\mathbb{Z}_n, +)$, the additive group of the ring of integers modulo n, and $\theta : \mathbb{Z} \to \mathbb{Z}_n$ is the map $\theta(k) = k + (n)$, then θ is a homomorphism and the kernel is the subgroup (n) of all integers divisible by n.

As in the case for homomorphisms of rings, the kernel determines when the homomorphism is one-to-one.

Theorem 3.2

If $\theta : G \to H$ is a homomorphism, then $\ker\theta$ *is a subgroup of G. Moreover,* $\ker\theta = \{e\}$, *the subgroup consisting of only the identity of G, if and only if θ is a one-to-one homomorphism.*

Proof: Let $K = \ker\theta$. The identity of H is $\theta(e)$. If $a, b \in K$, then

$$\theta(ab) = \theta(a)\theta(b) = \theta(e)\theta(e) = \theta(e),$$

and thus $ab \in K$ and K is closed with respect to multiplication. Moreover, if $a \in K$, then $\theta(a^{-1}) = \theta(a)^{-1} = \theta(e)^{-1} = \theta(e)$ and so $a^{-1} \in K$. By Theorem 1.2, K is a subgroup of G.

To prove the second statement, suppose first that θ is one-to-one. If $a \in \ker\theta$, then $\theta(a) = e_H$. We always have $\theta(e) = e_H$ so the one-to-one property implies $a = e$. Hence, $\ker\theta = \{e\}$. Conversely, let us assume that $\ker\theta = \{e\}$. If $a, b \in G$ are elements such that $\theta(a) = \theta(b)$, then

$$\theta(ab^{-1}) = \theta(a)\theta(b)^{-1} = \theta(b)\theta(b)^{-1} = e_H.$$

This shows that $ab^{-1} \in \ker\theta$ and so $ab^{-1} = e$ because e is the only element in the kernel. It follows that $a = b$ and θ is one-to-one. ∎

This result is often used in the following way: To show that a mapping of a group G into a group H is an isomorphism, it may be possible to show that it is a homomorphism of G onto H and then verify that its kernel consists of only the identity.

We conclude this section with a theorem that shows in some way why permutation groups are important and at the same time give an illustration of how a nontrivial isomorphism may be constructed. We prove the theorem first proved by the English mathematician Arthur Cayley (1821–1895). Cayley's theorem shows that every group is isomorphic to some group of permutations on some set. In fact, we will see later that a group may be isomorphic to different permutation groups and also there are homomorphisms of a group onto many permutation groups.

Theorem 3.3

CAYLEY'S THEOREM. *Every group is isomorphic to a group of permutations.*

Proof: Given a group G, we must begin the proof by finding a set, some of whose permutations can be associated to elements of G. We make the most obvious choice, namely, the set is G itself. To each element $a \in G$ we associate a permutation π_a of the elements of G as follows: π_a is the permutation sending an element $g \in G$ to the element ag. That is, π_a is the mapping that multiplies elements of G on the left by a. This is a permutation of G as we now show. To see that π_a is an onto map, select any element $c \in G$. Then $\pi_a(a^{-1}c) = a(a^{-1}c) = c$ so c is in the image of π_a. Next we show π_a is a one-to-one mapping. Suppose $\pi_a(g) = \pi_a(h)$ for $g, h \in G$. Then, by definition of π_a we have $ag = ah$. The cancelation law implies $g = h$ and so π_a is one-to-one. So far we have associated a permutation with each element of G. Next we show that the collection of permutations is a group and G is isomorphic to that group. Denote by $\pi(G)$ the collection

$$\pi(G) = \{\pi_a : a \in G\}.$$

We show that $\pi(G)$ is a group. It is sufficient to show that it is closed under multiplication and that the inverse of each element of $\pi(G)$ is in $\pi(G)$. For any $a, b \in G$ we show $\pi_a \pi_b = \pi_{ab}$. To prove these permutations are equal, we show that they have the same effect at every point. For $c \in G$ we have

$$(\pi_a \pi_b)(c) = \pi_a(\pi_b(c)) = \pi_a(bc) = a(bc).$$

On the other hand, we have

$$\pi_{ab}(c) = (ab)c.$$

Since $(ab)c = a(bc)$ it follows that $\pi_a \pi_b$ and π_{ab} have the same effect on every point $c \in G$ and hence they are equal functions. By the same kind of reasoning we obtain $(\pi_a)^{-1} = \pi_{a^{-1}}$. This step is left to the reader. Thus, $\pi(G)$ is a subgroup of the group of all permutations of the set G.

Finally, we show that the mapping $\theta : a \longrightarrow \pi_a$ is an isomorphism from G to $\pi(G)$. The equation $\pi_a \pi_b = \pi_{ab}$ shows that θ is a homomorphism. It is clearly onto by definition of $\pi(G)$. Let us show that it is one-to-one. Suppose $a, b \in G$ and $\pi_a = \pi_b$. Then the two functions have the same effect at every point, that is, $\pi_a(c) = \pi_b(c)$ for each $c \in G$. Evaluate at the identity of G:

$$\pi_a(e) = ae = a, \qquad \pi_b(e) = be = b.$$

Since $\pi_a(e) = \pi_b(e)$ it follows that $a = b$ and so θ is an isomorphism. ■

Here, we give a specific example that illustrates the proof of Cayley's theorem. For the group G we take the multiplicative group of the field $\mathbb{Z}_5$ so that $G = \{1, 2, 3, 4\}$ with multiplication taken modulo 5. Cayley's theorem shows how to associate a permutation of the set G with every element of G. For an element $g \in G$ we associate the function $\pi_g : x \to gx$. We write this explicitly for $g = 2$:

$$\pi_2(1) = 2 \cdot 1 = 2, \quad \pi_2(2) = 2 \cdot 2 = 4,$$
$$\pi_2(3) = 2 \cdot 3 = 1, \quad \pi_2(4) = 2 \cdot 4 = 3.$$

The permutation π_2 can be written in the double-row notation as

$$\pi_2 = \begin{pmatrix} 1 & 2 & 3 & 4 \\ 2 & 4 & 1 & 3 \end{pmatrix}.$$

We can perform the analogous computation to determine the permutations associated with the other group elements with the following result:

$$\pi_1 = \begin{pmatrix} 1 & 2 & 3 & 4 \\ 1 & 2 & 3 & 4 \end{pmatrix}, \quad \pi_3 = \begin{pmatrix} 1 & 2 & 3 & 4 \\ 3 & 1 & 4 & 2 \end{pmatrix}, \quad \pi_4 = \begin{pmatrix} 1 & 2 & 3 & 4 \\ 4 & 3 & 2 & 1 \end{pmatrix}.$$

The mapping $g \to \pi_g$ is an isomorphism of G with a group of permutations.

In view of Cayley's theorem, in order to prove a result about all groups it is sufficient to prove it for groups of permutations. Although the subject of group theory is not usually approached in this way, it will be seen in later chapters how the study of finite groups is facilitated by ingenious use of certain groups of permutations and the presentation of groups as permutations of selected sets.

EXERCISES

1. Let θ be a homomorphism from a group G to a group H. Suppose G has exactly n elements and H has exactly m elements. If there is a homomorphism from G *onto* H, show $n \geq m$. If there is a one-to-one homomorphism from G to H, show $n \leq m$. If G and H are isomorphic, show $n = m$.

2. Suppose G and H are groups with identities e_G and e_H, respectively, and $\theta : G \to H$ is an isomorphism. Prove the following facts:

(a) If there is an element $g \in G$ such that $g^n = e_G$, then there is an element $h \in H$ such that $h^n = e_H$.

(b) If G has exactly m subgroups (m some positive integer), then H has exactly m subgroups.

(c) If G has a pair of proper subgroups A and B such that $A \cap B = \{e_G\}$, then H has a pair of proper subgroups whose intersection is the identity of H.

3. It was shown in Example 3.3 that the multiplicative group of positive reals is isomorphic to the additive group of all real numbers. Show that the multiplicative group of all nonzero real numbers is not isomorphic to the additive group of all real numbers. [Hint: If there were an isomorphism, the element -1 in the multiplicative group would have to correspond to an element in the additive group with some impossible property.]

4. Let $\theta : G \to H$ be a homomorphism and $\phi : H \to G$ a homomorphism for two groups G and H. Suppose $\phi\theta = \epsilon_G$ (identity map on G) and $\theta\phi = \epsilon_H$. Prove that both θ and ϕ are isomorphisms. [By definition, the identity map satisfies $\epsilon_G(x) = x$, $x \in G$.]

4 CYCLIC GROUPS

If a is an element of an arbitrary group (with operation written as multiplication), then every power a^k is in G for any positive integer k. Also, a^0 is the identity e of G and, by definition, a^{-k} is the kth power of a^{-1}. It follows that the set $\{a^m : m \in \mathbb{Z}\}$ is a subgroup of G. We will consider the case in which this subgroup is the entire group. Accordingly, we make the following definition:

Definition 4.1

If the multiplicative group G contains an element a such that $G = \{a^k : k \in \mathbb{Z}\}$, we say that G is a **cyclic group** and that G is **generated by** $\boldsymbol{a}$ or that a is a generator of G.

Note that a cyclic group is necessarily abelian. Any pair of elements, g, h, may be written as $g = a^i$ and $h = a^j$ for some $i, j \in \mathbb{Z}$, so it follows that

$$gh = a^i \cdot a^j = a^{i+j} = a^{j+i} = a^j \cdot a^i = hg.$$

If G is an *additive* group, then G is a cyclic group with generator a if $G = \{ka : k \in \mathbb{Z}\}$ for some $a \in G$.

Example 4.1

Let G be the multiplicative group of units in the finite field $\mathbb{Z}_5$. The elements of G are written as $1, 2, 3, 4$ and multiplication is carried out modulo 5. Compute the powers of 2: $2^1 = 2$, $2^2 = 4$, $2^3 = 3$, and $2^4 = 1$. Thus, every element of G is a power of 2 and G is a cyclic group. The reader may verify that every element of G is also a power of 3. Thus, both 2 and 3 are generators of G. The only powers of 4 are $\{1, 4\}$, which is a cyclic subgroup of G.

Example 4.2

The additive group of integers is a cyclic group with 1 as generator. Every integer is a multiple $m1$ of 1 for an integer m. Note that -1 is also a generator of $\mathbb{Z}$. No other element of $\mathbb{Z}$ is a generator.

Example 4.3

The additive group of the ring $\mathbb{Z}_n$ of integers modulo n is cyclic with 1 as its generator. It may have other generators as well.

Even if a group is not cyclic, it will have cyclic subgroups; if $a \in G$, then the set of powers of a is a cyclic subgroup of G. The cyclic subgroup generated by a may have finitely many elements or infinitely many elements. We make the following definition to distinguish the cases.

Definition 4.2

(i) If G is a group with exactly n elements, for some positive integer n, we say G has **order n**. If no such integer exists, we say G has **infinite order**.

(ii) An element a of a group G has **order n** if the cyclic subgroup generated by a has n elements.

Using the language introduced in this definition we may say that the additive group of integers has infinite order, the symmetric group S_n has order $n!$, the additive group of the ring $\mathbb{Z}_n$ has order n, and the multiplicative group of the field $\mathbb{Z}_p$ has order $p-1$.

The next theorem gives an important characterization of the order of an element. To compute the order of an element, it is only necessary to find the lowest power of the element that gives the identity. This characterization is usually used to determine the order of an element.

We begin with a lemma that provides the most important step of the proof of the next theorem.

Lemma 4.1 *Let a be an element of the group G, and suppose that some power of a equals the identity; let n be the smallest positive integer such that $a^n = e$. If $k \in \mathbb{Z}$, then $a^k = e$ if and only if $k \equiv 0 \pmod{n}$. More generally, if $i, j \in \mathbb{Z}$, then $a^i = a^j$ if and only if $i \equiv j \pmod{n}$.*

Proof: By the division algorithm, we may write any integer k in the form $k = nq + r$ with integers q and r and $0 \leq r < n$. Using the equality $a^n = e$, we find

$$a^k = a^{nq+r} = (a^n)^q \cdot a^r = e^q \cdot a^r = a^r.$$

If $a^k = e$, then $a^r = e$. However, n was selected as the smallest positive integer such that $a^n = e$. Thus, $0 \leq r < n$ implies $r = 0$. The equality $a^k = e$ implies $k = nq$ or $k \equiv 0 \pmod{n}$. Conversely, if $k = nq$, then $a^k = (a^n)^q = e$. This proves the first part of the lemma. For the second part, suppose $a^i = a^j$. Then $a^{i-j} = e$ and, by what we have just proved, $i - j \equiv 0 \pmod{n}$ or $i \equiv j \pmod{n}$. ∎

We use this lemma to prove the next result.

Theorem 4.1 *If G is a group and $a \in G$ is an element of order n, then n is the least positive integer such that $a^n = e$, where e is the identity of G. If no such integer exists, then a has infinite order.*

Proof: Assume a has order n so the cyclic group generated by a has n distinct elements. There must exist $j < i$ with $a^i = a^j$ because there are only a finite number of distinct powers of a. Thus, $a^{i-j} = e$ and so there is a smallest positive integer N such that $a^N = e$; we must show $n = N$.

Since N is the least positive integer such that $a^N = e$, the N elements

$$e = a^0, a, a^2, \ldots, a^{N-1} \tag{6.5}$$

must all be different by the lemma just proved. Moreover, the previous lemma shows for any integer k, $a^k = a^r$ for some r with $0 \leq r < N$. Thus, every power of a appears in the list Eq. (6.5). Since n is the number of elements in the cyclic group of powers of a, it follows that $n = N$.

If there is no positive integer N such that $a^N = e$, then there are infinitely many distinct powers of a and a has infinite order. ∎

In any group, the identity has order 1. An element not equal to the identity either has infinite order or has finite order equal to some integer greater than 1.

If a group G has finite order, then every element of G has finite order. The converse is false; there exist infinite groups with every element of finite order. Next we examine some groups of finite order and determine the orders of some elements.

Example 4.4 Let us find the order of each element in the additive group of integers modulo 8. We make an observation about the order of an element $j + (8)$. If m is an integer such that $mj + (8) = 0 + (8)$, then the integer mj must be divisible by 8. The order of $j + (8)$ is the least positive integer such that mj is divisible by 8. So the four elements $1 + (8), 3 + (8), 5 + (8), 7 + (8)$ have order 8. The two elements $2 + (8), 6 + (8)$ have order 4, and $4 + (8)$ has order 2. The identity $0 + (8)$ has order 1. Note that the elements of this group with eight elements have orders $1, 2, 4$, and 8. It is not a coincidence that the order of an element divides the number of elements in the group. We will prove a more general fact using Lagrange's theorem.

Example 4.5 Let G be the multiplicative group of units in $\mathbb{Z}_{10}$. Then G consists of the cosets containing $1, 3, 7, 9$. Let us find the orders of each element. Since $9^2 = 81 \equiv 1$ we see that 9 has order 2. The powers of 3 are $3, 3^2 = 9, 3^3 = 7$, and $3^4 = 1$. Thus, 3 has order 4. We could compute the powers for 7, but since 7 is the inverse of 3 its order is also 4 because it is easily checked in general that an element and its inverse have the same order.

Example 4.6 The only subgroups of the additive group of $\mathbb{Z}_5$ are the trivial ones, $\{0\}$ and $\mathbb{Z}_5$. To see this let H be a subgroup other than $\{0\}$. Then H contains some element $a \neq 0$ and so H contains all of the multiples $a, 2a, 3a, 4a$, and $5a = 0$ of a. Since $a \neq 0$ it is easy to check that $\{a, 2a, 3a, 4a\} = \{1, 2, 3, 4\}$. For example, if $a = 2$, then

$$a = 2, \quad 2a = 4, \quad 3a = 1, 4a = 3.$$

Thus, H contains all five elements of $\mathbb{Z}_5$ and $H = \mathbb{Z}_5$. Note also that every nonidentity element of H is a generator of H.

It is easy to determine when two cyclic groups are isomorphic. Two cyclic groups are isomorphic if either they are both infinite or if both have order n. This follows from the next theorem.

Theorem 4.2

(i) *Every cyclic group of infinite order is isomorphic to the additive group of integers.*

(ii) *Every cyclic group of order n is isomorphic to the additive group of the ring $\mathbb{Z}_n$ of integers modulo n.*

Proof: Let G be a cyclic group with generator a and let $\theta : \mathbb{Z} \to G$ be the function defined by

$$\theta(k) = a^k, \qquad k \in \mathbb{Z}.$$

If $i, j \in \mathbb{Z}$, then

$$\theta(i + j) = a^{i+j} = a^i a^j = \theta(i)\theta(j);$$

thus, θ is a homomorphism from the additive group of $\mathbb{Z}$ onto G. It is onto because every element of G has the form a^k for some integer k. Now we consider two cases. Suppose first that G has infinite order. Then there is no positive power of a that equals the identity of G. Hence, there is no positive integer in the kernel of θ. Since $\ker\theta$ is a subgroup, there is no negative integer in the kernel of θ either. Thus, $\ker\theta = (0)$ and θ is one-to-one and onto. That is, θ is an isomorphism.

Now suppose G has finite order n. Then n is the smallest positive integer such that $a^n = e$. It follows that $a^k = e$ if and only if k is a multiple of n. This means that $\ker\theta = (n)$, the set of multiplies of n. The cosets $1 + (n)$, $2 + (n), \cdots$ correspond one-to-one with the elements $a^1, a^2, \cdots$. We define a mapping from the collection of cosets (i.e., from the additive group of $\mathbb{Z}_n$) to G by the rule

$$i + (n) \longrightarrow a^i.$$

This is well defined because $i + (n) = j + (n)$ implies $i \equiv j \pmod{n}$ and so $a^i = a^j$ by Lemma 4.1. Moreover, the correspondence is a homomorphism of $\mathbb{Z}_n$ onto G and it is one-to-one. This last assertion follows because $a^i = e$ if and only if $i + (n) = 0 + (n)$. It follows that G is isomorphic to the additive group n. ■

It may be useful to note that a homomorphism from $\mathbb{Z}_n$ to any group is completely determined by the image of the coset $1 + (n)$. By this we mean that if θ is a homomorphism defined on $\mathbb{Z}_n$ and if $\theta(1 + (n)) = b$ is a given image of $1+(n)$ in G, then any coset $k+(n)$ is an integer multiple $k[1+(n)]$ of $1+(n)$ and so

$$\theta(k+(n)) = \theta(k[1+(n)]) = \theta(1+(n))^k = b^k.$$

Hence, the image under θ of every element of $\mathbb{Z}_n$ is determined.

For example, consider the multiplicative group G of the field $\mathbb{Z}_5$. We have already seen that this is a cyclic group of order 4 with [2] as generator. We map the additive group of $\mathbb{Z}_4$ to G by sending $1+(4)$ to [2]; all other images are determined as follows:

$$1+(4) \to [2], \quad 2+(4) \to [2^2], \quad 3+(4) \to [2^3] = [3], \quad 0+(4) \to [2^0] = [1].$$

There is a second isomorphism between these groups obtained by using the generator [3] for G and the map $i+(4) \to [3^i]$.

4.1 Subgroups of Cyclic Groups

The final information about cyclic groups we will derive in this section is the determination of all the subgroups of a cyclic group. The description is slightly different depending on whether the group is finite or infinite. We first state a common feature of both cases.

Theorem 4.3 *Every subgroup of a cyclic group is also a cyclic group.*

Proof: Let G be a cyclic group with generator a, and let H be a subgroup of G. If $H = \{e\}$, the subgroup containing only the identity, then H is cyclic with generator e. Assume then that H contains some element other than the identity. Then it contains a power a^k with $k > 0$. Let m be the smallest positive integer such that $a^m \in H$. We will show that H is cyclic with generator a^m. Since $H \subseteq G$, every element of H has the form a^k for some integer k. By the division algorithm, we may write $k = mq + r$ with $0 \le r < m$. Hence,

$$a^k = a^{mq+r} = (a^m)^q \cdot a^r,$$

from which we conclude

$$a^r = a^k \cdot (a^m)^{-q}.$$

Since a^k and a^m are elements of H, it follows that $a^r \in H$. In view of m as the smallest positive integer with $a^m \in H$, and since $0 \le r < m$, the only choice is $r = 0$. Thus, $a^k = (a^m)^q$, which shows that every element of H is a power of a^m. ■

Now we give an explicit description of the subgroups.

Theorem 4.4

(i) *Let G be an infinite cyclic group with generator a. A subgroup H of G is a cyclic group generated by an element a^m, with m a nonnegative integer. The only other generator of H is a^{-m}.*

(ii) *Let G be a finite cyclic group of order n with generator a. A subgroup H of G is a cyclic group with generator a^m, where m is a positive integer dividing n. The number of elements in H is n/m. The generators of H are the elements a^r, where r is any integer multiple of m such that $m = (r, n)$.*

Proof: Let H be a subgroup of the infinite cyclic group G. If $H = \{e\}$, then $e = a^0$ is the only generator of H. Assume H has more than one element. By the proof Theorem 4.3 we know that a^m is a generator for H if m is the least positive integer such that $a^m \in H$. If a^k is another generator, then each is a power of the other; that is, $a^k = a^{mc}$ and $a^m = a^{kf}$ for some nonzero integers c and f. It follows that

$$a^{kf} = a^{mcf} = a^m.$$

This implies $a^{mcf-m} = e$. However, since a has infinite order, this can happen only if $mcf - m = 0$. Equivalently, $cf = 1$ and so $c = \pm 1$. Thus, the only generators for H are a^m and a^{-m}. This completes the proof of statement (i).

Now assume G is a cyclic group of order n and let H be a subgroup of G. If $H = \{e\}$, then H is generated by $a^n = e$ and this is the only generator of H. Suppose then that $H \neq \{e\}$. From the conclusion of Theorem 4.3, we know that H is a cyclic group and it must be generated by an element a^m, where m is the smallest positive integer such that $a^m \in H$. The proof of that theorem also shows that m is a divisor of n. Suppose $n = mt$. What is the order of a^m? Since n is the least power of a that equals e, the elements $a^m, a^{2m}, \ldots, a^{mt} = e$ are distinct and must consist of all the elements of H; thus, H has order t. What other elements of H can generate H? A generator of H must have the form a^{ms} for some integer s (because every element of H is a power of a^m) and it must have order $t = n/m$, which is the order of the cyclic subgroup generated by a^m. We argue that $m = (ms, n)$. Since $(ms, n) = (ms, mt) = m(s, t)$, it is sufficient to show $(s, t) = 1$. Suppose $d = (s, t)$ so that d divides both s and t. We show that a^{ms} has order at most t/d. This is shown as follows:

$$(a^{ms})^{t/d} = a^{mst/d} = (a^{mt})^{s/d} = (a^n)^{s/d} = e^{s/d} = e.$$

We have used the fact that s/d is an integer. Thus, the order of a^{ms} is at most t/d but must equal t. Thus, $d = 1$, as we wished to prove. ■

Next we consider an example to illustrate some of the ideas of this proof. The cyclic group of order 18, with generator a, has subgroups of orders $1, 2, 3, 6, 9, 18$, one for each divisor of 18. Which one of these is the subgroup generated by a^{10}? Since 10 does not divide 18, we examine the greatest common denominator (GCD) of 10 and 18, namely, 2. Then the group generated by a^{10} is contained in the group generated by a^2 since $a^{10} = (a^2)^5$. However, a^2 is also contained in the group generated by a^{10} because

$$(a^{10})^{-7} = a^{-70} = a^2$$

since $-70 \equiv 2 \pmod{18}$. Similarly, the reader may verify that the subgroup generated by a^{15} is also generated by a^3 since $3 = (15, 18)$.

Of course, these results apply equally well to subgroups of a cyclic group written additively. For example, the subgroups of the cyclic group $\mathbb{Z}_{14}$ are generated by the cosets containing divisors of 14: The coset $1 + (14)$ generates the whole group with 14 elements, $2 + (14)$ generates a group with $14/2 = 7$ elements, $7 + (14)$ generates a subgroup with $14/7 = 2$ elements, and $0 + (14)$ generates the trivial subgroup with 1 element. The coset $8 + (14)$ generates the same subgroup as the coset containing $(8, 14) = 2$, and so it has order 7.

EXERCISES

1. Find the order of each element in S_3.

2. Find elements of orders $2, 3, 4$, and 5 in S_5.

3. Verify that the multiplicative group of the field $\mathbb{Z}_p$ is cyclic of order $p - 1$ for the cases $p = 3, 5, 7$, and 11. In each case give a generator of the multiplicative group.

4. Find all the subgroups of the additive group of $\mathbb{Z}_{20}$.

5. Find all generators of the additive group of $\mathbb{Z}_{30}$.

6. Let G be a cyclic group of order n and k a divisor of n.

(a) Prove that G has exactly one subgroup of order k.

(b) Prove that there is a homomorphism from G onto a cyclic group of order k.

7. Determine all the isomorphisms from the additive group of $\mathbb{Z}_6$ with the multiplicative group of $\mathbb{Z}_7$.

8. Let G be a cyclic group of order n and H a cyclic group of order m.

(a) Prove that there is a homomorphism of G onto H if and only if m divides n.

(b) If $\theta : G \to H$ is a homomorphism, show that the order of $\theta(G)$ divides the GCD of n and m.

9. If a is an element of order n in a multiplicative group G, show that a^{-1} also has order n.

10. Show that the cyclic group of order 6 is the direct product of a cyclic group of order 2 and a cyclic group of order 3.

11. List the six elements of $GL(2, \mathbb{Z}_2)$ and find the order of each. Show that the subset consisting of the elements having a 0 in the upper right corner is a subgroup. Find two other subgroups of $GL(2, \mathbb{Z}_2)$ other than the two trivial subgroups.

12. Let p be any prime number and $\mathbb{Z}_p$ the field of p elements. Show that $g = \begin{bmatrix} 1 & 1 \\ 0 & 1 \end{bmatrix}$ is an element of order p in $GL(2, \mathbb{Z}_p)$. Let $a, b \in \mathbb{Z}_p$ be nonzero elements of multiplicative order r and s, respectively. Show that $\begin{bmatrix} a & 0 \\ 0 & b \end{bmatrix}$ is an element in $GL(2, \mathbb{Z}_p)$ and find its order.

13. Let $\varphi(n)$ be the Euler Phi Function defined in Chapter IV, Section 4 as the number of positive integers between 1 and n that are relatively prime to n.

(a) If A is a cyclic group of order n, show that A has $\varphi(n)$ generators.

(b) Prove the formula $n = \sum_{d|n} \varphi(d)$, where the sum is taken over all the positive integers d that divide n. [Hint: Let A be a cyclic group of order n; every element of order d in A generates a cyclic subgroup that has $\varphi(d)$ generators so the sum counts the number of elements in A.]

14. Let p be a positive prime integer and n a positive integer. Prove that $\varphi(p^n - 1)$ is divisible by n. [Hint: Consider the group of units in $\mathbb{Z}_N$ with $N = p^n - 1$. Its order is $\varphi(N)$ and it contains the element p since p is relatively prime to N. The order of p is n. Draw conclusions from this.]

5 COSETS AND LAGRANGE'S THEOREM

Let G be an arbitrary group and H a subgroup of G. If $a \in G$, we use the symbol aH to denote the set of all elements of the form ah, where $h \in H$. That is,

$$aH = \{ah : h \in H\}.$$

Definition 5.1

If H is a subgroup of the group G and $a \in G$, we call aH a **coset** of H in G.

It would be more precise to call aH a *left coset* to distinguish it from the closely related set $Ha = \{ha : h \in H\}$, which is called a *right coset*. We use the word "coset" to mean *left coset*. Of course, if the group G is abelian, then

$ah = ha$ for every pair of elements of G so there is no difference between a left coset and a right coset in this case.

One reason for studying cosets is that the collection of cosets of a given subgroup form a partition of G. We prove this now.

Lemma 5.1 *If H is a subgroup of the group G, then the following hold:*

(i) *For $a, b \in G$, either $aH = bH$ or $aH \cap bH = \emptyset$.*

(ii) *Every element of G lies in one, and only one, coset of H.*

Proof: Suppose the cosets aH and bH have a common element c. Then $c = ah_1 = bh_2$ for some $h_1, h_2 \in H$. It follows that $a = bh_2h_1^{-1}$. Any element ah of aH can be expressed as $ah = bhh_2h_1^{-1}h$; since $h_2h_1^{-1}h \in H$, it follows that $ah \in bH$. We have shown, therefore, that $aH \subseteq bH$. In a similar way we show that $bH \subseteq aH$, and it follows that $aH = bH$. Thus, if two cosets have one element in common, they are equal sets.

To prove (ii) take any element $a \in G$. Since H is a subgroup, it contains the identity element e and so aH contains $ae = a$. Thus, every element of G belongs to some coset of H. An element belongs to exactly one coset by part (i); that is, if $a \in aH$ and $a \in bH$, then $aH = bH$. ∎

As an example of cosets, let us take $G = S_3$, the symmetric group on three symbols whose multiplication table is given in Section 2. The element α has order 2 since $\alpha^2 = e$, so $H = \{e, \alpha\}$ is a subgroup. We write the cosets aH for the six elements $a \in S_3$:

$$\begin{array}{rclrcl} eH &=& \{e, \alpha\}, & \alpha H &=& \{\alpha, e\}, \\ \beta H &=& \{\beta, \alpha\beta^2\}, & \alpha\beta H &=& \{\alpha\beta, \beta^2\}, \\ \beta^2 H &=& \{\beta^2, \alpha\beta\}, & \alpha\beta^2 H &=& \{\alpha\beta^2, \beta\}. \end{array}$$

We see that of these six cosets, only three are different sets. For example, $\beta^2 H = \alpha\beta H$, which is predictable because $\alpha\beta \in \beta^2 H$. There are three cosets with two elements in each. Every element of S_3 lies in exactly one coset.

If the group G is an additive group, the notation for a coset of a subgroup G is

$$a + H = \{a + h : h \in H\}.$$

We have already discussed additive cosets in the context of rings and cosets of ideals. For a specific example, consider the cyclic, additive group of $\mathbb{Z}_{12}$ and the subgroup $H = \{[0], [3], [6], [9]\}$. There are three different cosets:

$$\begin{array}{rcl} [0] + H &=& \{[0], [3], [6], [9]\}, \\ [1] + H &=& \{[1], [4], [7], [10]\}, \\ [2] + H &=& \{[2], [5], [8], [11]\}. \end{array}$$

There are three cosets, each with four elements, and every element of $\mathbb{Z}_{12}$ appears in exactly one coset.

The number of cosets of a subgroup of a group is an important number to which we assign a name.

Definition 5.2

If G is a group and H a subgroup, then the number of cosets of H in G is called the **index** of H in G and is denoted by $[G : H]$.

If G is a finite group, then every subgroup has a finite number of cosets; even if G is an infinite group, the number of cosets of a subgroup may be finite. In these cases $[G : H]$ is a finite number. If H has infinitely many cosets, then its index in G is infinite.

If G is a finite group and $E = \{e\}$ is the subgroup consisting of only the identity element, then every coset of E contains one element so the index $[G : E]$ of E equals the number of elements in G. We abbreviate this by writing $|G|$ for the number of elements in G; that is, $|G| = [G : E]$.

We are now ready to prove the theorem of Lagrange, which is fundamental in the study of finite groups.

Theorem 5.1

LAGRANGE'S THEOREM *Suppose that G is a finite group and H is a subgroup. Then the order of H divides the order of G. More precisely, $|G| = |H| \cdot [G : H]$.*

Proof: We first observe that any coset of H has the same number of elements as H, namely, $|H|$. Let aH be any coset of H. Multiplication by a gives a correspondence $h \rightarrow ah$ that is a one-to-one function of H onto aH. Hence, the two sets have the same number of elements. Every coset has $|H|$ elements and there are $[G : H]$ cosets. Thus, there are exactly $|H| \cdot [G : H]$ elements in the cosets of H. Since every element of G lies in one, and only one, coset, it follows that $|H| \cdot [G : H] = |G|$, and we have proved the theorem. ∎

We give a few immediate applications of Lagrange's theorem. More applications are discussed in later sections. The order of an element in a group is the order of the cyclic subgroup generated by the element. We have the following corollary:

Corollary 5.1

The order of an element of a group of finite order is a divisor of the order of the group.

If the order of a group is a prime number p, then the only positive divisors of the order are 1 and p; thus, the group has only two subgroups, the group of order one and the entire group.

Corollary 5.2

A group of prime order is a cyclic group and every element, except the identity, is a generator of the group.

If a group G has order n and if $a \in G$ is an element of order m, then $n = mk$ for some positive integer k. It follows that $a^n = (a^m)^k = e$.

Corollary 5.3

If a is an element in a group of order n, then $a^n = e$.

This corollary may be used to give another proof of Fermat's little theorem (see Chapter IV, Theorem 4.4). For a prime p, the multiplicative group of the field $\mathbb{Z}_p$ has order $p - 1$. If a is an integer not divisible by p, then the class $[a] = a + (p)$ is nonzero and so by the corollary $[a]^{p-1} = [1]$. This is equivalent to saying that $a^{p-1} - 1$ is divisible by p as had been proved earlier. A proof of Euler's theorem (see Chapter IV, Theorem 4.3) may be given in a similar way.

The reader should notice that Lagrange's theorem does not assert the existence of any particular subgroups. It may be tempting, on the basis of examples of small order, to guess that if G has order n and d is a divisor of n, then there is a subgroup of G of order d. This is not true in general but it is true for certain classes of groups and for certain divisors. For example, if G is an abelian group, then there is a subgroup of order d for every divisor of $|G|$. If G is an arbitrary finite group and d is a prime power that divides $|G|$, there is a subgroup of order d in G. This is part of the content of Sylow's theorem to be studied later.

EXERCISES

1. Exhibit all the cosets of the subgroup $\{e, \alpha, \alpha^2, \alpha^3\}$ of the dihedral group D_8 whose multiplication table is given in Section 2 of this chapter.
2. Exhibit all the cosets of the subgroup $H = \{[0], [4], [8], [12], [16]\}$ of the additive group of $\mathbb{Z}_{20}$.
3. Let G be the multiplicative group of all nonzero real numbers and let H be the subgroup of all positive real numbers. List the representative elements from the distinct cosets of H in G and show that $[G : H] = 2$.
4. Let G be the additive group of all rational numbers and H the subgroup consisting of all rational numbers of the form a/b with b an odd integer and a an arbitrary integer. Prove that $[G : H]$ is infinite. [Hint: Show that no two of the numbers $1/2^i$, $i = 1, 2, \ldots$, lie in the same coset of H.]
5. Prove that a group of order n has a proper subgroup if and only if n is not prime.
6. Let p be a positive prime and let G be an abelian group containing elements a, b such that a is not contained in the cyclic group generated by b and a and b each have order p. Prove that $\{a^i b^j : 0 \leq i, j \leq p-1\}$ is a subgroup of order p^2.

7. Prove that an abelian group of order 6 must be cyclic. [Hint: The only possible orders of elements are 1, 2, 3 and 6. Use the previous exercise to argue that there is exactly one element of order 2 and two elements of order 3 and so there must be an element of order 6.]

8. Let G be a finite group and H and K subgroups of G such that $K \subseteq H$. Use Lagrange's theorem to prove the formula $[G : K] = [G : H][H : K]$.

9. Let H be a subgroup of a group G. If $a, b \in G$, let $a \sim b$ mean $b^{-1}a \in H$. Show that $\sim$ is an equivalence relation on G. For $a \in G$, show that the equivalence class containing a is the coset aH.

10. Let H be a subgroup of a group G. Show that there is a well-defined correspondence β from the set of left cosets of H to the set of right cosets of H that satisfies $\beta(aH) = Ha^{-1}$. Use this to show that the number of left cosets of H in G is the same as the number of right cosets of H in G.

11. Let G be a group of order n, and let H and K be subgroups of G with orders h and k, respectively. Let $D = H \cap K$ have d elements. Let HK denote the set of elements $\{xy : x \in H,\ y \in K\}$. Prove that the number of elements in HK is exactly hk/d. [Hint: If $h_1D, \ldots, h_mD$ is a complete set of cosets of D in H, then $h_1K, \ldots, h_mK$ is a partition of the set HK.]

12. Let H and K be subgroups of a group G. If $a \in G$, the set $HaK = \{hak : h \in H,\ k \in K\}$ is called a *double coset* of H and K in G.

(a) If $a, b \in G$ and $HaK \cap HbK \neq \emptyset$, show that $HaK = HbK$.

(b) If G is a finite group, show that the number of elements in HaK equals $|H| \cdot |K|/|H \cap aKa^{-1}|$. [Hint: The number of elements in HaK equals the number of elements in $H(aKa^{-1})$. Show that aKa^{-1} is a subgroup and apply the previous exercise.]

(c) Give an example to show that not all double cosets need have the same number of elements.

6 NORMAL SUBGROUPS AND FACTOR GROUPS

In the study of rings, the kernel of homomorphism from one ring to another is a special kind of subring which we called an ideal. In the study of groups, the kernel of a homomorphism from one group to another is a special subgroup which we now define.

Definition 6.1

A subgroup K of a group G is called a **normal subgroup** of G if $aK = Ka$ for every $a \in G$.

Note that K is a normal subgroup means that every left coset aK is equal to the right coset Ka. This does not mean that $ak = ka$ for each $k \in K$.

It merely states that the two *sets* aK and Ka are the same. Thus, for each $k \in K$ there is an element $k_1 \in K$ such that $ak = k_1 a$.

Clearly, every subgroup of an abelian group is a normal subgroup. For an example of a subgroup that is not normal, we may consider the symmetric group S_3. We provided its multiplication table in Section 2 of this chapter. The subgroup $H = \{e, \alpha\}$ of order 2 is not normal. For example, we may compute

$$\beta H = \{\beta, \beta\alpha\} = \{\beta, \alpha\beta^2\},$$

whereas

$$H\beta = \{\beta, \alpha\beta\}.$$

Thus, $\beta H \neq H\beta$. In the same group we can see that the subgroup $K = \{e, \beta, \beta^2\}$ is a normal subgroup. Rather than verify this with a direct analysis, we can prove a more general fact that includes this example. Note that K has order 3 and S_3 has order 6 so that $[S_3 : K] = 2$. We now show that in any group, a subgroup of index 2 is a normal subgroup.

Proposition 6.1 *If G is a group and K is a subgroup with $[G : K] = 2$, then K is a normal subgroup of G.*

Proof: To show that K is a normal subgroup, select any element $a \in G$. We must show $aK = Ka$. If $a \in K$, then $aK = K = Ka$. Therefore, assume $a \notin K$. Since $[G : K] = 2$, there are only two cosets of K which we may represent as K and aK since a is not in K. Now we have $G = K \cup aK$ and, of course, the two sets have no common element. Thus, aK is the set of all elements of G that do not lie in K. Now consider the right coset Ka. We also have $G = K \cup Ka$ and $K \cap Ka = \emptyset$. It follows that Ka is the set of all elements of G that do not lie in K. In other words, $aK = Ka$. ∎

There is an equivalent way of stating the condition in the definition of normal subgroup that is used more frequently than the definition itself. For an element $a \in G$ and a subgroup K of G we write aKa^{-1} for the set

$$aKa^{-1} = \{aka^{-1} : k \in K\}.$$

Note that aKa^{-1} is a subgroup of G when K is a subgroup, regardless of whether K is a normal subgroup or not. We easily check the closure:

$$(aka^{-1})(ak_1a^{-1}) = (ak)(a^{-1}a)k_1a^{-1} = akk_1a^{-1}.$$

The remaining properties required of a subgroup are easily verified. Thus, we have the following lemma:

Lemma 6.1 *K is a normal subgroup of G if and only if K is a subgroup of G and $K = aKa^{-1}$ for every $a \in G$.*

Proof: If K is a normal subgroup, then $aK = Ka$ for every $a \in G$. It follows, by multiplication on the right by a^{-1}, that $aKa^{-1} = K$. Conversely, the equation $aKa^{-1} = K$ implies $aK = Ka$.

Now we discuss the relation of homomorphisms and normal subgroups.

Theorem 6.1 *If θ is a homomorphism from a group G to a group H, then the kernel of θ is a normal subgroup of G.*

Proof: Let K denote the kernel of θ. An element $g \in G$ belongs to K if and only if $\theta(g) = e_H$, the identity of H. Suppose $g \in K$ and $a \in G$. Then

$$\theta(aga^{-1}) = \theta(a)\theta(g)\theta(a)^{-1} = \theta(a)e_H\theta(a)^{-1} = \theta(a)\theta(a)^{-1} = e_H.$$

Thus, $aga^{-1} \in K$ for every $a \in G$. This proves $aKa^{-1} \subseteq K$. To show equality, we need to prove $K \subseteq aKa^{-1}$. Pick any $k \in K$; then $a^{-1}ka \in K$ since a^{-1} is an element of G. Then

$$a(a^{-1}ka)a^{-1} = k$$

is an element of aKa^{-1} which proves that every element of K lies in aKa^{-1}. Hence, the kernel of θ is a normal subgroup. ∎

Now that we know the kernel of a homomorphism is a normal subgroup, we may ask if every normal subgroup is the kernel of some homomorphism. The answer is yes, as we now begin to verify.

6.1 Factor Groups

Let G be an arbitrary group and K a normal subgroup of G. Since K is normal, we need not distinguish between left and right cosets; therefore, we simply call them cosets and write them as left cosets.

Let G/K denote the collection of all cosets of K in G. We will define an operation of G/K which makes it into a group. We attempt the definition of multiplication of cosets in the form

$$(aK)(bK) = abK, \qquad a, b \in G. \tag{6.6}$$

In order to verify that this is a proper definition on the set of cosets, we must verify that the operation is well defined. If $aK = a_1K$ and if $bK = b_1K$, it is necessary to verify that $abK = a_1b_1K$. Therefore, we check this. The equation $aK = a_1K$ implies that $ak = a_1$ for some $k \in K$. Similarly, $bh = b_1$ for some $h \in K$. Then

$$a_1b_1 = (ak)(bh) = ab(b^{-1}kb)h.$$

By the alternative test for normality, we know that $b^{-1}kb \in K$ and so $(b^{-1}kb)h \in K$. It follows that $a_1b_1 \in abK$ and therefore $a_1b_1K = abK$. Thus, multiplication of cosets given in Eq. (6.6) is well defined.

With this step in hand, we now prove the following theorem:

Theorem 6.2 *Let K be a normal subgroup of a group G. With respect to the multiplication Eq. (6.6) of cosets, G/K is a group. Moreover, the mapping $\theta : G \to G/K$, defined by $\theta(a) = aK$, is a homomorphism of G onto G/K with kernel K.*

Proof: The set G/K of cosets has a well-defined operation and the identity is $eK = K$ since $(eK)(aK) = eaK = aK$ for every $a \in G$. The associative law for multiplication of cosets follows immediately from the associative law for multiplication in G:

$$(aKbK)cK = (abK)cK = (ab)cK = a(bc)K = aK(bcK) = aK(bKcK).$$

The inverse of aK is $a^{-1}K$ since $(aK)(a^{-1}K) = aa^{-1}K = eK$. Thus, G/K is a group. The definition of coset multiplication implies that θ is a homomorphism:

$$\theta(ab) = abK = (aK)(bK) = \theta(a)\theta(b), \qquad a, b \in G.$$

By definition of G/K we see that θ maps G onto G/K. Let $a \in G$ be an element of the kernel of θ. Then $\theta(a) = aK$ is the identity of G/K; that is, $aK = eK = K$. It follows that $a \in K$. Conversely, any element $a \in K$ is in the kernel of θ since $\theta(a) = aK = K$. Thus, $\ker \theta = K$ and this completes the proof. ■

The group G/K constructed here is called the *factor group* of G by K. The term *quotient group* is also used for G/K. If G has finite order, we may obtain a formula for the order of G/K by Lagrange's theorem, namely,

$$|G/K| = [G : K] = \frac{|G|}{|K|}.$$

If the group G is abelian with its operation written as addition, the operation of multiplication of cosets Eq. (6.6) is replaced by addition of cosets. Any subgroup K of G is normal and the cosets of K are $a + K$ with the addition of cosets defined by

$$(a + K) + (b + K) = (a + b) + K, \qquad a, b \in G.$$

The "zero" is $0 + K = K$ and the inverse of $a + K$ is $-a + K$. Of course, G/K is also an abelian group in this case.

We have shown that if K is a normal subgroup of an arbitrary group G, then there is a homomorphism of G, with kernel K, onto the factor group G/K. We now show that "essentially" all homomorphisms are of this type. More precisely, we show that any homomorphic image of G is isomorphic to a factor group G/K. We state the result in the following form. Recall that the image of a homomorphism θ defined on G is the set $\{\theta(g) : g \in G\}$.

Theorem 6.3 *Let θ and τ be homomorphisms defined on a group G such that* $\ker\theta = \ker\tau$. *Then the images $\theta(G)$ and $\tau(G)$ are isomorphic by an isomorphism that maps $\theta(g)$ to $\tau(g)$ for all $g \in G$.*

Proof: We attempt to define a map $\gamma : \theta(G) \to \tau(G)$ by the rule $\gamma : \theta(g) \to \tau(g)$ for $g \in G$. Realizing that there could be many different elements $x \in G$ with $\theta(x) = \theta(g)$ it is necessary to show that γ is well defined. That is, if $\theta(g) = \theta(x)$, it is necessary to show that $\tau(g) = \tau(x)$ since the two elements are each equal to $\gamma(\theta(g))$. The equation $\theta(g) = \theta(x)$ implies $\theta(x^{-1}g) = \theta(e)$ and so $x^{-1}g$ is in the kernel of θ. By assumption, the kernel of θ equals the kernel of τ so we have $\tau(x^{-1}g) = \tau(e)$; equivalently, $\tau(g) = \tau(x)$ as we wanted to show. Thus, γ is a well-defined map and it is easily verified that γ is a homomorphism from $\theta(G)$ to $\tau(G)$. One could have just as easily defined a map $\lambda : \tau(g) \to \theta(g)$ and verified that it is a homomorphism from $\tau(G)$ to $\theta(G)$. Then the composition $\gamma\lambda$ is the identity map on $\tau(G)$, whereas $\lambda\gamma$ is the identity map on $\theta(G)$. It follows that both γ and λ are isomorphisms. [Of course, one could have completed the proof by simply verifying that γ is one-to-one and onto.] ∎

We may restate the conclusion of this theorem in the form traditionally called the First Isomorphism Theorem.

Theorem 6.4 FIRST ISOMORPHISM THEOREM FOR GROUPS. *If $\theta : G \to H$ is a homomorphism of a group G onto a group H and if K is the kernel of θ, then H is isomorphic to the factor group G/K by an isomorphism $\alpha : G/K \to H$ defined by*

$$\alpha : aK \longrightarrow \theta(a).$$

Proof: The homomorphism θ has kernel K and the map $\tau : G \to G/K$, defined by $\tau(a) = aK$, also has kernel K. Since H is the image of θ, and G/K is the image of τ, the statements all follow from the previous theorem. ∎

In a certain sense we have described all possible homomorphic images of a group G; they are all isomorphic to some factor group G/K. Of course, it may be difficult, in any given example, to determine the normal subgroups of a group.

To illustrate the First Isomorphism Theorem, we consider the group D_8 of rigid motions of the square as discussed in Section 2.1 of this chapter. The rotation of the square through 90° is denoted by ρ and the "flip" through the horizontal line through its center is λ. The eight elements of the group are expressed as $\lambda^j \rho^i$ with $0 \leq j \leq 1$ and $0 \leq i \leq 3$. The full multiplication table is given in Section 2.

Now suppose that the square is painted white on one side and black on the other side. Assume that the square in its initial position has the white side up. After a rigid motion is performed, either the square will still have the white side up or the square will have been turned over and the black side will be up. We may think of the rigid motions as producing a permutation of the two element set $\{white, black\}$ which we abbreviate as $\{w, b\}$. For example, the flip λ produces the motion

$$\boxed{\begin{array}{ccc} 1 & & 4 \\ & \textbf{WHITE} & \\ 2 & & 3 \end{array}} \xrightarrow{\lambda} \boxed{\begin{array}{ccc} 2 & & 3 \\ & \textbf{BLACK} & \\ 1 & & 4 \end{array}}$$

which changes the color of the square and so produces a permutation which we may denote by

$$\tau = \begin{pmatrix} w & b \\ b & w \end{pmatrix}.$$

Let H be the group of order 2 $\{\epsilon, \tau\}$ containing the two permutations of the set $\{w, b\}$. We define a function from D_8 to H by the rule

$$\gamma \longrightarrow \begin{cases} \epsilon & \text{if } \gamma \text{ leaves the color of the square unchanged;} \\ \tau & \text{if } \gamma \text{ changes colors of the square.} \end{cases}$$

We leave it to the reader to verify that this correspondence is a homomorphism of D_8 onto H. The kernel is the subgroup of rigid motions that do not flip the square over; that is, the kernel is

$$K = \{e, \rho, \rho^2, \rho^3\}.$$

This is a normal subgroup of order 4 and index 2 in D_8. The First Isomorphism Theorem asserts that the map from D_8/K to H defined by

$$K \longrightarrow \epsilon, \qquad \lambda K \longrightarrow \tau = \begin{pmatrix} w & b \\ b & w \end{pmatrix}$$

is an isomorphism.

EXERCISES

1. Let a be an element of the group G and let $\phi : G \to G$ be the function defined by $\phi(g) = aga^{-1}$. Prove that ϕ is an isomorphism of G with itself.

2. Prove that the intersection of two (or more) normal subgroups of G is a normal subgroup of G.

3. If G is a group, prove that the set $Z(G) = \{a : a \in G \text{ and } ax = xa \text{ for every } x \in G\}$ is a normal subgroup of G. [$Z(G)$ is called the *center of* G.]

4. Verify that the center of D_8 (defined in the previous exercise) equals $\{e, \rho^2\}$.

5. If e is the identity of G, and H and K are normal subgroups of G such that $H \cap K = \{e\}$, prove that $hk = kh$ for any $h \in H$ and $k \in K$. [Hint: Argue that $hkh^{-1}k^{-1} \in H \cap K$.]

6. Prove that if every right coset of a subgroup H is also a left coset, then H is a normal subgroup. [Hint: If $aH = Hb$ show that $aH = bH$.]

7. Let G be a group that has a normal subgroup H that is cyclic of order 4. Prove that the subgroup of order 2 in H is also a normal subgroup of G.

8. Show that if H is a cyclic group of finite order that is a normal subgroup of G, then every subgroup of H is also a normal subgroup of G.

9. Let $\mathbb{Q}$ be the additive group of the field of rational numbers and $\mathbb{Z}$ the subgroup consisting of all integers. Show that every element of the factor group $\mathbb{Q}/\mathbb{Z}$ has finite order but that the group itself has infinite order.

10. Let $K = \{\pm 1\}$ be the subgroup of order 2 in the multiplicative group G of all nonzero real numbers. Prove that G/K is isomorphic to the multiplicative group of all positive real numbers.

11. Let G be any group and H a subgroup that contains every element $aba^{-1}b^{-1}$ with $a, b \in G$. Prove that H is a normal subgroup of G and that G/H is abelian.

12. Let K be a normal subgroup of the group G, and let A be a subgroup of the factor group G/K. We may consider A to be a collection of cosets of K in G. Prove that the union of these cosets is a subgroup of G.

13. Let H and K be subgroups of the group G, with K a normal subgroup of G. Prove each of the following:
    (a) $H \cap K$ is a normal subgroup of H.
    (b) If $HK = \{hk : h \in H,\ k \in K\}$, then HK is a subgroup of G.
    (c) K is a normal subgroup of the group HK.

14. Let H and K be as in the preceding exercise. Prove the following:
    (a) Every element in the factor group HK/K is expressible in the form $hK, h \in H$.
    (b) The mapping $\alpha : H \to HK/K$ defined by $\alpha(h) = hK$ for $h \in H$ is a homomorphism of H onto HK/K with kernel $H \cap K$.
    (c) $H/H \cap K$ is isomorphic to HK/K.

15. Let $\theta : G \to H$ be a homomorphism of G onto H with kernel K. For a subset A of G, $\theta(A)$ is the set $\{\theta(a) : a \in A\}$. If U is a subset of H, $\theta^{-1}(U)$ denotes the set $\{g : g \in G, \theta(g) \in U\}$ of all elements of G that are mapped into an element of U by θ. [This is not the "inverse" of θ as previously defined for isomorphisms.] Prove each of the following:

(a) If A is a subgroup of G, then $\theta(A)$ is a subgroup of H.

(b) If U is a subgroup of H, then $\theta^{-1}(U)$ is a subgroup of G which contains K.

(c) The mapping $A \to \theta(A)$ is a one-to-one mapping of the set of all subgroups of G which contain K and the set of all subgroups of H.

(d) If A is a normal subgroup of G, then $\theta(A)$ is a normal subgroup of H.

7 THE SYMMETRIC GROUP

We now further study permutations of a finite set $A = \{1, 2, \ldots, n\}$. The symmetric group S_n has already been defined as the group of all permutations of A. Throughout this section the word *permutation* will mean an element of S_n for some positive integer n. The elements of A are sometimes just called "symbols." We begin with a study of permutations of a special type.

Definition 7.1 An element α of S_n is called a **cycle of length k** if there exist elements $a_1, a_2, \ldots, a_k$, $k \geq 1$ of A such that

$$\alpha(a_1) = a_2, \quad \alpha(a_2) = a_3, \ldots, \alpha(a_{k-1}) = a_k \quad \alpha(a_k) = a_1,$$

and $\alpha(i) = i$ for each element i of A other than $a_1, a_2, \ldots, a_k$. The cycle is denoted by $\alpha = (a_1 a_2 \cdots a_k)$.

Notice that a cycle of length 1 is necessarily the identity permutation. It sometimes simplifies statements to consider the identity as a cycle, but we usually are interested in cycles of length greater than 1.

As an example of a cycle, suppose that β is the element of S_6 given by

$$\beta = \begin{pmatrix} 1 & 2 & 3 & 4 & 5 & 6 \\ 3 & 5 & 2 & 4 & 6 & 1 \end{pmatrix}.$$

Note that the action of β on elements of A may be described by

$$1 \xrightarrow{\beta} 3 \xrightarrow{\beta} 2 \xrightarrow{\beta} 5 \xrightarrow{\beta} 6 \xrightarrow{\beta} 1, \qquad 4 \xrightarrow{\beta} 4.$$

Then β is a cycle of length 5, and we may write $\beta = (13256)$. In a symbol such as (13256) the symbols appearing are permuted cyclically; that is, each symbol that is written down is mapped to the symbol immediately to the right except that the last is mapped to the first. The symbol 4 does not appear in the cycle because 4 is mapped to 4 by β. There are different ways to write the same permutation as a cycle. For example,

$$\beta = (13256) = (61325) = (56132) = (25613) = (32561),$$

where each cycle is obtained from the previous one by moving the last symbol to the first position. Each of these cycles represents the same permutation. There is a certain ambiguity in this notation. If we view β as an element of S_8 that leaves the symbols $4, 7, 8$ fixed, then we still write $\beta = (13256)$. One cannot tell from the cycle notation to which symmetric group S_n β belongs. This usually will not cause a problem because the particular group is usually fixed during the discussion. In fact, we usually adopt the view that $S_n \subset S_{n+1}$ when the symmetric groups are viewed as permutations of the symbols $1, 2, \ldots, n, n+1, \ldots$.

One of the advantages of the cycle notation is that one can easily determine the order of a cycle.

Theorem 7.1 *A cycle of length k is a permutation of order k.*

Proof: Let $\alpha = (a_1 a_2 \ldots a_k)$ be a cycle of length k. We compute the effect of powers of α on a_1: Clearly $\alpha(a_1) = a_2$. Suppose we have shown $\alpha^i(a_1) = a_{i+1}$ provided $i < k$. If $i \leq k - 2$, then

$$\alpha^{i+1}(a_1) = \alpha(\alpha^i(a_1)) = \alpha(a_{i+1}) = a_{i+2}.$$

When we reach $i = k - 1$ we have $\alpha^{k-1}(a_1) = a_k$ and then

$$\alpha^k(a_1) = \alpha(\alpha^{k-1}(a_1)) = \alpha(a_k) = a_1.$$

We have proved that $\alpha^k(a_1) = a_1$. However, the cycle could have been written with any of the a_j as the first element so that in fact $\alpha^k(a_j) = a_j$ for $j = 1, 2 \ldots, k$. For the other symbols i not equal to any a_j, we have $\alpha(i) = i$ so that $\alpha^k(i) = i$. Hence, α^k sends every symbol to itself and α^k is the identity permutation. Moreover, no smaller power of α has this property, so the order of α is k. ∎

We will soon show how this information can be used to determine the order of every permutation in S_n.

Two cycles, $(a_1 a_2 \ldots a_k)$ and $(b_1 b_2 \ldots b_l)$, are *disjoint* if the two sets of symbols $\{a_1, a_2, \ldots, a_k\}$ and $\{b_1, b_2, \ldots, b_l\}$ that they move have no elements in common. A set of more than two cycles is said to be disjoint if every pair of them are disjoint. Thus, (123) and (45678) are disjoint cycles, whereas (123) and (24567) are not disjoint cycles because they both move the symbol 2. The next result shows why cycles play an important role in the study of permutations.

Theorem 7.2 *Every element of S_n is either a cycle or a product of disjoint cycles.*

Proof: Let α be an element of S_n. We use α to produce a partition of the set $A = \{1, 2, \ldots, n\}$ as follows. For $i, j \in A$ we write $i \sim j$ if there is some power α^k of α with the property $\alpha^k(i) = j$. Then $\sim$ is an equivalence relation on A as we now show. Since $\alpha^0(i) = i$ we have $i \sim i$ for all $i \in A$. If $\alpha^k(i) = j$ then $\alpha^{-k}(j) = i$ so that $i \sim j$ implies $j \sim i$. Finally, suppose $i \sim j$ and $j \sim h$. If $\alpha^k(i) = j$ and $\alpha^t(j) = h$, then $\alpha^{k+t}(i) = \alpha^t(j) = h$, proving that $i \sim h$. Thus, A is the disjoint union of the equivalence classes with respect to this relation. Suppose $[a_1] = \{a_1, a_2, \ldots, a_m\}$ is the equivalence class of a_1. We first make the important remark that, for any i, $\alpha(a_i) \in [a_1]$. To see this we observe that $\alpha^k(a_1) = a_i$ for some k and hence $\alpha^{k+1}(a_1) = \alpha(a_i)$, proving that $a_1 \sim \alpha(a_i)$ and so $\alpha(i)$ is in the equivalence class $[a_1]$. Since each a_i is related to a_1, there is some power of α such that $\alpha^k(a_1) = a_i$. Let us select the numbering of the elements a_i so that $\alpha(a_1) = a_2$, $\alpha^2(a_1) = a_3$, and so on until $\alpha^{m-1}(a_1) = a_m$. Note that this is equivalent to writing

$$\alpha(a_1) = a_2, \ \ \alpha(a_2) = a_3, \ldots, \alpha(a_{m-1}) = a_m.$$

We almost have a cycle. Now we argue that $\alpha(a_m) = a_1$. Suppose that $\alpha(a_m) = a_i$ with $2 \leq i \leq m$. Then, since $a_i = \alpha^{i-1}(a_1)$, we must have

$$\alpha(a_m) = a_i = \alpha^{i-1}(a_1), \quad \text{and} \quad a_m = \alpha^{i-2}(a_1) = a_{i-1}.$$

We reach an impossible situation that $a_m = a_{i-1}$. Thus, the assumption $2 \leq i \leq m$ cannot stand and we have proved that the permutation α is a cycle $\xi = (a_1 a_2 \ldots a_m)$ when permuting the elements in the equivalence class of a_1. Note that $\xi(a_j) = \alpha(a_j)$.

Since A is the disjoint union of the equivalence classes, we have associated with each one a cycle and the collection of cycles is disjoint. We claim the product of these cycles equals α. To prove this we need only verify that the product of the cycles has the same effect on each symbol in A as α. If $i \in A$, then $i \to \alpha(i)$ is the effect of α. Since i belongs to one, and only one, equivalence class, there is one, and only one, cycle that moves i. This cycle carries i to $\alpha(i)$, so the theorem is proved. ■

Next, we show how to express some permutation as a product of cycles. For example, take

$$\alpha = \begin{pmatrix} 1 & 2 & 3 & 4 & 5 & 6 & 7 & 8 \\ 4 & 2 & 1 & 3 & 8 & 5 & 6 & 7 \end{pmatrix}.$$

We start with any symbol, e.g., 3 (just to be different), and compute the equivalence class of 3 by applying α. We write the computation symbolically as

$$3 \xrightarrow{\alpha} 1 \xrightarrow{\alpha} 4 \xrightarrow{\alpha} 3.$$

Thus, $\{3, 1, 4\}$ is an equivalence class and the cycle corresponding to it is (314). Next, take any symbol that does not appear in the class of 3, e.g., 5. Compute

$$5 \xrightarrow{\alpha} 8 \xrightarrow{\alpha} 7 \xrightarrow{\alpha} 6 \xrightarrow{\alpha} 5$$

so $\{5, 8, 7, 6\}$ is an equivalence class and the associated cycle is (5876). There is only one symbol left, 2, and $\alpha(2) = 2$. Thus,

$$\alpha = (314)(5876)$$

is the expression of α as a product of disjoint cycles.

The order of a cycle equals it length. The next step is to determine the order of a product of disjoint cycles. The first step is the following:

Lemma 7.1 *Let α and β be two cycles in S_n. If α and β are disjoint, then $\alpha\beta = \beta\alpha$. For any integer k we have $(\alpha\beta)^k = \alpha^k\beta^k$.*

Proof: In order to prove $\alpha\beta = \beta\alpha$, it is necessary to show that $\alpha\beta(i) = \beta\alpha(i)$ for every $i \in \{1, 2, \ldots, n\}$. This condition is obvious if i is left fixed by both α and β since in this case $\alpha\beta(i) = i = \beta\alpha(i)$. Next, suppose that $\beta(i) = j$ and $i \neq j$. The i and j both appear in the cycle description of β. Since α and β are disjoint, neither i nor j is moved by α; that is, neither i nor j appears in the cycle description of α. Thus, $\alpha(i) = i$ and $\alpha(j) = j$. Now we have

$$\alpha\beta(i) = \alpha(\beta(i)) = \alpha(j) = j,$$
$$\beta\alpha(i) = \beta(\alpha(i)) = \beta(i) = j.$$

The same reasoning shows that $\alpha\beta(i) = \beta\alpha(i)$ if i is moved by α. Therefore, we have covered all possible cases and $\alpha\beta = \beta\alpha$.

Now consider the last statement. We have, for k a positive integer,

$$(\alpha\beta)^k = \alpha\beta \cdot \alpha\beta \cdots \alpha\beta,$$

where there are k occurrences of α and of β on the right-hand side. Since $\beta\alpha = \alpha\beta$, each occurrence of β may be moved to the right and each occurrence of α may be moved to the left to achieve $(\alpha\beta)^k = \alpha^k\beta^k$, which proves our statement for positive exponents. For negative exponents, we use the equation

$$(\alpha\beta)^{-k} = (\alpha^{-1}\beta^{-1})^k = (\alpha^{-1})^k(\beta^{-1})^k = \alpha^{-k}\beta^{-k},$$

which follows because of what we have already proved for positive exponents and the fact that $\alpha\beta = \beta\alpha$ implies $\alpha^{-1}\beta^{-1} = \beta^{-1}\alpha^{-1}$. Thus, the formula holds for every integer exponent. ∎

Once we know this result for the product of two disjoint cycles, it is an easy application of mathematical induction to prove it for any collect of disjoint cycles, as follows:

Lemma 7.2 *Let $\alpha_1, \alpha_2 \ldots, \alpha_t$ be a collection of mutually disjoint cycles in S_n. Then for any integer k we have*

$$(\alpha_1\alpha_2\cdots\alpha_t)^k = \alpha_1^k\alpha_2^k\cdots\alpha_t^k.$$

This shows how to compute powers of products of disjoint cycles. For example, if $\alpha = (123)(45)(6789)$, then

$$\alpha^k = (123)^k(45)^k(6789)^k.$$

For a specific example we use $k = 4$ and note that $(45)^4 = \epsilon$ and $(6789)^4 = \epsilon$ because the cycle of length 2 has order 2 and The cycle of length 4 has order 4. The cycle of length 3 has order 3 so $(123)^4 = (123)$ and we find $\alpha^4 = (123)$.

Now we have the tools to describe the order of any given permutation in S_n.

Theorem 7.3 *Let σ be a permutation in the symmetric group and let $\sigma = \alpha_1\alpha_2\cdots\alpha_t$ be the product of disjoint cycles $\alpha_1, \alpha_2, \ldots, \alpha_t$. Then the order of σ is the smallest positive integer that is divisible by the length of each cycle α_i.*

Proof: Let k_i be the length of α_i so that k_i is the order of α_i. Let m be the smallest positive integer divisible by each k_i. Then $\alpha_i^m = \epsilon$ for each i and by Lemma 7.2 we have

$$\sigma^m = (\alpha_1\alpha_2\cdots\alpha_t)^m = \alpha_1^m\alpha_2^m\cdots\alpha_t^m = \epsilon.$$

This shows that the order of σ is a divisor of m. Now let h be a positive integer such that $\sigma^h = \epsilon$. Then

$$\epsilon = \sigma^h = \alpha_1^h\alpha_2^h\cdots\alpha_t^h.$$

The terms α_i^h might not be cycles but they are disjoint in the sense that no two of them move a common symbol. Thus, if each α_i^h is expressed as a product of disjoint cycles we obtain a decomposition of ϵ as a product of disjoint cycles. When ϵ is expressed as a product of disjoint cycles, every cycle must have length 1; that is, $\alpha_i^h = \epsilon$ for each i. It follows that h is divisible by the order of α_i, namely, by its length k_i. ∎

Note that the smallest positive integer divisible by each of the lengths is called the *least common multiple* of the lengths.

The permutation $\alpha = (12)(345)(6789)$ has order 12 because 12 is the smallest positive integer divisible by 2, 3, and 4. In the symmetric group S_{15} the permutation $\tau = (12345)(678)(9, 10, 11, 12, 13)$ has order 15 because 15 is the least common multiple of 5, 3, and 5.

7.1 Transpositions

The cycles of length 2 are of special interest, and we give them a name.

Definition 7.2

A cycle of length 2 is called a **transposition**.

The transposition (ij) interchanges the symbols i and j and leaves the other symbols unchanged. Since $(ij)(ij) = \epsilon$, *a transposition is its own inverse.*

Every cycle of length more than 2 is the product of transpositions. This is seen by carrying out the product

$$(a_{k-1}a_k)(a_{k-2}a_k)\cdots(a_2a_k)(a_1a_k) = (a_1a_2\ldots a_k).$$

Notice that the transpositions in this example all have a common symbol—in particular, they are not disjoint. In view of the fact that every permutation is a product of disjoint cycles, it follows that every permutation is a product of transpositions. However, it is easy to see that there is more than one way to express a permutation as a product of transpositions. Here is just one illustration:

$$(123)(45) = (13)(12)(45) = (23)(13)(45) = (34)(35)(24)(14)(34).$$

Since $(ij)(ij) = \epsilon$, we can insert as many pairs of identical transpositions as we wish and then regroup to produce other expressions. Clearly, a permutation can be expressed as a product of transpositions in many different ways. Even though a certain randomness appears to be present, there is something that remains constant in these representations. We will later prove that the number of transpositions appearing in the product for a given permutation will always be even or always be odd. In the previous example, we have three representations of $(123)(45)$ as a product of transpositions; there are three transpositions in two products and five in the other. No matter what representation you may find for this particular permutation, it can be expressed only as an odd number of transpositions. As a first step in the proof of this fact, we establish a lemma.

Lemma 7.3

The identity ϵ of S_n cannot be expressed as a product of an odd number of transpositions.

Proof: We assume the lemma is false and eventually reach an impossible situation. Suppose the identity is a product of an odd number of transpositions. From all possible such representations, let

$$\epsilon = \tau_1\tau_2\cdots\tau_k \tag{6.7}$$

be a representation in which each τ_i is a transposition and k is odd and as small as possible. Since k is odd, $k \neq 0$ and certainly $k \neq 1$, so we must have $k \geq 3$. Assume the right-most transposition is $\tau_k = (pq)$. We select one of the two symbols moved by τ_k at random; for example, we select p. There may be more than; one representation of ϵ as a product Eq. (6.7) in which the right-most transposition moves the symbol p. From all possible such products, assume that we have selected one in which the symbol p appears in the fewest possible number of transpositions τ_i. Since the product equals the identity, p must appear in at least one transposition other than τ_k. If this were not the case, then the product would move q to p, but the product is the identity so q cannot be moved. Let j be the largest index less than k such that $\tau_j(p) \neq p$ [e.g., $\tau_j = (pr)$], with $p \neq r$. Thus, none of the transpositions $\tau_{j+1}, \ldots, \tau_{k-1}$ between τ_j and τ_k move p. The next step is to arrange that $j = k - 1$. This is accomplished by altering the product without altering the conditions we have imposed upon our choice. We make use of two product formulas:

$$\begin{array}{llcll} \text{(i)} & (pr)(st) & = & (st)(pr) & \text{if } p, r, s, t \text{ are all different;} \\ \text{(ii)} & (pr)(tr) & = & (rt)(pt) & \text{if } p, r, t \text{ are all different.} \end{array} \tag{6.8}$$

These equations show that the nearest transposition to the left of τ_k which involves p may be moved to the right of each transposition not involving p. This involves a change of the original τ_i but it does not change k and does not change the number of occurrences of the symbol p. After these steps are done, we may assume that the two right-most transpositions in the product are

$$\tau_{k-1}\tau_k = (pm)(pq).$$

Now if $m = q$, then $(pq)(pq) = \epsilon$ and we have reached a product of $k - 2$ transpositions equal to ϵ, which is in conflict with the choice of k as the smallest number which could be used. Thus, $m \neq q$. Now we use the equation $(pm)(pq) = (mq)(pm)$ to replace the last two terms in Eq. (6.7) and thus obtain a product in which the number of occurrences of the symbol p is reduced by 1. We had assumed that this number was the least possible and so this is a conflict with the assumed existence of a product of an odd number of transpositions equal to the identity. Hence, the lemma is true. ■

The reader might wonder what happens if we apply this argument in the casé in which k is even. The result would be that the number of terms in the product would be reduced by 2 whenever $(pq)(pq)$ appears at the right end of the product, and eventually we end with $\epsilon = (pq)(pq)$, which is certainly true.

Now we prove one of the principal theorems about permutations.

Theorem 7.4 *Every permutation α can be expressed as a product of transpositions. Moreover, if α can be expressed as a product of r transpositions and also as a product of s transpositions, then either r and s are both even or they are both odd.*

Proof: We have already shown that every permutation is a product of transpositions. Suppose further that

$$\alpha = \beta_1\beta_2 \cdots \beta_r = \gamma_1\gamma_2 \cdots \gamma_s,$$

with each β_i and γ_j a transposition. Since $\gamma_i\gamma_i = \epsilon$, each γ_i satisfies $\gamma_i = \gamma_i^{-1}$. Thus,

$$\alpha^{-1} = \gamma_s^{-1}\gamma_{s-1}^{-1} \cdots \gamma_1^{-1} = \gamma_s\gamma_{s-1} \cdots \gamma_1.$$

We now have the equation

$$\epsilon = \alpha\alpha^{-1} = \beta_1\beta_2 \cdots \beta_r\gamma_s\gamma_{s-1} \cdots \gamma_1,$$

which expresses the identity as a product of $r + s$ transpositions. By the previous lemma, $r + s$ is an even integer and so either r and s are both even or they are both odd since the sum of an even integer and an odd integer is odd. ∎

7.2 The Alternating Group

Definition 7.3

A permutation is called an **even permutation** or an **odd permutation** according to whether it can be expressed as a product of an even number or an odd number of transpositions.

The set of all even permutations in S_n is denoted by A_n. This subset is a subgroup. If α is a product of k transpositions and β is the product of l transpositions, then their product $\alpha\beta$ can be expressed as the product of $k + l$ transpositions. Hence, if α and β are even permutations, then so is their product $\alpha\beta$. Since S_n is a finite group, we may conclude from Theorem 1.2 that A_n is a subgroup of S_n. The group A_n is called the *alternating group* on n symbols.

Theorem 7.5

The alternating group A_n is a normal subgroup of S_n having $[S_n : A_n] = 2$ and order $|A_n| = n!/2$.

Proof: We have already shown that A_n is a subgroup. Let us find the cosets of A_n in S_n. Let τ be any transposition; we will show that A_n has only two cosets, A_n and τA_n. Let σ be any permutation in S_n. If σ is an even permutation, then $\sigma \in A_n$; if σ is an odd permutation, the $\tau\sigma$ is an even permutation so $\tau\sigma \in A_n$. Thus, $\sigma \in \tau^{-1}A_n$. However, $\tau\tau = \epsilon$ so that $\tau^{-1} = \tau$; it follows that $\sigma \in \tau A_n$. Hence, $[S_n : A_n] = 2$. We have already seen that a subgroup of index 2 is normal, so A_n is a normal subgroup of S_n. The order of S_n is $n!$, so the order of A_n is half this number by Lagrange's theorem. ∎

A cycle of length k can be expressed as a product of $k - 1$ transpositions

$$(a_{k-1}a_k)(a_{k-2}a_k) \cdots (a_2a_k)(a_1a_k) = (a_1a_2 \ldots a_k),$$

as one verifies by evaluating the left side. Thus, a cycle of *odd length* is an *even permutation*. In particular, every cycle of length 3 lies in A_n. We show a more general property of A_n in the following theorem:

Theorem 7.6 *Every cycle of length 3 in S_n lies in A_n. Moreover, every element in A_n is a product of cycles of length 3.*

Proof: The formula $(abc) = (bc)(ac)$ shows that every cycle of length 3 is an even permutation and lies in A_n. Now we show that every element of A_n is a product of cycles of length 3. Since every element of A_n is a product of an even number of transpositions, it is sufficient to prove that the product of two transpositions is a product of cycles of length 3. We use formulas in which different letters are assumed to stand for different symbols being permuted. We have

$$\begin{array}{rrcl} \text{(i)} & (ab)(cd) & = & (abc)(bcd) \\ \text{(ii)} & (ab)(ac) & = & (acb), \\ \text{(iii)} & (ab)(ab) & = & (abc)^3 = \epsilon. \end{array}$$

It follows that the product of an even number of transpositions can be expressed as a product of cycles of length 3. ∎

This theorem is sometimes used to show that a subgroup of S_n must contain A_n by proving that the subgroup contains every cycle of length 3.

EXERCISES

1. In each of the following, γ is an element of S_7. Express it as a product of disjoint cycles and then determine its order.

(a) $\gamma(1) = 3$, $\gamma(2) = 4$, $\gamma(3) = 1$, $\gamma(4) = 7$, $\gamma(5) = 5$, $\gamma(6) = 6$, $\gamma(7) = 2$

(b) $\gamma(1) = 5$, $\gamma(2) = 3$, $\gamma(3) = 4$, $\gamma(4) = 7$, $\gamma(5) = 6$, $\gamma(6) = 1$, $\gamma(7) = 2$

(c) $\gamma = \begin{pmatrix} 1 & 2 & 3 & 4 & 5 & 6 & 7 \\ 3 & 4 & 1 & 2 & 6 & 7 & 5 \end{pmatrix}$

(d) $\gamma = \begin{pmatrix} 1 & 2 & 3 & 4 & 5 & 6 & 7 \\ 2 & 3 & 1 & 5 & 4 & 7 & 6 \end{pmatrix}$

2. Express the following products as a product of *disjoint* cycles and then determine the order of each element:

(a) $(123)(16543)$

(b) $(213456)(172)$

(c) $(4215)(3426)(5671)$

(d) $(1234)(124)(3127)(56)$

3. Prove that the order of S_6 is divisible by 8 but that there is no element in S_6 with order 8.

4. List all the possible orders of elements in the symmetric group S_6, and find examples of elements with each order.

5. Prove that S_{12} does not have an element of order 13 and it does have an element of order 60.

6. Compute $\theta\tau\theta^{-1}$ in the following cases:
(a) $\tau = (12345)$, $\theta = (136)$
(b) $\tau = (123)$, $\theta = (13524)$
(c) $\tau = (123)(45)(67)$, $\theta = (23)(15)(47)$

7. Let $\tau = (12\ldots r) \in S_n$ be a cycle of length r and let θ be any element of S_n (where $n \geq r$). Prove that $\theta\tau\theta^{-1}$ is a cycle of length r. [Hint: Show that the product cyclically permutes the elements $\theta(1)$, $\theta(2), \ldots, \theta(n)$ and fixes all other symbols.]

8. Let K be a normal subgroup of S_n and suppose that K contains a cycle τ of length r. Use the previous exercise to show that K contains *every* cycle of length r.

9. Let p be a positive prime number. Find the number of subgroups of order p in S_p. [Hint: Count the number of cycles of length p and then use the fact that each subgroup of order p has exactly $p - 1$ elements of order p.]

10. Define a function $\theta : S_n \to \mathbb{Z}_2$, the additive group of integers modulo 2, by the rules $\theta(\alpha) = 0$ if α is an even permutation, $\theta(\alpha) = 1$ if α is an odd permutation. Show that θ is a homomorphism of S_n onto $\mathbb{Z}_2$.

11. List the elements of the alternating group A_3 and of the alternating group A_4.

12. Show that A_4 is a group of order 12 but has no subgroup of order 6. This shows that the converse of Lagrange's theorem is false. [Hint: If $N \subset A_4$ is a subgroup of order 6, then $[A_4 : N] = 2$ and so N is a normal subgroup of index 2. Argue that this would imply $\alpha^2 \in N$ for every $\alpha \in A_4$. This would force every cycle of length 3 to lie in N.]

13. Let G be a subgroup of S_n and suppose that G contains an odd permutation. Show that the set of even permutations in G forms a normal subgroup of G having index 2 in G.

8 SIMPLE GROUPS AND THE ALTERNATING GROUP

Every group G has the two trivial subgroups, $\{e\}$ and G, both of which are normal subgroups. If G has some other normal subgroup K, then one may form the factor group G/K so that, in the case in which G is a finite group, one may hope to obtain information about G from the two smaller groups K and G/K. One might say loosely that G is built from the smaller groups, although this building may not be easy to describe. If the only normal subgroups of G are the two trivial subgroups, then no such strategy is possible and the group must be studied in some other way. Such groups are given a special name.

Definition 8.1

A group G is called a **simple group** if the only normal subgroups are the trivial subgroups $\{e\}$ and G.

As examples, we may begin with the cyclic groups of prime order. The group $\mathbb{Z}_p$, for a prime p, has no subgroups other than the two trivial subgroups. Therefore, in particular, it has no nontrivial normal subgroups; $\mathbb{Z}_p$ is simple. These are *abelian* simple groups and are the only abelian simple groups.

The nonabelian, finite simple groups are essentially known; (some details still have to be written completely). The classification of all finite simple groups is one of the greatest achievements of mathematics in the twentieth century. It is not possible to describe all the finite simple groups here, but we do give one important infinite class of simple groups, namely, the alternating groups.

Theorem 8.1

If $n \neq 4$, the alternating group A_n is a simple group.

Proof: The group A_2 has order 1, and the group A_3 has order 3; these are simple groups. We exclude $n = 4$ so we may now assume that $n \geq 5$. Suppose K is a normal subgroup of A_n and that $|K| \neq 1$. We must prove that $K = A_n$. The method used to carry out the proof is to show that K contains a cycle of length 3. Once we know that K contains a cycle of length 3, then it follows that K contains all cycles of length 3 and Theorem 7.6 implies $K = A_n$.

Since $K \neq \{\epsilon\}$, we may select a nonidentity element τ in K. We proceed by considering various possibilities of the form of τ when it is written as a product of disjoint cycles.

Case 1. K contains an element τ that is expressed as a product of disjoint cycles, at least one of which has length ≥ 4.

Let $\tau = (1234\ldots r)\beta$, where $r \geq 4$ and β is a product of cycles, none of which involve $1, 2, \ldots, r$. Note: In this and later parts of the argument, we use the symbols $1, 2, \ldots, r$ in a generic sense. If the particular element involves other symbols $p, q, s, \ldots$, we carry out the proof using $p, q, s \ldots$ in place of $1, 2, 3, \ldots$.

Let $\sigma = (123)$; then

$$\sigma\tau\sigma^{-1} = \sigma(1234\ldots r)\beta\sigma^{-1} = (\sigma(1)\sigma(2)\sigma(3)\ldots\sigma(r))\beta = (2314\ldots r)\beta$$

and this is an element of K because K is a normal subgroup. Note that in this computation we used the fact that $\sigma\beta = \beta\sigma$, which holds because σ and β are disjoint. Since $\tau^{-1} \in K$, we may form the product

$$(\sigma\tau\sigma^{-1})\tau^{-1} = (2314\ldots r)\beta\beta^{-1}(r, r-1, \ldots 321) = (124),$$

which is a cycle of length 3 in K.

Now we may assume that no element of K has any cycle of length 4 or more in its disjoint-cycle product representation.

Case 2. K contains an element τ that is expressed as a product of disjoint cycles, at least two of which have length 3.

Let $\tau = (123)(456)\beta$, where β does not involve any of the symbols 1, 2,. . .,6, be an element of K. Let $\sigma = (124)$. Then

$$\sigma\tau\sigma^{-1} = (243)(156)\beta$$

is an element of K and so is

$$\sigma\tau\sigma^{-1}\tau^{-1} = (243)(156)(132)(465) = (12534).$$

Thus, in this case, K contains a cycle of length greater than 4, and by Case 1 K contains a cycle of length 3.

Case 3. K contains and element τ that is expressed as a product of disjoint cycles, one of which has length 3 and all others have length 2. In this case, we have $\tau = (123)\beta$, where β is a product of disjoint transpositions. In particular, $\beta^2 = \epsilon$. Thus, $\tau^2 = (132)\beta^2 = (132)$ and so K contains a cycle of length 3.

Case 4. K contains an element τ that is expressed as a product of disjoint transpositions.

Since $K \subseteq A_n$, the product of transpositions must involve an even number of disjoint transpositions. Let $\tau = (12)(34)\beta$, where β does not involve $1, 2, 3, 4$. Let $\sigma = (123)$; we make the computation, by now familiar,

$$\sigma\tau\sigma^{-1}\tau^{-1} = [(23)(14)\beta]\beta^{-1}(12)(34) = (13)(24)$$

to get the element $(13)(24)$ in K. Now, since $n \geq 5$ there is an element in $\{1, 2, \ldots, n\}$ different from $1, 2, 3, 4$; therefore, we may consider $\gamma = (135) \in A_n$. Then we have

$$(13)(24)\gamma(13)(24)\gamma^{-1} = (13)(24)(35)(24) = (135)$$

is an element of K. Thus, in all cases, K contains a cycle of length 3.

Now let us show that K contains every cycle of length 3. Assume $(123) \in K$ and let a, b, c be any three symbols in $\{1, 2, \ldots, n\}$ and let σ be some permutation such that

$$\sigma(1) = a, \quad \sigma(2) = b, \quad \sigma(3) = c.$$

If σ is not in A_n, then σ is an odd permutation. Let (pq) be a transposition that does not involve $1, 2, 3$. Such p and q exist because $n \geq 5$. Then the product $\sigma(pq)$ is an even permutation that has the same effect on $1, 2, 3$ as does σ. Hence, we may assume that $\sigma \in A_n$ since we could replace it by $\sigma(pq)$ if that were not the case. Then

$$\sigma(123)\sigma^{-1} = (\sigma(1)\sigma(2)\sigma(3)) = (abc)$$

is an element of K because K is normal in A_n. Since K contains every cycle of length 3, $A_n \subseteq K$ by Theorem 7.6. Thus, $K = A_n$ and A_n is a simple group. ∎

It is worth noting that the previous proof used the restriction $n \geq 5$ only at the very end to rule out Case 4. Thus, in the case of A_4, a nontrivial normal subgroup must consist only of elements that are products of disjoint transpositions. This case actually arises. The elements

$$V = \{\epsilon, (12)(34), (13)(24), (14)(23)\}$$

form a normal subgroup of A_4. This case is an exception that must be stated in the theorem.

As a corollary of the simplicity of A_n, for $n \geq 5$ we may obtain the following about normal subgroups of S_n:

Theorem 8.2 *If $n \geq 5$, then a normal subgroup of S_n must be $\{\epsilon\}$, A_n, or S_n.*

Proof: Let N be a normal subgroup of S_n with $N \neq \{\epsilon\}$. Then $N \cap A_n$ is a normal subgroup of A_n so it must equal $\{\epsilon\}$ or A_n by the simplicity of A_n. If $N \cap A_n = A_n$, then $A_n \subseteq N$ so either $N = A_n$ or $N = S_n$. This last statement follows because $[S_n : A_n] = 2$, so N contains either one or both cosets of A_n.

Now consider the possibility that $N \cap A_n = \{\epsilon\}$. Then the only even permutation in N is the identity. Suppose $\sigma, \tau \in N$ with neither one equal to the identity. Then σ and τ are odd permutations and their product $\sigma\tau$ is an even permutation in N; that is, $\sigma\tau = \epsilon$. Thus, $\tau = \sigma^{-1}$ and it follows that N contains only one element other than the identity; every nonidentity element is equal to σ^{-1}. If $N = \{\epsilon, \sigma\}$ with $\sigma^2 = \epsilon$, we may write $\sigma = (12)\beta$, with β a product of disjoint transpositions not involving 1 or 2. The assumption that N is normal in S_n implies $\gamma\sigma\gamma^{-1} = \sigma$ for every $\gamma \in S_n$ because σ is the only nonidentity element in N. Take $\gamma = (13)$ to see that $\gamma\sigma\gamma^{-1}$ does not equal σ. Hence, this case cannot occur and the proof is complete. ∎

EXERCISES

1. Verify that the subgroup of order 4 in A_4 is normal in A_4 but is not normal in S_4. [Note: This shows that the relation "is a normal subgroup of," defined for subgroups of a group, is not transitive.]

2. If G is a group and H a subgroup of G, then the *normalizer of* H in G is $N_G(H) = \{g \in G : gHg^{-1} = H\}$. Show that $N_G(H)$ is a subgroup of G containing H as a normal subgroup. In fact, $N_G(H)$ is the largest subgroup of G in which H is a normal subgroup.

3. If G is a simple group and H a proper subgroup of G, then $N_G(H)$ is a proper subgroup of G. [See previous exercise.]

4. Let $G = A_5$ so that G is a nonabelian simple group. Let H be the cyclic subgroup of G generated by (123). Then $N_G(H)$ is not all of G. Find the elements of $N_G(H)$. [Hint: First show that the normalizer of H actually lies in S_3.]

5. Find the normalizer in A_5 of the cyclic subgroup generated by (12345). [Hint: It has order 10.]

9 THE 15-PUZZLE

The 15-puzzle is a children's game (for children of all ages) to which we apply the theory of the symmetric group in order to make some adult observations.

The 15-puzzle is a 4×4 frame holding movable squares, numbered from 1 to 15 with one space left vacant (Fig. 6.1). The squares may be moved horizontally or vertically into the vacant space, but no square may be lifted out of its plane. A *simple move* consists of sliding one square into the blank space, thus in effect moving the blank space to an adjacent position either horizontally or vertically.

1	2	3	4
5	6	7	8
9	10	11	12
13	14	15	

FIGURE 6.1 Starting position.

This puzzle was first introduced in the 1870s by its creator Sam Loyd, who was famous for many other puzzle creations and for his chess problems. The task he proposed was to reverse the positions of blocks 14 and 15 by a sequence of simple moves and return the other squares to their starting position (Fig. 6.2). Loyd wrote that his creation kept many people occupied for hours

1	2	3	4
5	6	7	8
9	10	11	12
13	14	15	

$\longrightarrow$

1	2	3	4
5	6	7	8
9	10	11	
13	14	15	12

FIGURE 6.2 A simple move.

12	14	5	7
10		3	9
8	2	1	11
6	4	15	13

Position I

8	10	7	9
6	12	5	11
4	14	3	13
2		1	15

Position II

1	2	3	4
8	7	6	5
9	10	11	12
	15	14	13

Position III

FIGURE 6.3

but no person succeeded in claiming the prize he offered for the solution. Well, no surprise! We will prove no solution is possible. More generally, one may ask, What positions can be achieved by applying a sequence of simple moves to the starting position? For example, given the three positions in Fig. 6.3, we will be able to tell rather quickly that only one of the positions may be reached from the starting position by a series of simple moves.

Now we set up the notation which permits us to solve this problem. Use the numbers 1–16 to indicate the *locations* in the frame which hold the squares. In the starting position, square 1 is in location 1, square 2 is in location 2, and so on; the blank space is in location 16 and we assign the number 16 to the "blank" square. The location numbers remain fixed; the numbered squares in the locations may change. After a series of simple moves, the numbered squares will be in different locations and the new position may be viewed as a permutation of the set $\{1, 2, \ldots, 16\}$.

Here, we describe the way in which the symmetric group enters the problem. Any permutation $\sigma \in S_{16}$ may be applied to an arrangement A of the puzzle to produce a new arrangement, $\sigma(A)$, by the following rule: The square in location i of A is moved to the location $\sigma(i)$ in $\sigma(A)$ for each $i = 1, 2, \ldots, 16$. For example, when the transposition $(12, 16)$ is applied to the starting position, the square in location 12, which is the square numbered 12, is moved into the location of the blank square, location 16, and the blank square is moved to location 12, with the result shown in Fig. 6.2. In this case, the permutation $(12, 16)$ is accomplished by a simple move. If we apply the transposition $(12, 16)$ to the arrangement indicated as position I in Fig. 6.3, then the squares numbered 11 and 13 would be interchanged because they occupy locations 12 and 16. It is not evident that this could be accomplished by a series of simple moves. We are not assuming that every permutation of the starting position can be reached by a series of simple moves. In fact, the problem is precisely to determine which permutations can be achieved by a series of simple moves. At this point we are simply describing the connection between permutations in S_{16} and arrangements of the squares in the puzzle frame.

For any arrangement A of the squares in the frame there is a unique permutation $\sigma \in S_{16}$ such that $\sigma(\text{SP}) = A$, where SP is the starting position. This gives the correspondence of arrangements with group elements. We now describe the permutations in S_{16} that correspond to arrangements that arise after a series of simple moves. We make a restriction by insisting that the blank square be returned to position 16.

Theorem 9.1

Let H be the subset of S_{16} consisting of all permutations that correspond to arrangements of the 15-puzzle that are obtained from the starting position, SP, by a series of simple moves which ends with the blank square in location 16. Then H is a subgroup of S_{16} and H consists of all the even permutations which leave 16 fixed; that is, $H = A_{15}$.

Proof: The proof requires many steps.

(A) H is a subgroup of S_{16}.

Every simple move corresponds to a transposition (ab), where either a or b is the location of the blank square just before the move is applied to a position. Any series of simple moves which is applied to a position in which the blank is at location 16 before and after the series is represented by a permutation of the form

$$\sigma = (16, x_{t-1})(x_{t-1}, x_{t-2}) \cdots (x_2, x_1)(x_1, 16). \quad (6.9)$$

In this product, two adjacent transpositions must have a common symbol; $(ab)(bc)$ occurs in the product if c is the location of the blank before this part of the product is applied. The blank is then moved to location b and then to a. The 16 must appear in the first and last transpositions because the only arrangements we consider are those that start and end with the blank in location 16. There are restrictions on the entries a, b, c as well since b must be one of the locations horizontally or vertically adjacent to the location of the blank at c; similarly, a is adjacent to b. When all these conditions are met, then σ describes a series of simple moves. Clearly, the product of two such permutations is again a permutation that meets this description. Hence, H is closed under multiplication of permutations and, since S_{16} is a finite group, H is a subgroup.

(B) $H \subseteq A_{15}$.

Every element $\sigma \in H$ is a product of the form given in Eq. (6.9); let us prove it is an even permutation. The description given shows that σ is the product of t transpositions, so it is necessary to prove that t is even. We think of σ as moving the blank space from location 16 to x_1, x_2, and so on until the blank returns to 16. Every move of the blank is left, right, up, or down. Let l, r, u, and d stand for the number of left moves, right moves, up moves, or down moves, respectively. Then $t = l + r + u + d$ is the total number of moves of the blank square. Since the blank returns to its original position, the number of left moves must equal the number of right moves, that is, $l = r$; similarly, the number of up moves equals the number of down moves, that is, $u = d$. Thus, we have $t = 2r + 2d$, which is an even number. Thus, $\sigma \in A_{16}$. However, the permutation of the locations given by σ leaves 16 fixed and so $\sigma \in A_{16} \cap S_{15}$. It remains to be shown that this intersection is actually A_{15}. Clearly, $A_{15} \subset A_{16}$ by definition of alternating groups as even permutations. Thus, $A_{15} \subseteq A_{16} \cap S_{15} = K \subseteq S_{15}$. We know that A_{15} has only two cosets in S_{15} so either both cosets are in K and $K = S_{15}$ or only one coset is in K and $K = A_{15}$. If $K = S_{15}$, then K contains a transposition, which is an odd permutation, whereas we have seen that $K \subseteq A_{16}$ contains only even permutations. Thus, $A_{15} = A_{16} \cap S_{15}$ and H is a subgroup of A_{15}.

We interrupt the flow of the proof to make an observation. We already can show that certain arrangements of the squares cannot be obtained by a series of simple moves. For instance, it is impossible to exchange the 14 and 15 squares leaving the others in their original position because this corresponds to a transposition $(14, 15)$ which is not an element of A_{15}. The remainder of the proof requires that we prove H actually equals A_{15} so that we can assert that arrangements corresponding to even permutations can be reached by simple moves.

The next steps are designed to show that $H = A_{15}$. This will be accomplished by proving that every cycle (x, y, z) of length 3 is in H and then invoking Theorem 7.6, which asserts that every element of A_n is a product of cycles of length 3 in S_n.

(C) H contains the elements α, β, γ, and σ defined by

$$\begin{aligned}\alpha &= (16,12)(12,8)(8,4)(4,3)(3,2)(2,1)(1,5)(5,9)\\ &\quad\cdot(9,13)(13,14)(14,15)(15,16)\\ &= (1,5,9,13,14,15,12,8,4,3,2),\end{aligned}$$

$$\begin{aligned}
\beta &= (16,15)(15,14)(14,10)(10,6)(6,7)(7,8)(8,12)(12,16)\\
&= (6,7,8,12,15,14,10),\\
\gamma &= (16,12)(12,11)(11,15)(15,16)\\
&= (11,15,12),\\
\sigma &= (16,15)(15,14)(14,10)(10,6)(6,2)(2,3)(3,4)(4,8)(8,12)(12,16)\\
&= (4,8,12,15,14,10,6,2,3).
\end{aligned}$$

Each element is presented as a product of simple moves which are applied to the starting position to achieve the indicated permutation. The effect of each move is illustrated in Fig. 6.4.

5	1	2	3
9	6	7	4
13	10	11	8
14	15	12	

α

1	2	3	4
5	10	6	7
9	14	11	8
13	15	12	

β

1	2	3	4
5	6	7	8
9	10	15	11
13	14	12	

γ

1	6	2	3
5	10	7	4
9	14	11	8
13	15	12	

σ

FIGURE 6.4

(D) H contains every cycle (x, y, z) in S_{15}.

We have one cycle of length 3 in H, namely, $\gamma = (11, 15, 12)$. We get others as follows: For any $\theta \in S_{16}$ and any cycle (x, y, z) of length 3, we have the relation

$$\theta(x,y,z)\theta^{-1} = (\theta(x), \theta(y), \theta(z)).$$

This is easily verified by computing the effect of the two sides on a general symbol. In particular, if $\theta \in H$, then $\theta\gamma\theta^{-1}$ is a cycle of length 3 in H. For example, $\beta \in H$, so

$$\beta^5\gamma\beta^{-5} = (\beta^5(11), \beta^5(15), \beta^5(12)) = (11, 8, 7) \in H.$$

Now notice that both α and σ fix 11 and 7 so that if τ is a power of either α or σ, then

$$\tau\gamma\tau^{-1} = (\tau(11), \tau(8), \tau(7)) = (11, \tau(8), 7).$$

Every x, with $1 \leq x \leq 15$ and $x \neq 7, 11$, is equal to either $\alpha^i(8)$ or $\sigma^j(8)$ for a suitable i or j. It follows that every cycle of length 3 having the form $(11, x, 7)$ is in H. Let x and y denote any integers between 1 and 15 other than 7 or 11. Then

$$(11, x, 7)(11, y, 7)(11, x, 7)^{-1} = (x, y, 11);$$

therefore, every element of this form lies in H. Finally, we use the product

$$(11, z, 7)(x, y, 11)(11, z, 7)^{-1} = (x, y, z)$$

to conclude that every cycle (x, y, z) is in H if none of x, y, z are equal to 7 or 11. However, we already know that $(x, y, 11)$ is in H and so $\tau(x, y, 11)\tau^{-1} = (x, y, 7)$ is in H if $\tau = (11, z, 7)^2$. Hence, every cycle of length 3 from S_{15} is in H and this proves $H = A_{15}$. ∎

Now we return to the challenge posed at the beginning of this section and show that position I cannot be obtained by a series of simple moves from the SP. We begin by applying a small number (four in this case) of simple moves to move the blank square to location 16. Call this position I′ (Fig. 6.5)

12	14	5	7
10	2	3	9
8	4	1	11
6	15	13	

FIGURE 6.5 Position I′.

Then position I′ corresponds to the permutation

$$\sigma = \begin{pmatrix} 1 & 2 & 3 & 4 & 5 & 6 & 7 & 8 & 9 & 10 & 11 & 12 & 13 & 14 & 15 \\ 11 & 6 & 7 & 10 & 3 & 13 & 4 & 9 & 8 & 5 & 12 & 1 & 15 & 2 & 14 \end{pmatrix},$$

which can be written in cycle form as

$$\sigma = (1, 11, 12)(2, 6, 13, 15, 14)(3, 7, 4, 10, 5)(8, 9).$$

If σ is applied to the starting position, then position I′ is obtained. The first three cycles in the product have odd length so they are even permutations and lie in A_{15}. However, the transposition $(8, 9)$ is odd and so the product σ is not in A_{15} and position I′ cannot be reached by a series of simple moves from the SP. If it were possible to reach position I by simple moves from the SP, then position I′ could also be so obtained. Hence, position I cannot be obtained from the SP.

The theory developed here allows one to determine if a given position can be obtained, but it does not give very useful information for actually carrying out the moves when an arrangement can be obtained. Interested readers with a puzzle in their hands probably should use the theory to determine if an arrangement is possible and then apply trial and error to attain those that are possible.

EXERCISES

1. Determine which of positions II and III can be obtained from the starting position by a series of simple moves.
2. Show that A_{15} is the smallest subgroup of S_{15} that contains α, β, and γ as defined in C; in other words, the permutation σ was not needed to complete the proof of the theorem.
3. How many arrangements of the squares in the 15-puzzle can be reached by applying a series of simple moves to the SP? [Do not assume the blank square remains in location 16.]
4. Show that the subset of permutations in S_{16} consisting of all elements that correspond to arrangements of the 15-puzzle which are obtained from the SP by a series of simple moves ending with the blank square in location i forms a coset of the subgroup H defined in Theorem 9.1, for any i, $1 \leq i \leq 16$.
5. Develop a theory of the 8-puzzle, which consists of a 3×3 frame of squares numbered 1–8 with one space left blank, and give a method to decide which arrangements can be obtained from a given SP by a series of simple moves.

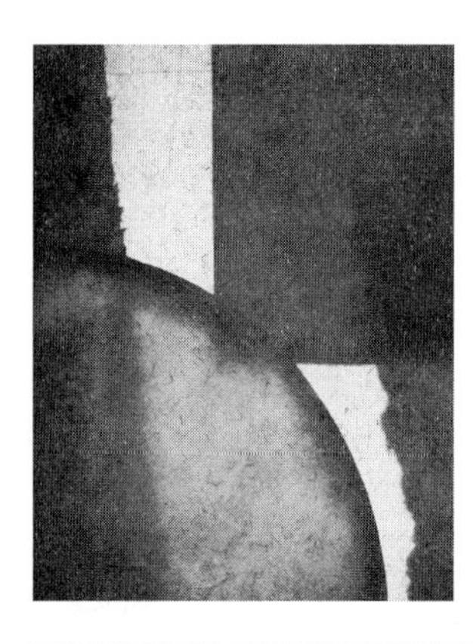

VII FINITE ABELIAN GROUPS

The problem of describing all finite groups is very difficult and, in fact, is an unsolved problem. However, it is possible to classify all finite abelian groups. What does it mean to "classify" finite abelian groups? It means that we give a description of each group in terms of familiar data in such a way that it is possible to tell from the data when two groups are isomorphic. The familiar data used to describe finite abelian groups will be cyclic groups and direct sums. The purpose of this chapter is to give a precise description of all finite abelian groups along with a statement indicating when two such groups are isomorphic.

Let G be a finite abelian group. Throughout this chapter we use addition as the operation in G. Of course, everything could just as well be stated in terms of multiplication as the operation. Let us recall some essential facts that will be used frequently. Since G has finite order, each element a of G has finite order. If a has order n, then $na = 0$ and n is the least positive integer with this property. Moreover, $ka = 0$ if an only if $n|k$. The cyclic group generated by a is isomorphic to the additive group $\mathbb{Z}_n$.

1 DIRECT SUMS OF SUBGROUPS

Let $G_1, G_2, \ldots, G_r$ be subgroups of the abelian group G. We define the *sum*

$$G_1 + G_2 + \cdots + G_r$$

of these subgroups to be the collection of all elements of G which can be expressed in the form

$$g_1 + g_2 + \cdots + g_r, \qquad g_i \in G_i, \quad i = 1, 2, \ldots, r.$$

This set is seen to be a subgroup of G that contains each subgroup G_i. In fact, the sum is the smallest subgroup of G that contains each G_i.

We will be interested in *direct sums* of subgroups. The direct sum of a collection of groups was considered in Chapter 6, Section 6. We review the definition. The *direct sum* of the groups $G_1, \ldots, G_r$ is the set

$$G_1 \oplus G_2 \oplus \cdots \oplus G_r = \{(g_1, g_2, \ldots, g_r) : g_i \in G_i,\ i = 1, 2, \ldots, r\}.$$

The operation is coordinatewise addition:

$$(g_1, g_2, \ldots, g_r) + (g_1', g_2', \ldots, g_r') = (g_1 + g_1', g_2 + g_2', \ldots, g_r + g_r').$$

Note that this definition does not assume that the groups G_i are contained in some common group. However, when all the G_i are contained in a larger group G then the *sum* of the G_i is defined as previously shown. There is a connection between these two types of "sums."

Lemma 1.1

Let $G_1, G_2, \ldots, G_r$ be subgroups of the abelian group G. Then there is a homomorphism θ from the direct sum $G_1 \oplus G_2 \oplus \cdots \oplus G_r$ onto the sum $G_1 + G_2 + \cdots + G_r$ given by

$$\theta : (g_1, g_2, \ldots, g_r) \longrightarrow g_1 + g_2 + \cdots + g_r.$$

Proof: We need only verify that θ preserves the operations since visibly θ is an onto map. Thus,

$$\begin{aligned}
\theta[(g_1, g_2, \ldots, g_r) + (g_1', g_2', \ldots, g_r')] &= \theta[(g_1 + g_1', g_2 + g_2', \ldots, g_r + g_r')] \\
&= (g_1 + g_1') + (g_2 + g_2') + \ldots + (g_r + g_r') \\
&= (g_1 + g_2 + \cdots + g_r) + (g_1' + g_2' + \cdots + g_r') \\
&= \theta(g_1, g_2, \cdots, g_r) + \theta(g_1', g_2', \cdots, g_r'),
\end{aligned}$$

which proves θ is a homomorphism. ■

This homomorphism need not be an isomorphism in all cases. For example, consider the simple case in which $G_1 = G_2 = G$ so that $G_1 + G_2 = G + G = G$. Then the map $\theta : G \oplus G \rightarrow G$ is given by

$$\theta(a, b) = a + b, \qquad a, b \in G.$$

The kernel is the set of elements $\{(a, -a) : a \in G\}$.

The case in which θ is an isomorphism is an important one which we distinguish using the following terminology:

Definition 1.1

Let $G_1, G_2, \ldots, G_r$ be subgroups of the abelian group G. The sum $G_1 + G_2 + \cdots + G_r$ is called an internal **direct sum** if the homomorphism θ in

Lemma 1.1 is an isomorphism. We indicate that this is the case by writing $G_1 \oplus G_2 \oplus \cdots \oplus G_r$ for the internal direct sum.

We refer to this as an internal direct sum to distinguish it from the "direct sum"; when the meaning is clear from the context, we may simply refer to the sum of the G_i as a direct sum if θ is an isomorphism.

The following are some important properties of internal direct sums:

Theorem 1.1 *Let $G_1, G_2, \ldots, G_r$ be subgroups of the abelian group G and let $H = G_1 + G_2 + \cdots + G_r$, the sum of these subgroups. The following are equivalent statements:*

(i) *H is the internal direct sum of the subgroups G_i, $i = 1, 2, \ldots, r$.*

(ii) *If $g_i \in G_i$, $i = 1, 2, \ldots, r$, are such that*

$$g_1 + g_2 + \cdots + g_r = 0,$$

then $g_i = 0$ for each $i = 1, 2, \ldots, r$.

(iii) *If $g_i, h_i \in G_i$, $i = 1, 2, \ldots, r$ are elements such that*

$$g_1 + g_2 + \cdots + g_r = h_1 + h_2 + \cdots + h_r,$$

then $g_i = h_i$ for $i = 1, 2, \ldots, r$.

Proof: Assume (i) holds and we prove (ii). The homomorphism θ defined in Lemma 1.1 is an isomorphism. The equation in (ii) implies $(g_1, g_2, \ldots, g_r)$ is in the kernel of θ; the kernel consists only of $(0, 0, \ldots, 0)$, so $g_i = 0$ for each i.

Assume (ii) holds and we now prove (iii). The condition given in (iii) is equivalent to the equation

$$(g_1 - h_1) + (g_2 - h_2) + \cdots + (g_r - h_r) = 0,$$

which implies $g_i - h_i = 0$ by (ii). Thus, $g_i = h_i$ for all i and (iii) holds.

Assume (iii) holds and we try to prove (i). It is necessary to show that θ is one-to-one since we know it is an onto homomorphism. If

$$\theta(g_1, g_2, \ldots, g_r) = 0 = \theta(h_1, h_2, \ldots, h_r),$$

then, by definition of θ,

$$g_1 + g_2 + \cdots + g_r = h_1 + h_2 + \cdots + h_r$$

and so (iii) implies $(g_1, g_2, \ldots, g_r) = (h_1, h_2, \ldots, h_r)$. ■

The equivalence of statements (i) and (iii) is often expressed by saying that $G_1 + G_2 + \cdots + G_r$ is a direct sum if and only if each element of the sum is *uniquely* expressible in the form

$$g_1 + g_2 + \cdots + g_r, \qquad g_i \in G, i = 1, 2, \ldots, r.$$

We make one further remark about the direct sum $H = G_1 \oplus G_2 \oplus \cdots \oplus G_r$; if each G_i is a finite group of order $n_i = |G_i|$, then the direct sum has order $|H| = n_1 n_2 \cdots n_r$.

1.1 Primary Subgroups

One goal of this chapter is to show that any finite abelian group is a direct sum of cyclic groups. We begin by making a first reduction to abelian groups with prime power orders.

Let G be a finite abelian group and p a positive prime number. Denote by $G(p)$ the set of all elements in G with order a power of p; that is,

$$G(p) = \{x : x \in G, p^k x = 0, \text{ for some nonnegative integer } k\}.$$

We verify that $G(p)$ is a subgroup of G; if $a, b \in G(p)$ with $p^k a = p^m b = 0$, then $p^s(a+b) = 0$ with s equal to the larger of k and m. We call $G(p)$ the ***p*-primary** subgroup of G. The only primes p such that $G(p) \neq 0$ are the prime divisors of the order of G. This follows because if there is an element of order p then p divides the order of G. At this point, we do not know the converse, which would assert that for every prime divisor of the order of G, the subgroup $G(p)$ is not the identity. This will follow from the next theorem, which provides the first step in our classification.

Theorem 1.2 PRIMARY DECOMPOSITION. *Let G be an abelian group of finite order n and let $p_1, p_2, \ldots, p_k$ be the distinct prime divisors on n. If $G(p_i)$ denote the subgroup of G consisting of the elements of order a power of p_i, then*

$$G = G(p_1) \oplus G(p_2) \oplus \cdots \oplus G(p_k).$$

Proof: Let n have the factorization

$$n = p_1^{e_1} p_2^{e_2} \cdots p_k^{e_k}$$

and let $n_j = n/p_j^{e_j}$, $1 \leq j \leq k$, so that n_j is the largest factor of n that is not divisible by p_j. The main step of the proof is to show that there exist integers m_j such that

$$m_1 n_1 + m_2 n_2 + \cdots + m_k n_k = 1. \tag{7.1}$$

To prove this consider the additive subgroup M of $\mathbb{Z}$ consisting of all integers of the form

$$a_1 n_1 + a_2 n_2 + \cdots + a_k n_k, \qquad a_i \in \mathbb{Z},\ i = 1, 2, \ldots, k.$$

We know all the subgroups of $\mathbb{Z}$ from Theorem 4.3; there is an integer d such that M is the set of all integer multiples of d. In particular, since $n_j \in M$

for every j, d must be a factor of every n_j. However, there is no prime that divides every n_j; the only candidates are the divisors p_i of n and p_i does not divide n_i. Hence, $d = 1$ and so, in particular, $1 \in M$ and there must exist integers m_i such that Eq. (7.1) holds. Next, we make use of this property.

Take any element $g \in G$ and write

$$g = 1 \cdot g = (m_1 n_1 + m_2 n_2 + \cdots m_k n_k)g = g_1 + g_2 + \cdots + g_k,$$

where $g_j = m_j n_j g$. We assert that $g_j \in G(p_j)$. To see this we multiply g_j by $p_j^{e_j}$ to get

$$p_j^{e_j} g_j = m_j (p_j^{e_j} n_j) g = m_j \cdot ng = 0$$

because $ng = 0$ for every $g \in G$. Thus, every element g is expressed as a sum of elements in the primary subgroups and we have proved that

$$G = G(p_1) + G(p_2) + \cdots + G(p_k).$$

Next we must show that this is a direct sum. Let $g_i \in G(p_i)$ for each i and suppose

$$0 = g_1 + g_2 + \cdots + g_k.$$

We will prove that each $g_i = 0$ so that Theorem 1.1(ii) can be invoked to assert that the sum is direct. First note that every element in $G(p_j)$ has order a power of p_j and the order must divide the order of G. Hence, every element in $G(p_i)$ has order dividing $p_i^{e_i}$; in other words, $p_i^{e_i} a_i = 0$ for every $a_i \in G(p_i)$.

The integer n_1 is divisible by $p_i^{e_i}$ for every $i \geq 2$, so $n_1 g_i = 0$ for $i \geq 2$ and

$$0 = n_1(g_1 + g_2 + \cdots + g_k) = n_1 g_1 + n_1 g_2 + \cdots + n_1 g_k = n_1 g_1.$$

This implies that the order of g_1 divides n_1; however, the order of g_1 is also a power of p_1. Since the only power of p_1 dividing n_1 is $p_1^0 = 1$, it follows that g_1 has order 1 and $g_1 = 0$ as 0 is the only element of order 1. Similarly, we show $g_j = 0$ for all j and we have shown that the sum is direct. ∎

Definition 1.2 For a prime p, a group G is called a **p-group** if the order of each of its elements is a power of p.

Let us immediately prove that an abelian p-group has order a power of p.

Theorem 1.3 *If G is a finite abelian p-group for some prime p, then the order of G is a power of p.*

Proof: If the theorem is false, then the collection of finite abelian p-groups for which the theorem does not hold must have a group with smallest order. Let G be such a group. Then G is not just the identity because that would have order a power of p, namely, $p^0 = 1$. Thus, G has some nonidentity element a whose order must be a power of p by definition of a p-group. The cyclic subgroup $\langle a \rangle$ generated by a has order p^k for some integer k. Since G is abelian, every subgroup of G is normal so we may consider the factor group $A = G/\langle a \rangle$ and the homomorphism $\theta : G \to A$. Every element of G has order a power of p, so therefore every element in $\theta(G) = A$ has order a power of p. Thus, A is a p-group and the order of A is $[G : \langle a \rangle] = |G|/p^k < |G|$. By the minimal choice of G, the theorem must be true for A and so the order of A is a power of p. We may now apply Lagrange's theorem to get

$$|G| = [G : \langle a \rangle]|\langle a \rangle| = |A| \cdot |\langle a \rangle|.$$

Both $|A|$ and $|\langle a \rangle|$ are powers of p, as is $|G|$. This conflicts with the assumption that there is some p-group for which the theorem fails, and so the theorem holds for all finite abelian p-groups. ∎

Corollary 1.1 *If G is an abelian group of order $n = p^e m$ for an integer m not divisible by the prime p, then the set $G(p)$ of elements of p-power order form a subgroup of order p^e.*

Proof: If the order of G is $|G| = p_1^{e_1} p_2^{e_2} \cdots p_r^{e_r}$, then the primary decomposition theorem implies $|G| = |G(p_1)| \cdot |G(p_2)| \cdots |G(p_r)|$ because G is the direct sum of the groups $G(p_i)$. Moreover, by the theorem just proved, the order $|G(p_i)|$ is a power of p_i, so by the unique factorization of integers it follows that $|G(p_i)| = p_i^{e_i}$. ∎

So far we have proved that a finite abelian group is a direct sum of p-groups, with p restricted to be a prime divisor of the order of G. The next step is to determine the structure of a finite abelian p-group. We know that an abelian p-group has order a power of p. Our goal is to show that a finite abelian p-group is a direct sum of cyclic p-groups.

Toward this end we begin with the following technical lemma which, in fact, involves the difficult part of the proof of our main result.

Lemma 1.2 *Let H be a finite abelian p-group that is a direct sum of cyclic p-groups and let θ be a homomorphism defined on H. Then the image $\theta(H)$ of θ is a direct sum of cyclic p-groups.*

Before giving the proof, we illustrate one of the subtle points in dealing with direct sums. Suppose $H = C \oplus D$ and let θ be a homomorphism defined on H. Then the image $\theta(H)$ of θ is surely the sum of its two subgroups $\theta(C)$ and $\theta(D)$ since every element in H is a sum $c + d$, with $c \in C$ and $d \in D$. The equation $\theta(c + d) = \theta(c) + \theta(d)$ shows that $\theta(C \oplus D) = \theta(C) + \theta(D)$. However, this latter sum need **not** be a direct sum. We give a concrete example.

Let $C = \mathbb{Z}_{p^2}$, $D = \mathbb{Z}_p$, and $H = C \oplus D$. The elements of H are the ordered pairs (a, b) with a an integer taken modulo p^2 and b an integer taken modulo p. Consider the cyclic subgroup K of H generated by $x = (p, 1)$. Then $px = (p^2, p) = (0, 0) = 0$ so K has order p. Let θ be the canonical homomorphism of H onto H/K.

We identify $c \in C$ with the element $(c, 0)$ in H and the element $d \in D$ with $(0, d) \in H$. In this way we view C and D as subgroups of H. Let us show that C and $\theta(C)$ are isomorphic, as are D and $\theta(D)$. The kernel of the map $C \to \theta(C)$ is the set of elements in C that lie in the kernel K of θ. Elements of C look like $(c, 0)$, whereas those in K look like $mx = m(p, 1) = (mp, m)$ for some $m \in \mathbb{Z}$. If $(c, 0) = (mp, m)$, then $m \equiv 0 \pmod{p}$ and so $p|m$. It follows that $mp \equiv 0 \pmod{p^2}$ and $m(p, 1) = (0, 0) = 0$. Hence, $c = 0$. In a similar way, argue that D and K have only 0 in common.

Then $\theta(C)$ is a cyclic subgroup of H/K and it has order p^2; $\theta(D)$ is a cyclic subgroup of H/K of order p. If the sum $\theta(C) + \theta(D)$ were direct, then H/K would have order $|\theta(C)||\theta(D)| = p^3$. However, the order of H/K is $[H : K] = |H|/|K| = p^2$ so the sum is not direct. This does not present a contradiction to the assertion of the lemma, however. The lemma asserts that H/K is a direct sum of cyclic groups. The only conclusion to be drawn from this example is that H/K is not a direct sum of the particular cyclic groups $\theta(C)$ and $\theta(D)$. We have to look elsewhere for the cyclic direct summands of H/K. In fact, this is quite easy in this example. Since $\theta(C)$ is cyclic of order p^2 and H/K has order p^2, we must have $H/K = \theta(C)$ so H/K is a cyclic group. The obstacle in the general case is that the kernel of θ need not lie in any of the particular direct summands. When this occurs, it is necessary to change to a different set of summands. In this example the summand D can be replaced by the cyclic group $\langle x \rangle$ and still retain the direct sum property; that is,

$$H = C \oplus \langle x \rangle.$$

With this change, the kernel of θ is contained in one of the cyclic summands (and even equals one of the summands). Thus, it is seen by inspection that $\theta(H) = \theta(C)$. The general situation is modeled on this example.

Now we begin the proof of Lemma 1.2. Let

$$H = C_1 \oplus C_2 \oplus \cdots \oplus C_r,$$

with C_i a cyclic group of order p^{e_i}.

We will use specific notation here; identify C_i with the cyclic group $\mathbb{Z}_{p^{e_i}}$ and write the elements of H as rows $(m_1, m_2, \ldots, m_k)$ with m_i an integer taken modulo p^{e_i}. We still use C_i to denote the subgroup of rows having 0 in all coordinates except the ith coordinate.

The homomorphism θ is defined on H with image $\theta(H)$. If θ is an isomorphism, then there is nothing more to prove because $\theta(H)$ is the direct sum of the r subgroups $\theta(C_i)$. We may assume that θ is not an isomorphism.

First consider the case in which $K = \ker\theta$ is a subgroup of order p. Let x be a generator of K. Using the fact that $px = 0$ and that $x \in H$, we know there must exist integers t_i such that

$$x = (p^{e_1-1}t_1, p^{e_2-1}t_2, \ldots, p^{e_r-1}t_r). \tag{7.2}$$

It is possible that some of the t_j are divisible by p so that the jth coordinate of x equals 0. From among the indices j for which p does not divide t_j, select one for which e_j is as small as possible. To be specific, suppose that index is $j = 1$; that is, $p^{e_1-1}t_1 \not\equiv 0 \pmod{p^{e_1}}$, and if i is some index such that $p^{e_i-1}t_i \not\equiv 0 \pmod{p^{e_i}}$ then $e_1 \leq e_i$.

We want to "factor out" p^{e_1-1} from Eq. (7.2) and write x as p^{e_1-1} times some element y of H. This is possible because of our choices. Define the coordinates of y as follows: If j is an index such that $t_j \not\equiv 0 \pmod{p}$, let $b_j = p_j^{e_j-e_1}t_j$, which makes sense because $e_1 \leq e_j$ by our choice of index $j = 1$. If $t_j \equiv 0 \pmod{p}$, let $b_j = 0$. Then we set $y = (b_1, b_2, \cdots, b_r)$ so that

$$x = p^{e_1-1}(b_1, b_2, \cdots, b_r) = p^{e_1-1}y.$$

We will now show that the cyclic group $\langle y\rangle$ generated by y can replace C_1, with the other groups C_j remaining unchanged, and still have a direct sum equal to H. Note first that the order of y is exactly p^{e_1} because $p^{e_1-1}y = x$ is an element of order p.

Let $B = C_2 \oplus \cdots \oplus C_r$ so that $H = C_1 \oplus B$. We claim that $\langle y\rangle \oplus B$ is a direct sum equal to H. If the sum were not direct, there would be a pair of nonzero elements $sy \in \langle y\rangle$ and $b \in B$ with $sy + b = 0$. Now examine the expression for sy in terms of the b_j:

$$sy = s(b_1, b_2, \cdots, b_r) = s(t_1, p^{e_2-e_1}t_2, \cdots)$$

and this element equals $-b$ in B. Notice that every element of B has 0 in the first coordinate and so sy must have a 0 in the first coordinate; that is, $st_1 \equiv 0 \pmod{p^{e_1}}$. However, the choice of the index $j = 1$ was such that t_1 is not divisible by p. It follows that s must be divisible by p^{e_1} and so $sy = 0$ (as y has order p^{e_1}) and also $b = 0$, as $-b = sy$. Thus, the sum $\langle y\rangle \oplus B$ is a direct sum contained in H. In fact, it equals H by considering the orders; $|H| = |C_1| \cdot |B|$. Since $|\langle y\rangle| = |C_1|$, it follows that the order of $\langle y\rangle \oplus B$ equals the order of H and so $H = \langle y\rangle \oplus B$.

Now we have achieved a new representation of H:

$$H = \langle y\rangle \oplus C_2 \oplus \cdots \oplus C_r \tag{7.3}$$

as a direct sum with an additional property that we did not have at the start, namely, the kernel $\langle x\rangle$ of θ is contained in one of the summands, $\langle y\rangle$. This will allow us to complete this case. We claim

$$\theta(\langle y\rangle) + \theta(C_2) + \cdots + \theta(C_r)$$

is a direct sum of cyclic groups that equals $\theta(H)$. To see that this is a direct sum, we suppose

$$\theta(my) + \theta(c_2) + \cdots + \theta(c_r) = 0 \tag{7.4}$$

for some integer m and elements $c_i \in C_i$. We must show each element $\theta(my)$ and $\theta(c_j)$ equals 0. Since θ is a homomorphism, we have

$$\theta(my + c_2 + \cdots + c_r) = 0.$$

This means $my + c_2 + \cdots + c_r$ is in the kernel of θ. The kernel of θ is contained in the subgroup generated by y and so $c_2 = \cdots = c_r = 0$ because the sum in Eq. (7.3) is direct. Then $\theta(my) = 0$ because my is in $\ker\theta$ and so each term is 0 and the sum in Eq. (7.4) is direct. Since the sum visibly equals $\theta(H)$, we have proved $\theta(H)$ is a direct sum of cyclic groups in the case that $\ker\theta$ has order p.

Now we discuss the general case. We give a proof by contradiction. We suppose the lemma is not true for every p-group H that equals a direct sum of cyclic groups and all homomorphisms defined on them. From the collection of all pairs (H, θ) of finite abelian p-groups H that equal a direct sum of cyclic groups, and homomorphisms θ defined on H such that $\theta(H)$ is not a direct sum of cyclic groups, select a pair (H, θ) with $|H|$ as small as possible. Since H is a direct sum of cyclic groups, and $\theta(H)$ is not, θ cannot be an isomorphism. Thus, the kernel K of θ is not the identity and must contain an element of order p. Select $x \in K, x \neq 0$ so that the cyclic subgroup $K_1 = \langle x\rangle$, generated by x, has order p. By the case just proved, H/K_1 is a direct sum of cyclic groups because it is a homomorphic image of H with kernel of order p. Now define a homomorphism θ' from H/K_1 to $\theta(H)$ by the rule

$$\theta' : h + K_1 \longrightarrow \theta(h).$$

This map is well defined because for any $b \in K_1$ we have $\theta(h+b) = \theta(h) + \theta(b) = \theta(h)$ because $K_1 \subseteq \ker\theta$. We have a pair $(H/K_1, \theta')$, where H/K_1 is a finite abelian group that is a direct sum of cyclic subgroups and a homomorphism θ' defined on H/K_1. For this pair, $|H/K_1| < |H|$, and by our choice of H as a smallest group for which the lemma fails, the lemma must hold for $(H/K_1, \theta')$. Thus, we conclude $\theta'(H/K_1)$ is a direct sum of cyclic groups. However, we see that $\theta'(H/K_1) = \theta(H)$ because for any $h \in H$, $\theta(h) = \theta'(h + K_1)$. It follows that $\theta(H)$ is a direct sum of cyclic subgroups after all. In other words, there is no counterexample to the lemma and it holds in all cases. ∎

This lemma will now be used to prove the main result about finite abelian p-groups.

Theorem 1.4 *A finite abelian p-group is isomorphic to a direct sum of cyclic p-groups.*

Proof: Let G be a finite abelian p-group. We begin by selecting elements $a_1, a_2, \ldots, a_n$ in G such that

$$G = \langle a_1 \rangle + \langle a_2 \rangle + \cdots + \langle a_n \rangle.$$

This is a sum of subgroups of G but is not necessarily a direct sum. Note that such a choice of a_i is possible. For example, one could take $n = |G|$ so that every element of G is one of the a_i. Then G is contained in the right side of the sum and equality must hold. (This is a very wasteful choice, but it works.) Let a_i have order p^{e_i} so that $\langle a_i \rangle$ is a cyclic group of order p^{e_i} and is isomorphic to $\mathbb{Z}_{p^{e_i}}$. The isomorphism from $\mathbb{Z}_{p^{e_i}} \to \langle a_i \rangle$ may be expressed as $m \to ma_i$ for an integer m taken modulo p^{e_i}. Now form the direct sum

$$H = \mathbb{Z}_{p^{e_1}} \oplus \mathbb{Z}_{p^{e_2}} \oplus \cdots \oplus \mathbb{Z}_{p^{e_n}}.$$

Define a mapping $\theta : H \to G$ by

$$\theta(m_1, m_2, \ldots, m_n) \longrightarrow m_1 a_1 + m_2 a_2 + \cdots + m_n a_n.$$

Since each mapping $m_i \to m_i a_i$ is a homomorphism, θ is also a homomorphism. Since H is a direct sum of cyclic p-groups, Lemma 1.2 may be applied to conclude that every homomorphic image of H is also a direct sum of cyclic p-groups. We have just shown that G is a homomorphic image of H so G is a direct sum of cyclic p-groups, proving our theorem. ∎

We now know that a finite abelian p-group G is isomorphic to a direct sum

$$\mathbb{Z}_{p^{e_1}} \oplus \mathbb{Z}_{p^{e_2}} \oplus \cdots \oplus \mathbb{Z}_{p^{e_r}} \tag{7.5}$$

for some set $\{e_1, e_2, \ldots, e_r\}$. The next task is to show the uniqueness of this set of numbers. That is, we must prove that if G is also isomorphic to

$$\mathbb{Z}_{p^{f_1}} \oplus \mathbb{Z}_{p^{f_2}} \oplus \cdots \oplus \mathbb{Z}_{p^{f_s}}, \tag{7.6}$$

then $r = s$ and the set of numbers $\{f_1, f_2, \ldots, f_s\}$ equals the set $\{e_1, e_2, \ldots, e_r\}$. One of the subtle points about this problem is that an isomorphism γ from the direct sum of Eq. (7.5) onto the direct sum of Eq. (7.6) will usually cause the coordinates to cross. That is, $\gamma(m, 0, 0, \ldots, 0)$ will, in general, not be of the form $(m', 0, 0, \ldots, 0)$ but will usually have many nonzero coordinates. In other words, γ will not carry a given cyclic summand of the first group onto one of the given summands of the second group.

Therefore, we must look for another approach.

2 ELEMENTARY DIVISORS

Suppose G is a finite abelian p-group and

$$\theta : G \longrightarrow \mathbb{Z}_{p^{e_1}} \oplus \mathbb{Z}_{p^{e_2}} \oplus \cdots \oplus \mathbb{Z}_{p^{e_r}}$$

is an isomorphism. It is assumed that $e_i \geq 1$ for each i. The numbers $p^{e_1}, \ldots, p^{e_r}$ are called the **elementary divisors** of G. It is our goal to show that G has a unique set of elementary divisors and that they determine G up to isomorphism.

One property is evident: *The product of the elementary divisors of G equals the order of G.*

We will use this several times in the arguments that follow.

Theorem 2.1 *If G and H are isomorphic finite abelian p-groups, then the set of elementary divisors of G and H are the same, except for the ordering in which they are listed.*

Proof: We use an argument by induction. The theorem is clearly true if G has order p because in this case there is only one elementary divisor and it equals p. Then also $|H| = p$ because G and H are isomorphic.

Suppose that the theorem is true for groups of order p^m for $m < n$ and let $|G| = p^n$. We show that it is also true for G.

Let G have elementary divisors $\{p^{e_1}, \ldots, p^{e_r}\}$, let H have elementary divisors $\{p^{f_1}, \ldots, p^{f_s}\}$, and let $\theta : G \to H$ be an isomorphism between the two groups.

We make use of the subgroup that we denote by pG and define as

$$pG = \{pg : g \in G\}.$$

It is quite easy to verify that pG is a subgroup of G since $pa + pb = p(a + b)$. If we view pG as an abelian group, then it has a set of elementary divisors. We can determine these easily from the elementary divisors of G. First, note that if $A = \mathbb{Z}_{p^e}$ is a cyclic group of order p^e, then pA is the cyclic subgroup of A generated by p taken mod p^e; thus, pA is a cyclic subgroup of A with order p^{e-1} if $e > 1$, and pA equals (0) when $e = 1$. Thus, the decomposition of pG is given by

$$pG = p\mathbb{Z}_{p^{e_1}} \oplus \cdots \oplus p\mathbb{Z}_{p^{e_r}},$$

where the subgroups $p\mathbb{Z}_{p^{e_i}}$ are omitted if $e_i = 1$ since the group is 0 in that case. Let us write the elementary divisors of G in decreasing order so that $e_1 \geq e_2 \geq \cdots \geq e_t > 1$ and $e_{t+1} = \cdots = e_r = 1$. (This amounts to simply a change of the numbering of the e_i.) Thus, there are $r - t$ elementary divisors of G that equal p. We conclude that the elementary divisors of pG are the numbers $\{p^{e_1-1}, p^{e_2-1}, \ldots, p^{e_t-1}\}$.

With this as preparation, we are ready to consider the isomorphism θ mapping G onto H. The subgroup pG is mapped by θ to the subgroup $\theta(pG) = p\theta(G) = pH$. Moreover, pG has smaller order than G and so theuniqueness of the elementary divisors holds for pG by induction. Thus, the elementary divisors of pG and pH are the same numbers. If we arrange the elementary divisors of H in the same order, namely, with $f_1 \geq f_2 \geq \cdots \geq f_a > 1$ and $f_{a+1} = \cdots = f_s = 1$, then the elementary divisors of pH are $p^{f_j - 1}$ for $1 \leq j \leq a$. It follows that $a = t$ and $e_j - 1 = f_j - 1$ for $1 \leq j \leq t$. Thus, $e_i = f_i$ for $1 \leq i \leq t$ and so the first t elementary divisors of G and of H are the same. For both groups the remaining elementary divisors all equal p. What remains to be shown is that the number of elementary divisors equal to p is the same for both groups. Let G have k elementary divisors equal to p ($k \geq 0$) and let H have l elementary divisors equal to p ($l \geq 0$). The equality $k = l$ will be seen as a consequence of the formula for the orders of the groups. Let $E = e_1 + \cdots + e_t$. The product of the elementary divisors of G equals p^{E+k}, whereas the product of the elementary divisors of H is p^{E+l}. Since the order of G equals the order of H and these orders equal the product of the elementary divisors of the respective groups, it follows that $E + k = E + l$ and $k = l$. Thus, the number of elementary divisors equal to p is the same for both groups and the set of all elementary divisors of G equals the set of elementary divisors of H. ■

3 THE CLASSIFICATION THEOREM

Let G be an abelian group of finite order $n = p_1^{a_1} p_2^{a_2} \cdots p_k^{a_k}$. The primary subgroup $G(p_i)$ is a p_i-group of order $p_i^{a_i}$ and has a set of elementary divisors that determines $G(p_i)$ up to isomorphism. We define the *elementary divisors* of G to be the union of the sets of elementary divisors of the groups $G(p_1)$, $G(p_2), \ldots, G(p_k)$.

Therefore, for example, the cyclic group of order 100 has elementary divisors $\{2^2, 5^2\}$.

Theorem 3.1 THE CLASSIFICATION OF FINITE ABELIAN GROUPS. *Every finite abelian group is isomorphic to a direct sum of cyclic groups of prime power order, and the orders of the cyclic groups in a given direct sum decomposition determine the group up to isomorphism.*

Proof: All the work for the proof of this theorem has already been done. Let G and H be isomorphic finite abelian groups by an isomorphism θ that carries G onto H. For each prime p, the set of elements of p-power order is mapped onto the set of elements of p-power order in H; thus, $\theta(G(p)) = H(p)$. For each prime p dividing the order of G, the elementary divisors of $G(p)$ are the same as the elementary divisors of $H(p)$. Hence, the elementary divisors of G and H are the same. ■

3.1 Examples

We now give a few examples that illustrate the theory developed here.

Let G be a cyclic group of order n. We know that G is isomorphic to $\mathbb{Z}_n$. Let n have the factorization

$$n = p_1^{e_1} p_2^{e_2} \cdots p_r^{e_r}.$$

For each i, the subgroup $G(p_i)$ is cyclic (all subgroups of a cyclic group are cyclic) of order $p_i^{e_i}$. By the primary decomposition theorem we have

$$G \cong \mathbb{Z}_n \cong \mathbb{Z}_{p_1^{e_1}} \oplus \mathbb{Z}_{p_2^{e_2}} \oplus \cdots \oplus \mathbb{Z}_{p_r^{e_r}}$$

and we conclude the elementary divisors of G are $\{p_1^{e_1}, p_2^{e_2}, \cdots, p_r^{e_r}\}$. Here we write "$G \cong H$" as shorthand for "$G$ is isomorphic to H."

Cyclic groups are characterized by the following:

Theorem 3.2 *A finite abelian group G is cyclic if and only if for each prime p dividing $|G|$, G has exactly one subgroup of order p.*

Proof: We know by the characterization of subgroups of a cyclic group that G has only one subgroup of order d for each divisor d of its order. Hence, this holds when d is a prime divisor of the order.

Conversely, suppose G is a finite abelian group with just one subgroup of order p for each prime p dividing its order. For such a p, the subgroup $G(p)$ has only one subgroup of order p. If $G(p)$ has two or more elementary divisors, p^m and p^s, each greater than 1, then $G(p)$ has a subgroup isomorphic to $\mathbb{Z}_{p^m} \oplus \mathbb{Z}_{p^s}$. Then each summand has a subgroup of order p so $G(p)$ has more than one subgroup of order p. Hence, $G(p)$ has just one elementary divisor and it follows that G has the same elementary divisors as the cyclic group of order $|G|$. By the classification theorem, G must be isomorphic to the cyclic group of order $|G|$. ■

As another application of the classification theorem, let us determine all nonisomorphic abelian groups of order 24. Since $24 = 2^3 \cdot 3$, and since the product of the elementary divisors must equal 24, we may make a list of all possible sets of elementary divisors; the possible sets are

$$\{3, 2^3\}, \quad \{3, 2^2, 2\}, \quad \{3, 2, 2, 2\}$$

corresponding to the groups

$$\mathbb{Z}_3 \oplus \mathbb{Z}_8, \quad \mathbb{Z}_3 \oplus \mathbb{Z}_4 \oplus \mathbb{Z}_2, \quad \mathbb{Z}_3 \oplus \mathbb{Z}_2 \oplus \mathbb{Z}_2 \oplus \mathbb{Z}_2.$$

Therefore, there are three, and only three, abelian groups of order 24 when isomorphic groups are considered the same.

More generally, to make a list of all possible abelian p-groups of order p^n, one must make a list of the possible elementary divisors p^{e_j} whose product equals p^n; equivalently, this requires that the sum of the e_j equals n. For example, there are five nonisomorphic abelian groups of order p^4 corresponding to the equations

$$4 = 4, \quad 1 + 3 = 4, \quad 2 + 2 = 4, \quad 1 + 1 + 2 = 4, \quad 1 + 1 + 1 + 1 = 4.$$

These correspond to the five groups

$$\mathbb{Z}_{p^4}, \quad \mathbb{Z}_p \oplus \mathbb{Z}_{p^3}, \quad \mathbb{Z}_{p^2} \oplus \mathbb{Z}_{p^2}, \quad \mathbb{Z}_p \oplus \mathbb{Z}_p \oplus \mathbb{Z}_{p^2}, \quad \mathbb{Z}_p \oplus \mathbb{Z}_p \oplus \mathbb{Z}_p \oplus \mathbb{Z}_p,$$

which are all the nonisomorphic abelian groups of order p^4.

EXERCISES

1. Let $G_1, G_2, \ldots, G_r$ be subgroups of a finite abelian group and suppose the order of the sum $G_1 + G_2 + \cdots + G_r$ equals $|G_1| \cdot |G_2| \cdots |G_r|$. Prove the sum is a direct sum.
2. Let $G = \mathbb{Z}_p \oplus \mathbb{Z}_p = \{(x, y) : x, y \in \mathbb{Z}_p\}$.
   (a) Show that there are $p + 1$ different subgroups of G with order p.
   (b) Let C be a subgroup of G with order p. Show that there are p subgroups D of G such that $C + D$ is a direct sum equal to G.
   (c) If C is the cyclic group generated by $(1, 1)$, give a generator of each of the different subgroups D of order p for which $C + D$ equals G.
3. Show that the two groups $\mathbb{Z}_{12} \oplus \mathbb{Z}_{18}$ and $\mathbb{Z}_6 \oplus \mathbb{Z}_{36}$ are isomorphic.
4. List all the nonisomorphic abelian groups of order 100.
5. Prove that if two abelian groups have the same order and that order is not divisible by the square of any prime, then the groups are isomorphic (and therefore cyclic).
6. If G is an abelian group of order p^m (p a prime) and if $0 \leq t \leq m$, prove that G has a subgroup of order p^t.
7. State a condition on the elementary divisors of an abelian p-group G that ensures G has a subgroup isomorphic to $\mathbb{Z}_{p^2} \oplus \mathbb{Z}_{p^3}$.
8. Using the theory of elementary divisors, state a necessary and sufficient condition that an abelian p-group has a homomorphism onto $\mathbb{Z}_{p^2} \oplus \mathbb{Z}_{p^3}$.

9. If $G = \mathbb{Z}_{p^3} \oplus \mathbb{Z}_{p^2} \oplus \mathbb{Z}_p$, count the number of elements in G of order p. Do the same for orders p^2 and p^3. Check your results by taking the sum of the three counts, which must equal $p^6 - 1$.

10. Let G be an abelian p-group with elementary divisors $p^{e_1}, \ldots, p^{e_r}$. Give a formula for the number of elements of order p in G.

11. With G as in the previous exercise, give a formula for the number of elements in G that have order exactly p^2.

4 THE MULTIPLICATIVE GROUP OF A FINITE FIELD

Now that we have proved results about finite abelian groups, we are able to complete some of the ideas which were studied in Chapter V, Section 6 regarding finite fields. We saw there that for every prime p and integer n there is a field with exactly p^n elements and that such a field could be constructed as a factor ring $\mathbb{Z}_p[x]/(g(x))$ if there is an irreducible polynomial $g(x)$ of degree n in $\mathbb{Z}_p[x]$. We prove that such a polynomial exists. In addition, we show that the multiplicative group of a finite field is cyclic.

Theorem 4.1 *Let F be a finite field. Then the multiplicative group of nonzero elements of F is a cyclic group.*

Proof: The field F has r elements and the multiplicative group of F has $r - 1$ elements. Let q be any prime divisor of $r - 1$. Every element of multiplicative order q is a root of the polynomial $x^q - 1$. However, a polynomial with coefficients in a field has no more roots than its degree. Thus, there are at most q solutions of the equation $x^q - 1 = 0$ in F, one of which is $x = 1$. There are at most $q - 1$ others. This means that there is at most one subgroup of order q because if there were two such groups we would obtain more than $q - 1$ elements of multiplicative order q. By Theorem 3.2, it follows that the multiplicative group of F is cyclic. ∎

We know that the mapping $n \to n \cdot 1_F$, with 1_F the multiplicative identity of F, is a homomorphism from $\mathbb{Z}$ onto an integral domain so its kernel is an ideal (p) generated by a prime number p. The image of the map is the field $\mathbb{Z}_p$, which is a subfield of F.

Theorem 4.2 *Let p be a prime and F a finite field containing $\mathbb{Z}_p$. Then there is an element $\alpha \in F$ such that $\mathbb{Z}_p[\alpha] = F$. The number of elements in F is p^n, where n is the smallest degree of a polynomial in $\mathbb{Z}_p[x]$ having α as a root.*

Proof: Let α be a generator of the multiplicative group of F. Every nonzero element of F is a power of α, so necessarily $\mathbb{Z}_p[\alpha] = F$. The evaluation map $\mathbb{Z}_p[x] \to F$ given by $f(x) \to f(\alpha)$ has kernel $(g(x))$ generated by an irreducible polynomial $g(x)$ of degree n, for some integer n. By the results in Chapter V, Section 6, we know that the image of the evaluation map has p^n elements. However, the image is all of F so F has p^n elements and the irreducible polynomial $g(x)$ having α as a root may be used to construct F. ■

Corollary 4.1

For any prime p and any positive integer n, there is an irreducible polynomial of degree n in $\mathbb{Z}_p[x]$.

Proof: We have seen in Chapter V, Section 6 that there is a field with p^n elements. By the theorem just proved, that field is obtained as a factor ring $\mathbb{Z}_p[x]/(g(x))$ with $g(x)$ irreducible of degree n in $\mathbb{Z}_p[x]$. ■

As a last result about finite fields, we show that there is only one field, up to isomorphism, with p^n elements. Here is the idea of the proof. Let F be a field with p^n elements and let $g(x)$ be any irreducible polynomial of degree n in $\mathbb{Z}_p[x]$. Suppose we can show that there is an element $\alpha \in F$ such that $g(\alpha) = 0$. (This is the key step.) Then there is a homomorphism $\gamma : \mathbb{Z}_p[x] \to F$ given by $f(x) \to f(\alpha)$, and the kernel is the ideal $(g(x))$ of $\mathbb{Z}_p[x]$ since $g(x)$ is irreducible. It then follows that the image of γ is a subfield of F isomorphic to $\mathbb{Z}_p[x]/(g(x))$. However, we know that this factor ring has p^n elements because $g(x)$ has degree n and so we conclude that γ maps onto F and F is isomorphic to $\mathbb{Z}_p[x]/(g(x))$. This proves the uniqueness of F. That is, just select one irreducible polynomial $g(x)$ of degree n and conclude that any field with p^n elements is isomorphic to $\mathbb{Z}_p[x]/(g(x))$.

Therefore, we turn to the key step. Let $h(x) = x^{p^n} - x$. Since the multiplicative group of F has order $p^n - 1$, every nonzero element is a root of the polynomial $x^{p^n-1} - 1$ and so every element of F, including zero, is a root of $h(x)$. Since this statement is true for *every* field with p^n elements, it is also true for the field $\mathbb{Z}_p[x]/(g(x))$. In particular, since $g(x)$ has a root in $\mathbb{Z}_p[x]/(g(x))$, and that root must also be a root of $h(x)$, it follows that $g(x)$ divides $h(x)$. The factorization of $h(x)$ over F is given by

$$h(x) = x^{p^n} - x = \prod_{i=1}^{p^n} (x - \alpha_i),$$

where $\{\alpha_1, \alpha_2, \ldots, \alpha_{p^n}\}$ are the elements of F. Since $g(x)$ divides this polynomial, at least one of the α_j in F is a root of $g(x)$. More precisely, n of the distinct elements of F are roots of $g(x)$.

Thus, we have proved the uniqueness.

Theorem 4.3

For any positive prime p and positive integer n there is a field with p^n elements and any two such fields are isomorphic.

EXERCISES

1. Let $F = \mathbb{Z}_3[x]/(x^2 + 1)$, a field with nine elements.

(a) Find all the elements in the multiplicative group of F with order 8.

(b) Evaluate the product of all the terms $(x - \alpha)$ as α runs through the four elements of F with multiplicative order 8. [Hint: It might be easier to work backwards: What polynomial of degree 4 is satisfied by all the elements of order 8?]

2. There are two irreducible polynomials of degree 3 in $\mathbb{Z}_2[x]$, $g_1(x) = x^3 + x + 1$ and $g_2(x) = x^3 + x^2 + 1$. By the proof of the uniqueness of the field with 2^3 elements, the field $\mathbb{Z}_2[x]/(g_1(x))$ has three elements $\alpha_1, \alpha_2, \alpha_3$ such that $g_2(x) = (x - \alpha_1)(x - \alpha_2)(x - \alpha_3)$. Find these α_i.

3. If F is a finite field containing $\mathbb{Z}_p$ as a subfield and if $f(x)$ is an irreducible polynomial in $\mathbb{Z}_p[x]$ such that $f(x)$ has one root in F, then $f(x)$ factors into a product of degree 1 factors in $F[x]$.

4. Prove that the number of monic irreducible polynomials of degree n in $\mathbb{Z}_p[x]$ is at least $\varphi(p^n - 1)/n$, where φ is the Euler function defined in Chapter IV, Section 4. [Hint: Use the facts that a cyclic group of order m has $\varphi(m)$ generators and a monic irreducible polynomial of degree n has n roots in a field with p^n elements. Note that $\varphi(p^n - 1)/n$ is an integer (see the exercises in Chapter VI, Section 4).]

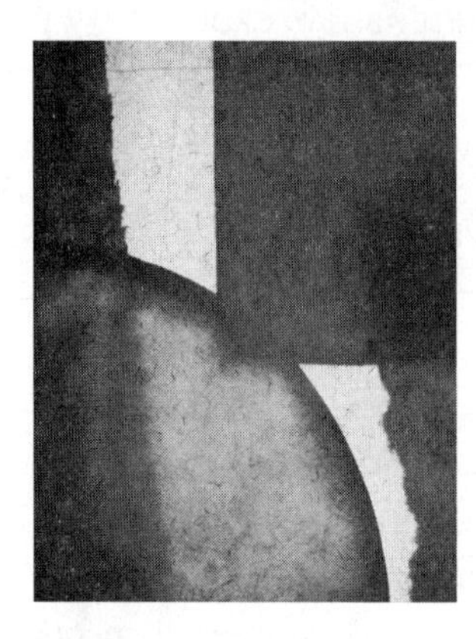

VIII FINITE GROUPS

A great deal more is known about finite groups than is known about groups in general. One important reason for this is the ability to use certain types of counting arguments to study finite groups that are not available when studying infinite groups. A common technique is to construct some finite set closely related to the group and count group elements that have some special properties. This is accomplished by finding homomorphisms from a finite group into the group of permutations of some finite set. This theme will be exploited throughout this chapter. One of the main results is Sylow's theorem, which asserts the existence of a subgroup of order p^a whenever p^a divides the order of the group. We will apply Sylow's theorem and other ideas to classify groups of small order.

1 GROUPS ACTING ON SETS

Let G be a group and X a set. We say G *acts on* X if, for each element $g \in G$, there is assigned a permutation θ_g of X in such a way that the equation $\theta_g\theta_h = \theta_{gh}$ for all $g, h \in G$. Another way to state this is to say that the correspondence $g \to \theta_g$ is a homomorphism from G into the group of all permutations of X. We have seen examples in which subgroups of the symmetric group S_n act on the set $X = \{1, 2, \cdots, n\}$; our goal here is to find actions of more general finite groups acting on some set. In other words, we want to find homomorphisms of some group G into the group of permutations of a set.

In order to simplify the notation, the homomorphism θ of G into the permutations of X is not written explicitly. Instead of writing $\theta_g(x)$, we may simply write $g(x)$ or even gx. The condition required by the definition of an action takes a simpler form; that is, we require $g(hx) = (gh)x$ for all $g, h \in G$

and $x \in X$. Since θ is a homomorphism, θ_e must be the identity permutation if e is the identity of G. This means $x = \theta_e(x)$ or $x = ex = e(x)$ for all $x \in X$. Next, we give some examples of actions of a group on a set. We begin with one of the familiar examples.

Example I.1

Let G be a subgroup of the symmetric group S_n of permutations of $X = \{1, 2, \ldots, n\}$. Then G acts on X in accordance with the definition of permutation; for $x \in X$ and $\sigma \in G$, $\sigma(x)$ is the value of the function σ at the point x. In this case, the homomorphism $\theta \to S_n$ is just the inclusion map that sends $\sigma \in G$ to itself, as an element of S_n.

Next we consider one of the most important actions of a group on itself.

Example I.2

CONJUGATION. We define an action of an arbitrary group G on itself; that is, $X = G$. For each $g \in G$ let ι_g be the function defined by

$$\iota_g : x \longrightarrow gxg^{-1}.$$

We first check that ι_g is indeed a permutation of G. It is an onto function because for any $x \in G$, $\iota_g(g^{-1}xg) = x$. It is one-to-one because $gxg^{-1} = gyg^{-1}$ implies $x = y$ as we see by multiplying on the left by g^{-1} and on the right by g. Next we must show that the correspondence $g \to \iota_g$ is a homomorphism. For $g, h \in G$ and any $x \in G$ we have

$$\begin{aligned} \iota_{gh}(x) &= (gh)x(gh)^{-1} = g(hxh^{-1})g^{-1} \\ &= \iota_g(hxh^{-1}) = \iota_g(\iota_h(x)) = (\iota_g\iota_h)(x). \end{aligned}$$

Thus, $\iota_{gh} = \iota_g\iota_h$, proving that this defines an action of G on G.

This is an important example in the study of nonabelian groups. We call an element gxg^{-1} a *conjugate* of x and the action of G is called *conjugation*. There is another important property of this action. Each permutation ι_g is an isomorphism of the group G onto itself. This is easily checked directly from the definition

$$\iota_g(xy) = g(xy)g^{-1} = gxeyg^{-1} = gxg^{-1}gyg^{-1} = \iota_g(x)\iota_g(y).$$

An isomorphism of a group with itself is called an **automorphism**. Thus, conjugation by any element of G produces an automorphism of G.

Example I.3

Let G be any group and let X be the collection of all subgroups of G. For each $g \in G$ and each subgroup $H \in X$, let γ_g be a function defined by

$$\gamma_g(H) = gHg^{-1} = \{ghg^{-1} : h \in H\}.$$

It is necessary to verify that this defines an action of G on X. Since the conjugation action, ι_g, defines an isomorphism of the group G, it is clear that

$\gamma_g(H)$ is a subgroup of G and hence is an element of X. Moreover, γ_g has an inverse, namely, $\gamma_{g^{-1}}$, so γ_g is a permutation of X. Just as in the previous example, we compute $\gamma_{gh} = \gamma_g\gamma_h$ for $g, h \in G$.

By analogy with the previous example, we call gHg^{-1} a *conjugate* of the subgroup H. The conjugate gHg^{-1} is a subgroup isomorphic to H.

Example 1.4 Let $G = S_3$ be the symmetric group of permutations of $\{1, 2, 3\}$. We will describe the conjugation action of G acting on itself explicitly. For each $g \in G$, ι_g is the permutation of the six elements of G given by the rule $\iota_g(x) = gxg^{-1}$. For simplicity of notation, let us number the elements of G as follows:

$$\begin{array}{lll} \sigma_1 = \epsilon, & \sigma_2 = (12), & \sigma_3 = (13), \\ \sigma_4 = (23), & \sigma_5 = (123), & \sigma_6 = (132). \end{array}$$

We work out the action of the element (12) on G by computing $\iota_{(12)}(\sigma_i)$ for $1 \le i \le 6$:

$$\begin{aligned} \iota_{(12)}(\sigma_1) &= (12)\epsilon(12)^{-1)} = \epsilon = \sigma_1, \\ \iota_{(12)}(\sigma_2) &= (12)(12)(12)^{-1)} = (12) = \sigma_2, \\ \iota_{(12)}(\sigma_3) &= (12)(13)(12)^{-1)} = (23) = \sigma_4, \\ \iota_{(12)}(\sigma_4) &= (12)(23)(12)^{-1)} = (13) = \sigma_3, \\ \iota_{(12)}(\sigma_5) &= (12)(123)(12)^{-1)} = (132) = \sigma_6, \\ \iota_{(12)}(\sigma_6) &= (12)(132)(12)^{-1)} = (123) = \sigma_5. \end{aligned}$$

If we use the double-row notation for this permutation, then

$$\iota_{(12)} = \begin{pmatrix} \sigma_1 & \sigma_2 & \sigma_3 & \sigma_4 & \sigma_5 & \sigma_6 \\ \sigma_1 & \sigma_2 & \sigma_4 & \sigma_3 & \sigma_6 & \sigma_5 \end{pmatrix}.$$

In the cycle notation we write $\iota_{(12)} = (\sigma_3, \sigma_4)(\sigma_5, \sigma_6)$. In the analogous manner we could determine that $\iota_{(13)} = (\sigma_2, \sigma_4)(\sigma_5, \sigma_6)$. All of the other permutations ι_g can be determined from the calculations already made. Every element in S_3 is a product of transpositions and, since $(23) = (12)(13)(12)$, every element of S_3 is a product formed using only the transpositions (12) and (13). Using the fact that $g \to \iota_g$ for any $g \in S_3$ is a homomorphism, ι_g can be determined by expressing g as a product using only (12) and (13). For example, using $(12)(13) = (132)$ yields

$$\iota_{(132)} = \iota_{(12)}\iota_{(13)} = (\sigma_3, \sigma_4)(\sigma_5, \sigma_6)(\sigma_2, \sigma_4)(\sigma_5, \sigma_6) = (\sigma_2, \sigma_3, \sigma_4).$$

We have thus obtained a homomorphism of S_3 into the group of permutations of six elements σ_i. One may verify that this is a one-to-one homomorphism.

1.1 Equivalence Classes of an Action

Before giving further examples of actions of groups, we develop some general principles that will apply to all examples. These will be applied later in concrete situations.

Let G be a group acting on the set X. Define a relation on X by the rule $x \sim y$ if there is some $g \in G$ such that $g(x) = y$. Let us verify that this is an equivalence relation. Since $e(x) = x$, it follows that $x \sim x$. If $g(x) = y$, then $x = g^{-1}(y)$ so $x \sim y$ implies $y \sim x$. Finally, if $g(x) = y$ and $h(y) = z$, then $(hg)(x) = h(y) = z$ so that $x \sim y$ and $y \sim z$ imply $x \sim z$. Therefore, this is an equivalence relation and X is partitioned by the equivalence classes which are defined by this relation:

$$[x] = \{y : y \in X \text{ and } y = g(x) \text{ for some } g \in G\}.$$

The equivalence classes are certain subsets of X defined by reference to a certain group G acting on X. It may sometimes be convenient to emphasize the role of G in this definition so we may call the set $[x]$ a *G-equivalence class.*

Next we use the action of G on X to define certain subgroups of G. For any $x \in X$ let

$$G_x = \{g : g \in G \text{ and } g(x) = x\}.$$

Then G_x is a subgroup because if $g, h \in G_x$, then $(gh)(x) = g(h(x)) = g(x) = x$, showing that $gh \in G_x$. Furthermore, $g(x) = x$ implies $x = g^{-1}(x)$ so that $g^{-1} \in G_x$. Thus, G_x is a subgroup of G and it is called the *subgroup fixing* x since it consists of all the group elements that do not move the point x.

The following is an important connection between the equivalence class $[x]$ and the subgroup fixing x. This is the first of many "counting" theorems:

Theorem 1.1 *Let G be a group acting on the finite set X. For each $x \in X$, the number of elements in the equivalence class $[x]$ equals the index $[G : G_x]$ of the subgroup fixing x.*

Proof: The proof is carried out by establishing a one-to-one correspondence between the collection of cosets gG_x and the elements of $[x]$, thereby showing the two sets have the same number of elements. Define a correspondence ψ by the rule

$$\psi(gG_x) = g(x).$$

Note that this correspondence is well defined because $gG_x = hG_x$ implies $g = ha$ with $a \in G_x$. It follows that $g(x) = ha(x) = h(a(x)) = h(x)$. Then $\psi(gG_x) = \psi(hG_x)$, proving ψ has the same value no matter which element is used from the coset. Thus, ψ associates with each coset of G_x, an element of $[x]$. The correspondence is onto $[x]$ because if $y \in [x]$ then $y = g(x)$ for some $g \in G$ (by definition of $[x]$) and so $\psi(gG_x) = g(x) = y$.

Now let us show that ψ is one-to-one. Suppose $\psi(gG_x) = \psi(hG_x)$. Then $g(x) = h(x)$ and so $x = g^{-1}h(x)$. This means $g^{-1}h \in G_x$, by definition of G_x. Then $h = g(g^{-1}h)$ is in the coset gG_x and it follows that $gG_x = hG_x$. Thus, the number of cosets of G_x in G equals the number of elements in $[x]$. ■

For emphasis, we state an immediate consequence of this theorem.

Corollary 1.1 *If the finite group G acts on the finite set X, then the number of elements in a G-equivalence class is a divisor of the order of G.*

1.2 Application to Binomial Coefficients

The following is an immediate application of this counting theorem: Let $Y = \{1, 2, \ldots, n\}$ and let X be the collection of all k element subsets of Y for some positive integer k at most n. (By a k element set, we mean simply a set with k distinct elements.) We show how to determine the number of elements in X. Let $G = S_n$ be the symmetric group of all permutations of Y. Then G acts on X in the obvious way: If $A = \{a_1, a_2, \ldots, a_k\} \in X$ and $\sigma \in S_n$, then

$$\sigma(A) = \{\sigma(a_1), \sigma(a_2), \ldots, \sigma(a_k)\}.$$

Clearly σ transforms a k element set into another k element set. Let $E = \{1, 2, \ldots, k\}$. For any k element set A, there is a $\sigma \in G$ with $\sigma(E) = A$; that is, for any k distinct elements $a_1, \ldots, a_k$ of Y there is a permutation $\sigma \in G$ with $\sigma(i) = a_i$, $i = 1, 2, \ldots, k$. Thus, the G-equivalence class of E is all of X; by Theorem 1.1, the number of elements $|X|$ in X equals the index of the subgroup fixing E. Let us determine this index. If $\sigma(E) = E$, then also $\sigma(E') = E'$, where $E' = \{k+1, k+2, \ldots, n\}$. An element that fixes E can permute the k elements of this set in any way; similarly, an element fixing the set E' can permute its elements arbitrarily. If we take any permutation τ of E and any permutation γ of E', then $\tau\gamma$ is an element of G fixing E. There are $k!$ possibilities for τ and $(n-k)!$ possibilities for γ so the subgroup G_E fixing E has $k!(n-k)!$ elements. Thus, we conclude

$$|X| = [G : G_E] = \frac{|G|}{|G_E|} = \frac{n!}{k!(n-k)!}.$$

This number is the binomial coefficient denoted earlier by $C(n, k)$. Thus, we have proved that the number of k element subsets of an n element set is $C(n, k)$. This also gives an independent proof that $C(n, k)$ is an integer.

1.3 The Class Equation

If G acts on a finite set X, then X is partitioned by the G-equivalence classes. Let

$$X = [x_1] \cup [x_2] \cup \cdots \cup [x_r]$$

be this partition. Since every element of X lies in one, and only one, of the $[x_i]$, the number of elements in X is the sum of the number of elements in the equivalence classes $[x_i]$. We denote the number of elements in X by $|X|$ and apply Theorem 1.1 to conclude

$$|X| = [G : G_{x_1}] + [G : G_{x_2}] + \cdots + [G : G_{x_r}], \tag{8.1}$$

where $x_1, x_2, \ldots, x_r$ are elements of X, one taken from each of the distinct G-equivalence classes. This gives a formula for the number of elements of X as a sum of certain divisors of $|G|$.

There is a special instance of this that is important. We first illustrate it with a concrete example.

Example 1.5 Let $G = S_3$ and $X = S_3$ with the action of G on X given by conjugation as in Example 1.4. The equivalence relation $x \sim y$ now means $x = gyg^{-1}$ for some $g \in G$. The equivalence classes are

$$[\epsilon] = \{\epsilon\}, \quad [(12)] = \{(12), (13), (23)\}, \quad [(123)] = \{(123), (132)\}.$$

In this case the formula Eq. (8.1) reads

$$\begin{aligned} |X| &= [G : G_\epsilon] + [G : G_{(12)}] + [G : G_{(123)}] \\ 6 &= 1 + 3 + 2, \end{aligned}$$

which expresses the order of G as a sum of divisors of G.

The number of elements in the equivalence class $[(12)]$ is the index of the subgroup of elements that fixes (12). In this case an element $g \in G$ fixes (12) if and only if $(12) = g(12)g^{-1}$. Since $g(12)g^{-1} = (g(1), g(2))$ we see that the only possibilities for g are $g = \epsilon, (12)$. Thus, $G_{(12)} = \{\epsilon, (12)\}$ and this is a subgroup of index 3, verifying the fact that the number of elements in the equivalence class of (12) is the index of $G_{(12)}$ in G.

We can put this example into a more general context.

1.4 Conjugate Classes

If G is any finite group and G acts on G by conjugation, the equivalence class of an element $x \in G$ is the set

$$[x] = \{gxg^{-1} : g \in G\}.$$

This set is called the **conjugate class** of x. Since G is the disjoint union of these classes, we have a count of the elements of G as the sum of the numbers of elements in the distinct conjugate classes:

$$|G| = |[x_1]| + |[x_2]| + \cdots + |[x_r]|, \tag{8.2}$$

where the x_i are elements of G selected, one from each of the distinct conjugate classes.

The subgroup G_x is the set of all $g \in G$ such that $x = gxg^{-1}$. This is equivalent to the condition $xg = gx$. When $x, g \in G$ satisfy $xg = gx$, we say x and g *commute* as a short form of saying that the "commutative law holds for x and g." The set of elements g that commute with x is usually called the **centralizer** of x in G and is denoted as $C_G(x)$. Thus, the number of elements in the conjugate class of x is the index of the centralizer of x in G, namely,

$$|[x]| = [G : C_G(x)],$$

and so Eq. (8.2) may be written in the form

$$|G| = [G : C_G(x_1)] + [G : C_G(x_2)] + \cdots + [G : C_G(x_r)].$$

The class $[e]$ containing the identity has just one element, that is, $C_G(e) = G$. Other classes may also contain only one element. If the group is abelian, then every class contains one element. For a group that may not be abelian, the conjugate class of x contains only x precisely when $xg = gx$ holds for all $g \in G$. The collection of such x forms a subgroup $Z(G)$ called the **center** of G.

We should verify that this set is a subgroup. Suppose $x, y \in Z(G)$. Then for every element $g \in G$ we have $gx = xg$ and $gy = yg$. Now to show that the product is in $Z(G)$ we compute

$$g(xy) = (gx)y = (xg)y = x(gy) = x(yg) = (xy)g.$$

Thus, $xy \in Z(G)$. Similarly, $x^{-1} \in Z(G)$ so $Z(G)$ is a subgroup of G.

Note that the elements of $Z(G)$ are exactly the elements whose conjugate class has only one element. If we let $x_1, \ldots, x_s$ be elements taken from distinct conjugate classes having more than one element, Eq. (8.2) takes the form

$$|G| = |Z(G)| + [G : C_G(x_1)] + \cdots + [G : C_G(x_s)]. \tag{8.3}$$

This equation is called the **class equation** for G since it arises from a counting of elements of G by arranging them in their conjugate classes. It is the basis for the proofs of several basic results about subgroups of finite groups.

EXERCISES

1. Let G be the cyclic subgroup of S_5 consisting of powers of $\alpha = (123)(45)$. Then G has order 6 and acts on the set $X = \{1, 2, 3, 4, 5\}$. Give the G-equivalence classes of X, select one element x from each, and determine G_x. Verify Eq. (8.1) for this example.

2. Let $G = S_3$ be the symmetric group of all permutations of $\{1, 2, 3\}$. Let X be the set of all ordered pairs (i, j) with $i, j \in \{1, 2, 3\}$. Then G acts on X with the action $\gamma(i, j) = (\gamma(i), \gamma(j))$, $\gamma \in G$. Explicitly give the G-equivalence classes of X and obtain the decomposition of $|X|$ given in Eq. (8.1).

3. Repeat Exercise 2 using X as given there and $G = \{e, (123), (132)\}$ in place of S_3.

4. Let $G = S_3$ be the symmetric group of all permutations of $\{1, 2, 3\}$. Let Y be the set of all unordered pairs $\{i, j\}$ with $i \neq j \in \{1, 2, 3\}$. Then G acts on Y with the action $\gamma\{i, j\} = \{\gamma(i), \gamma(j)\}$, $\gamma \in G$. Explicitly give the G-equivalence classes of Y and obtain the decomposition of $|Y|$ given in Eq. (8.1).

5. Let $\mathbb{Z}_p$ denote the field with p elements for a prime p, and let G be the group of all two-by-two matrices of the form

$$g = \begin{bmatrix} a & b \\ c & d \end{bmatrix}, \qquad a, b, c, d \in \mathbb{Z}_p,\ ad - bc \neq 0.$$

Let X denote the set of all "columns" $z = \begin{bmatrix} u \\ v \end{bmatrix}$ with $u, v \in \mathbb{Z}_p$.

(a) Show that the rule

$$gz = \begin{bmatrix} a & b \\ c & d \end{bmatrix} \begin{bmatrix} u \\ v \end{bmatrix} = \begin{bmatrix} au + bv \\ cu + dv \end{bmatrix}$$

defines an action of G on X.

(b) Show that X has just two G-equivalence classes and one of them consists only of one element. [Hint: Show that if x and y are not both zero, there is an element $g \in G$ having a first column equal to $\begin{bmatrix} x \\ y \end{bmatrix}$. Conclude that the equivalence class of $\begin{bmatrix} 1 \\ 0 \end{bmatrix}$ contains every element of X except $\begin{bmatrix} 0 \\ 0 \end{bmatrix}$.]

(c) Show that the subgroup G_z, for $z = \begin{bmatrix} 1 \\ 0 \end{bmatrix}$, is the set of all $g \in G$ with $a = 1$, $c = 0$, and $d \neq 0$. Conclude that $|G_z| = p(p - 1)$.

(d) Use the counting equation $|X| = 1 + [G : G_z]$ to prove that the order of G is $p(p-1)(p^2-1)$.

6. Let G be a finite group that acts on a set X and let $x, y \in X$ be elements in the same G-equivalence class. Prove that $|G_x| = |G_y|$. In fact, even more is true. If $g \in G$ is any element such that $g(x) = y$, then $gG_xg^{-1} = G_y$. Hence, the isomorphism of G to itself given by ι_g, conjugation with g, carries G_x onto G_y and so G_x is isomorphic to G_y.

7. Let G be the cyclic group of order 7 and let X be a set with five elements. Prove that the only action of G on X is the one in which $g(x) = x$ for all $x \in X$.

8. Let G be a cyclic group of prime order p and X a set with fewer than p elements. Prove that the only action of G on X is the "trivial" action defined by $g(x) = x$ for all $g \in G$ and $x \in X$.

9. If σ in S_n is the product of disjoint cycles of lengths $l_1, l_2, \ldots, l_t$, including the cycles of length 1. Define the *cycle shape* of σ to be the set of integers $\{l_1, l_2, \ldots, l_t\}$. Thus, $l_1 + l_2 + \ldots + l_t = n$. Prove that two elements of S_n are conjugate if and only if they have the same cycle shape. Prove that any set $\{l_1, l_2, \ldots, l_t\}$ of positive integers having a sum equal to n is a cycle shape so the number of conjugate classes in S_n equals the number of different sets of positive integers having a sum equal to n.

10. Use the previous exercise to determine the number of conjugate classes in each of the symmetric groups S_3, S_4, and S_5.

11. Show that the center of S_n consists of only the identity if $n \geq 3$.

12. Let p be a positive prime and $G = S_p$ the symmetric group of permutations of $\{1, 2, \ldots, p\}$. Let X be the collection of subgroups of G that have order p. If H is a subgroup of order p and $g \in G$, then gHg^{-1} is also a subgroup of order p. Thus, G acts by conjugation on X. Prove that the *normalizer* of H,

$$N_G(H) = \{g : g \in G, \text{ and } gHg^{-1} = H\},$$

is a subgroup of order $p(p-1)$. There is more than one solution to this exercise. The following is one possible outline:

(a) Show that there is only one G-equivalence class in X.

(b) The number of elements in X is $(p-2)!$

(c) Deduce the order of $N_G(H)$ from the fact that the subgroup G_H of G fixing H is $N_G(H)$ and then use the relation $|X| = [G : N_G(H)]$.

13. Let K be a normal subgroup of a finite group G and let $x \in K$. Show that the G conjugate class of x lies in K. Give an example to show that the K conjugate class of a may be different than the G conjugate class of a. [Hint: Find an example in which K is an abelian normal subgroup.]

2 GROUPS OF PRIME POWER ORDER

If a finite group has order p^n, for some prime number p, then every divisor of the order of G is a power of p and so is either equal to 1 or is divisible by p. This simple fact has surprising consequences which will be explored in this section. For brevity, we use the following terminology:

Definition 2.1

A finite group G with order a power of a prime p is called a ***p*-group**.

If we say G is a p-group, it is to be understood that p is some prime number.

Theorem 2.1

Let G be a finite p-group. If G acts on a finite set X, then either

(i) *the number of elements in X is divisible by p*

or

(ii) *there is some $x \in X$ such that $g(x) = x$ for every $g \in G$.*

Proof: Suppose that condition (ii) does not hold. We must show that (i) does hold. For each $x \in X$, there is some $g \in G$ such that $g(x) \neq x$ because (ii) does not hold. Thus, the G-equivalence class of x has more than one element. The subgroup G_x fixing x is not all of G so the index $[G : G_x]$ is a power of p and not equal to 1. Now refer to Eq. (8.1) for the action of G on X to conclude that $|X|$ is a sum of numbers $[G : G_{x_i}]$, each of which is divisible by p. Thus, $|X|$ is divisible by p, as required by statement (i). ∎

When G acts on a set X, a point $x \in X$ with the property $g(x) = x$ for all $g \in G$ is called a *fixed point* of G or, more precisely, a *G-fixed point.* The theorem states that whenever G has prime power order, any set on which it acts either has a fixed point or the number of elements in the set is divisible by p. We improve this theorem slightly by giving some information about the number of fixed points.

Theorem 2.2

Let G be a finite p-group and X a finite set on which G acts. If f is the number of G-fixed points in X, then $f \equiv |X| \pmod p$.

Proof: The element $x \in X$ is a fixed point if and only if the G-equivalence class $[x]$ has exactly one element. All classes containing more than one element must contain p^s elements for various integers $s > 0$ because the number of elements in such a class is a divisor of $|G| = p^n$. When we count the number of elements in X using Eq. (8.1), we see that $|X|$ is the sum of f ones and other numbers which are powers of p greater than 1. It follows that $|X| \equiv f \pmod p$. ∎

Recall that the center of a group G is the set of all elements x such that $xg = gx$ for every $g \in G$. Many groups, such as the symmetric groups S_n with $n \geq 3$, have only the identity element in the center. However, this is not the case when G is a p-group.

Theorem 2.3 *If G is a p-group of order $p^n > 1$, then the center $Z(G)$ of G has order greater than 1.*

Proof: Let G act on G by conjugation so that for each $g \in G$ the action of g is $g(x) = gxg^{-1}$. An element x is a fixed point of G if and only if $x = gxg^{-1}$ for each $g \in G$. This is equivalent to the condition that $xg = gx$ for all $g \in G$. There is certainly one fixed point, namely, the identity e of G. If f denotes the number of fixed points, then $f > 0$. Theorem 2.2 implies $f \equiv |G| \pmod{p}$. This means that p divides f and so $f \geq p$. The center of G has order f so $|Z(G)| > 1$. ■

We now apply this result to prove something about groups of order p^2. We know that a group of order p must be cyclic. We show that a group of order p^2 must be abelian.

Theorem 2.4 *A group of order p^2 is abelian.*

Proof: Let G be a group of order p^2. If G contains an element of order p^2, then G is cyclic and hence abelian. Assume then that there is no element of order p^2. Since the order of an element in G must divide the order of G, each nonidentity element then has order p. The center of G is not the identity so we may select an element x in the center with x of order p. Since G has order p^2 there must exist an element y in G with y not in the cyclic group $\langle x \rangle$ generated by x. Since x is in the center, $xy = yx$. We claim that the elements $x^i y^j$ with $0 \leq i, j \leq p$ are distinct. Suppose $x^i y^j = x^a y^b$. Then $x^{i-a} = y^{b-j}$ and this element lies in the intersection $\langle x \rangle \cap \langle y \rangle$ of the two cyclic groups of order p. If the intersection were not the identity, then $\langle x \rangle = \langle y \rangle$ because the only subgroup, other than the identity of the cyclic group of order p, is the entire group. By our choice of y this cannot happen. The intersection is the identity and so $x^{i-a} = y^{b-j} = e$, and it follows that $x^i = x^a$ and $y^b = y^j$. Thus, the p^2 elements of G are exactly the elements $x^i y^j$. From the fact that $xy = yx$ it is easy to verify that $(x^i y^j)(x^a y^b) = (x^a y^b)(x^i y^j)$ for all exponents and hence the commutative law holds for every pair of elements in G. ■

Thus, groups of order p or p^2 are abelian. We cannot go any further. There are nonabelian groups of order p^3. In fact, the dihedral group D_8 introduced in Chapter VI, Section 2 is a nonabelian group of order $8 = 2^3$. See the exercises following this section for additional examples.

We show one more property of p-groups in the following theorem:

Theorem 2.5 *Let G be a p-group of order p^n. Then G has a subgroup of order p^i for $0 \leq i \leq n$.*

Proof: The theorem is certainly true if the order of G is p. We give an inductive argument based on the assumption that the theorem is true for p-groups of order smaller than the order of G.

The center $Z(G)$ of G is an abelian p-group which contains a nonidentity element g of order p^t for some $t \geq 1$. The cyclic subgroup generated by g contains a subgroup of order p, so we may select an element $a \in Z(G)$ with order of a equal to p. The cyclic subgroup $A = \langle a \rangle$ generated by a is a normal subgroup of G because $gag^{-1} = a$ for every $g \in G$. Thus, we may form the factor group G/A which is a group of order p^{n-1}. Let $\theta : G \to G/A$ be the canonical homomorphism defined by $\theta(g) = gA$.

Since the order of G/A is less than the order of G, G/A has a subgroup of order p^t for $0 \leq t \leq n-1$. Let H be a subgroup of G/A with order p^t and let K be the set of elements of G that map into H under θ; that is,

$$K = \{g : g \in G, \theta(g) \in H\}.$$

We may regard H as a collection of cosets of A and K is the union of these cosets. The number of elements in K is the order of H times p since each coset has p elements; that is, $|K| = p|H| = p^{t+1}$. Now we argue that K is a subgroup of G. If $u, v \in K$, then $\theta(uv) = \theta(u)\theta(v) \in H$ and $\theta(u^{-1}) \in H$. Thus, uv and u^{-1} lie in K and K is a subgroup of G. Thus, G has a subgroup of order p^i for every i, $1 \leq i \leq n$. The identity alone is a subgroup of order $p^0 = 1$. ∎

EXERCISES

1. Let D_8 denote the dihedral group described in Chapter VI, Section 2.
   (a) Find the center of D_8.
   (b) Find a subgroup of D_8 with order 2^i for $0 \leq i \leq 3$.

2. Let $\alpha = (123)(456)$ and $\beta = (347)$ be permutations in S_7. Both α and β have order 3. Use Theorem 2.2 to argue that α and β are not contained in a subgroup of order 3^n in S_7. [One could prove this in several ways, but the use of Theorem 2.2 does not require the multiplication of any permutations.]

3. Let G be a group of order p^n. Suppose that the center of G has order at least p^{n-1}. Use the ideas from the proof of Theorem 2.4 to show that G must be abelian and so the center of G equals G.

4. Suppose that a group contains the elements g, h, and gh, all of which have order 2. Prove that $\{e, g, h, gh\}$ is an abelian group of order 4.

5. Let $\alpha = (123456789)$ and $\beta = (258)(396)$ be elements of S_9. Show that the set of all elements $\alpha^i\beta^j$ with $0 \le i \le 8$ and $0 \le j \le 2$ forms a nonabelian group of order 3^3. [Hint: It will be most efficient to first prove that $\beta\alpha\beta^{-1} = \alpha^4$.]
6. Let p be a prime number and let X be the ring of integers modulo p^2. We define permutations of X as follows: $\sigma(x) = x+1$ and $\tau(x) = x(1+p)$, where the addition and multiplication are taken modulo p^2.
   (a) Verify that σ is a permutation of X having order p^2 and that τ is a permutation of X having order p.
   (b) Prove the relation $\tau\sigma\tau^{-1} = \sigma^{1+p}$.
   (c) Conclude that the set of all powers $\sigma^i\tau^j$ is a finite nonabelian group $G = \langle\sigma, \tau\rangle$ of order p^3.
   (d) Show that σ^p lies in the center of G and use Exercise 3 to conclude that $Z(G)$ is the cyclic group generated by σ^p.

3 SYLOW'S THEOREMS

In this section we apply the results of the previous sections to prove the existence of subgroups of order p^n in a finite group whose order is divisible by p^n, where p is a prime number. In addition, the subgroups of the largest prime power order possible are shown to all be conjugates of any one of them and the number of such subgroups is shown to be congruent to 1 modulo p. We state these in separate parts.

Theorem 3.1

FIRST SYLOW THEOREM. *Let G be a finite group of order $|G| = p^a m$ with p a prime number and m an integer not divisible by p. Then G has a subgroup of order p^a.*

Proof: The proof is trivially true if $m = 1$. We carry out the proof by induction; assume the statement of the theorem is true for groups of order $p^a k$ with $k < m$.

Let X denote the collection of all subsets of G having p^a elements. The number of elements in X is the binomial coefficient $C(p^a m, p^a)$. This number is not divisible by p as we show at the end of this section. Let G act on X by the following rule: If $A \in X$ and $A = \{a_1, \ldots, a_{p^a}\}$ and if $g \in G$, then

$$g(A) = gA = \{ga_1, \ldots, ga_{p^a}\},$$

where ga_i is the product of elements in G. Thus, gA is the subset obtained by multiplying each element of A on the left by g. The set X is the disjoint union of the G-equivalence classes and the total number of elements in X is not divisible by p. Hence, there must be at least one G-equivalence class $[A]$ with $|[A]|$ not divisible by p. Let A be one such choice. The number of elements in the G-equivalence class $[A]$ is the index $[G : G_A]$, where G_A is the subgroup of all $g \in G$ such that $gA = A$. Our choice of A is such that $[G : G_A]$ is not divisible by p. Since we know

$$[G : G_A] = \frac{|G|}{|G_A|} = \frac{p^a m}{|G_A|}$$

is not divisible by p, it follows that $|G_A| = p^a k$ for some integer k that divides m. If $k < m$, then G_A is a group with order divisible by p^a and smaller than $|G|$. By induction, G_A has a subgroup of order p^a. Since any subgroup of G_A is a subgroup of G, we are done in this case. Therefore, we assume that $k = m$ and $G = G_A$. Then every element $g \in G$ satisfies $gA = A$. Pick one element $a \in A$; then every element ga is in A. If $g \neq h$, then $ga \neq ha$, so the number of elements in G equals the number of elements of the set $\{ga : g \in G\}$ and all these elements are in A, which has only $|A| = p^a$ elements. Thus, $|G| \leq p^a$. Since $|G| = p^a m$, this case can only arise if $m = 1$ and $|G| = p^a$. Thus, the theorem is also true in this case and we are finished (except for the statement about the binomial coefficient which will be proved in Section 3.1). ■

Corollary 3.1 *If G is a finite group of order $p^a m$ with p a prime number not dividing m, then G has a subgroup of order p^i for each integer i, $0 \leq i \leq a$.*

Proof: We have just proved that G contains a subgroup S of order p^a. Then S is a p-group which has subgroups of order p^i for $0 \leq i \leq a$ by Theorem 2.5. ■

Definition 3.1 If G is a group of order $p^a m$ with p a prime number not dividing m, a subgroup S of G with order p^a is called a ***p*-Sylow subgroup** (or a Sylow subgroup).

If S is a p-Sylow subgroup of G and $g \in G$, then the conjugate gSg^{-1} is a subgroup of G with the same order as S; hence, gSg^{-1} is also a p-Sylow subgroup of G. The next theorem shows that every p-Sylow subgroup of G is obtained as a conjugate of any one of them.

Before giving a proof of this, we need a computation of a fairly general nature that gives information about constructing larger subgroups from given subgroups.

Lemma 3.1 *Let P and A be subgroups of a finite group H. Assume that $gPg^{-1} = P$ for every $g \in A$. Then the set*

$$AP = \{xy : x \in A,\ y \in P\}$$

is a subgroup of H having order $[A : A \cap P]|P|$.

Proof: The set AP is the union of the cosets of P of the form aP, $a \in A$. Two cosets aP and bP are equal if an only if $a^{-1}b \in P$. Of course, $a^{-1}b$ is in A because both a and b are in A. Thus, $aP = bP$ if and only if $a^{-1}b \in A \cap P$. This is equivalent to the assertion that $a(A \cap P) = b(A \cap P)$. Therefore, the number of distinct cosets aP in G equals the number of distinct cosets

$a(A \cap P)$ in A and this number is $[A : A \cap P]$. Thus, the number of elements in AP equals $[A : A \cap P]|P|$, proving part of the lemma. Now we have left to show that AP is a subgroup. The condition $gPg^{-1} = P$ is equivalent to $gP = Pg$. It follows that

$$gPhP = ghP, \qquad g, h \in A.$$

We may interpret this as saying that the product of two elements—one from gP and one from hP—lies in ghP for all $g, h \in A$. Since every element of AP lies in a coset of P, this is the same as saying that AP is closed under multiplication. Since H is finite, AP is a subgroup. ∎

Corollary 3.2 *Let A and P be subgroups of the finite group H. Suppose that each of A and P have order a power of the prime p and in addition that $gPg^{-1} = P$ for each $g \in A$. The AP is a subgroup of H which has p power order. Moreover, if A is not contained in P, then AP has order larger than the order of P.*

Proof: We know from the previous theorem that AP is a subgroup of H having order $[A : A \cap P]|P|$. Since both $|A|$ and $|P|$ are powers of p and $[A : A \cap P]$ is a divisor of $|A|$, it follows that $|AP|$ is a power of p. The only way $|AP|$ can fail to be larger than $|P|$ is for $A = A \cap P$, which is the same as saying that $A \subseteq P$. Thus, if A is not contained in P, then AP has order larger than the order of P. ∎

Now we provide additional information about the Sylow subgroups of G.

Theorem 3.2 SECOND SYLOW THEOREM. *Let G be a finite group of order $p^a m$ with p a prime number not dividing m.*

(i) *If P and Q are two subgroups of G having order p^a, then there exists $g \in G$ with $P = gQg^{-1}$.*

(ii) *The number of subgroups of G with order p^a has the form $1 + kp$ for an integer k and $1 + kp$ is a divisor of m.*

Proof: By the first Sylow theorem we know that there is a subgroup Q of G with order p^a. Let X denote the collection of all G conjugates of Q; that is,

$$X = \{gQg^{-1} : g \in G\}.$$

Then G acts on X by conjugation and there is just one G-equivalence class. Consider the conjugation action of Q on X. Then X is the disjoint union of the Q-equivalence classes and Q is a p-group. Thus, we may apply the counting result (Theorem 2.2) to conclude $|X| \equiv f \pmod{p}$, where f is the number of fixed points of the action of Q. There is at least one fixed point, namely, Q, since $xQx^{-1} = Q$ for every $x \in Q$. We show that this is the only fixed point. Suppose P is a fixed point of the Q action. Then $xPx^{-1} = P$

for every $x \in Q$. It follows from Corollary 3 that the set QP is a subgroup of order a power of p. The largest power of p that divides $|G|$ is p^a; therefore, by Lagrange's theorem, the order of QP divides p^a. However, $P, Q \subset QP$ are subgroups of order p^a contained in QP, so the only possibility is that $P = Q = QP$. Hence, Q is only one fixed point and the number $|X|$ of subgroups of order p^a satisfies $|X| \equiv 1 \pmod{p}$; equivalently, this number equals $1 + pk$ for some integer k.

By considering the action of one Sylow subgroup Q on X, we determined that there is exactly one fixed point. Now suppose that P is another p-Sylow subgroup of G so that $|P| = p^a$. Since all of G acts on X, we may consider the action of P on X. Once again, P is a p-group and so the formula $|X| \equiv f \pmod{p}$ holds, where f now means the number of fixed point of P. Since $|X| = 1 + kp$, we conclude that $1 + kp \equiv 1 \equiv f \pmod{p}$. Hence, there is at least one fixed point under the action of P. Let $R = gQg^{-1}$ be an element of X fixed by P. By the same argument used in the previous case, RP is a p-group and $R = P = RP$. Thus, $P = gQg^{-1}$ for some $g \in G$ as we had to show to prove statement (i). All that remains in the statement is the assertion that $1 + kp$ divides m. For this we consider the action of all of G on X. The number of elements, $1 + kp$, in X equals $[G : N_G(Q)]$ since X is a G-equivalence class containing Q and $N_G(Q)$ is the subgroup of G that leaves Q fixed. Thus, $1 + kp$ is a divisor of the order of $G = p^a m$ and $1 + kp$ is relatively prime to p. Thus, $1 + kp$ must divide m. This completes the proof of all parts the theorem. ■

The two Sylow's theorems assert that a finite group of order $p^a m$, with the prime number p not dividing m, has a p-Sylow subgroup (i.e., a subgroup of order p^a) and any two such subgroups are conjugate. There also exist subgroups of order p^i for all $i \leq a$ but two of the same p power order p^i need not be conjugate, or even isomorphic if $i < a$.

For a subgroup P of a group G, the **normalizer** of P is the set

$$N_G(P) = \{g : g \in G, \text{ and } gPg^{-1} = P\}.$$

The normalizer of any subgroup P is a subgroup of G containing P. When G operates on the set of subgroups of G by conjugation, then the normalizer of P is the subgroup fixing P. The second Sylow theorem implies that if P is a p-Sylow subgroup of G, then the number of p-Sylow subgroups is the index $[G : N_G(P)]$ since all p-Sylow subgroups are G conjugate to P and $N_G(P)$ is the subgroup fixing P in this action.

Let us illustrate some of the general facts that appear in the proof of the Sylow theorem in a concrete case. Consider $G = A_4$, the alternating group of order 12. A_4 consists of all the even permutations of $\{1, 2, 3, 4\}$. Except for the identity, there are two types of even permutations: The three elements of type $(ab)(cd)$ have order 2, and eight elements of the type (abc) have order 3. The 2-Sylow subgroup must have order 4 and consists of all the elements of order 2 along with the identity. The 2-Sylow subgroup is a normal subgroup and so there is only one of these.

A 3-Sylow subgroup must have order 3; it will contain two elements of order 3 along with the identity. There are eight cycles of order 3, so there must be four subgroups of order 3:

$$\begin{aligned} P_1 &= \{e, (234), (243)\}, \\ P_2 &= \{e, (134), (143)\}, \\ P_3 &= \{e, (124), (142)\}, \\ P_4 &= \{e, (123), (132)\}. \end{aligned}$$

The total number of 3-Sylow subgroups is $4 = 1+3\cdot 1$, providing an illustration of statement (ii) of the second Sylow theorem. To illustrate that these four subgroups are conjugate, we must do some computation. Let $\sigma = (123)$, an element of P_4. According to the proof, the action of P_4 on the set of A_4 conjugates of one Sylow group should have exactly one fixed point. σ will fix P_4 and permute the other three P_i. We check that this is the case:

$$\sigma P_1 \sigma^{-1} = P_2, \quad \sigma P_2 \sigma^{-1} = P_3, \quad \sigma P_3 \sigma^{-1} = P_1.$$

Similarly, using $\tau = (134)$ we find $\tau P_2 \tau^{-1} = P_2$ and

$$\tau P_1 \tau^{-1} = P_3, \quad \tau P_3 \tau^{-1} = P_4 \quad \tau P_4 \tau^{-1} = P_1.$$

From these we may find some element in A_4 to conjugate any P_i to another P_j.

3.1 Binomial Coefficients

We provide a proof of the assertion used in the proof of Sylow's theorem that the binomial coefficient $C(p^a m, p^a)$ is not divisible by p if p is a prime number not dividing m. The idea of the proof is to determine an expression for the exact power of p that divides $n!$ for any integer n and then apply it to determine the exact power of p dividing $C(p^a m, p^a)$. The proof makes use of the *greatest integer function* $[x]$ defined for all real numbers x by the rule $[x]$ is the largest integer not exceeding x. For example, $[3.2] = 3$, $[-3.2] = -4$, and $[22/7] = 3$.

Lemma 3.2 *If n and k are positive integers, then $[n/k]$ is the number of integers between 1 and n that are divisible by k.*

Proof: Suppose $k \leq n$. The multiples of k that are less than or equal to n are $1 \cdot k, 2 \cdot k, \ldots, r \cdot k$, where r is the largest integer such that $r \cdot k \leq n$. This is equivalent to saying that r is the greatest integer less than or equal to n/k;

in other words, $r = [n/k]$ and r is the number of multiples of k which are at most equal to n. If $k > n$, then there are 0 integer multiples of k between 1 and n and $0 = [n/k]$ because $0 < n/k < 1$. ∎

Lemma 3.3 *Let p denote a prime number. For any positive integer n, the exact power of p dividing $n!$ is p^M, where*

$$M = \sum_{i=1}^{\infty} \left[\frac{n}{p^i}\right].$$

Proof: Note that the sum is really a finite sum because $[n/p^i] = 0$ for every i such that $n < p^i$.

By definition, $n! = 1 \cdot 2 \cdot 3 \cdots (n-1) \cdot n$; every multiple kp of p with $kp \leq n$ contributes a factor p to the total power of p dividing $n!$. There are $[n/p]$ such multiples which together contribute p^t, $t = [n/p]$ to the total. The term $[n/p]$ counts every multiple of p just once. There may be multiples of p less than or equal to n that are also multiples of p^2. Thus, further powers of p come from the multiples of p^2 that are at most n; there are $[n/p^2]$ of these. Continuing on in this way, there are $[n/p^3]$ multiples of p^3 that are at most n, each of which contributes an additional power of p. It should be clear that the sum $\sum_{i=1}^{\infty}[n/p^i]$ counts the exact power of p dividing $n!$. ∎

Theorem 3.3 *For a prime p, a positive integer a, and a positive integer m not divisible by p, the binomial coefficient*

$$C(p^a m, p^a) = \frac{(p^a m)!}{(p^a)!(p^a m - p^a)!}$$

is an integer not divisible by p.

Proof: The exact power of p dividing each of the factorial terms is computed using the previous lemma. The power of p in the numerator $(p^a m)!$ is p^{N_1}, where

$$N_1 = \sum_{i=1}^{\infty} \left[\frac{p^a m}{p^i}\right] = (p^{a-1} + \cdots + p + 1)m + \sum_{j=1}^{\infty} \left[\frac{m}{p^j}\right].$$

The exact power of p dividing $(p^a)!$ is p^{N_2}, where

$$N_2 = \sum_{i=1}^{\infty} \left[\frac{p^a}{p^i}\right] = (p^{a-1} + \cdots + p + 1).$$

The exact power of p dividing $(p^a m - p^a)! = \big(p^a(m-1)\big)!$ is p^{N_3}, where

$$N_3 = \sum_{i=1}^{\infty} \left[\frac{p^a(m-1)}{p^i}\right] = (p^{a-1} + \cdots + p + 1)(m-1) + \sum_{j=1}^{\infty} \left[\frac{m-1}{p^j}\right].$$

The exact power of p dividing $C(p^a m, p^a)$ is $p^{N_1-N_2-N_3}$. There are many cancelations when we simplify this exponent and we are left with

$$N_1 - N_2 - N_3 = \sum_{j=1}^{\infty} \left[\frac{m}{p^j}\right] - \sum_{j=1}^{\infty} \left[\frac{m-1}{p^j}\right].$$

The term $[m/p^j]$ counts the number of integer multiples of p^j that lie between 1 and m. Since p does not divide m, any multiple of p^j that is less than or equal to m is actually less than m. In other words, every multiple of p^j that lies on the interval from 1 to m actually lies on the interval from 1 to $m - 1$. That is,

$$\left[\frac{m}{p^j}\right] = \left[\frac{m-1}{p^j}\right].$$

It follows that $N_1 - N_2 - N_3 = 0$ and the theorem holds. ■

EXERCISES

1. List all of the p-Sylow subgroups of S_3 for $p = 2$ and $p = 3$.

2. Let G be a group of order $p^a m$ with p a prime number not dividing m. Suppose $m < p$. Prove that there is only one p-Sylow subgroup of G and it is a normal subgroup of G.

3. Let X be the set of 2-Sylow subgroups of the symmetric group S_4, and let S_4 act on X by conjugation. Show that $|X| = 3$ and so S_4 acts as a permutation group on a set of three elements. Number the elements of X as 1,2,3 so the action of S_4 may be viewed as permutations of $\{1, 2, 3\}$; thus, the action provides a homomorphism from S_4 into S_3. Determine the kernel and the image of this homomorphism.

4. Show that a 2-Sylow subgroup of S_4 has order 2^3 and contains two nonisomorphic subgroups of order 2^2. Conclude that two subgroups of order 2^2 need not be conjugate.

5. Let $G = GL(2, \mathbb{Z}_p)$ be the group of two-by-two matrices as described in the exercises following Section 1 of this chapter. Then the order of $|G| = p(p^2 - 1)(p - 1)$.

(a) Show that the cyclic group P generated by

$$g = \begin{bmatrix} 1 & 1 \\ 0 & 1 \end{bmatrix}$$

is a p-Sylow subgroup of G.

(b) Show that the normalizer of the subgroup P consists of all elements in G of the form

$$g = \begin{bmatrix} a & b \\ 0 & d \end{bmatrix}, \qquad ad \neq 0.$$

(c) Use a and b to determine the number of p-Sylow subgroups of G.

6. Let K be a normal subgroup of a finite group G. Suppose $|G|$ is divisible by the prime number p and suppose also that the index $[G : K]$ is not divisible by p. Prove that every p-Sylow subgroup of G is contained in K.

7. For which primes p is it true that every p-Sylow subgroup of S_7 is also a p-Sylow subgroup of S_8?

8. Without the use of Theorem 3.3, show that the binomial coefficient $C(2m, 2)$ is odd when m is odd.

9. Compute the exact power of p in $n!$ in each of the following cases:

(a) $p = 2$, $n = 10$.

(b) $p = 3$, $n = 15$.

(c) $p = 5$, $n = 100$.

10. Find the complete factorization $(20)!$ as a product of primes using the formulas of this section to determine the powers of each prime factor.

11. Compute the exact power of 2 and of 3 that divides the binomial coefficient $C(18, 6)$.

12. Prove that 2^M is the exact power of 2 dividing the binomial coefficient $C(2n, n)$, where $M = n - \sum_{i=1}^{\infty}[n/2^i]$ and n is any positive integer.

4 APPLICATIONS OF SYLOW'S THEOREMS

We will give a very small number of the many possible applications of Sylow's theorems. For a first application we show how the theorem may be used to prove under certain conditions that a group has only one p-Sylow subgroup and it is a normal subgroup.

If a finite group G has a p-Sylow subgroup P and g is an element of G, then gPg^{-1} is also a p-Sylow subgroup of G. If it is known in advance that G has only one p-Sylow subgroup, then $P = gPg^{-1}$ for every $g \in G$. That is, P is a normal subgroup of G. Conversely, if some p-Sylow subgroup P is a normal subgroup of G, then there is only one p-Sylow subgroup as every p-Sylow subgroup has the form gPg^{-1} and this is equal to P when P is a normal subgroup.

Consider a group of order $45 = 3^2 \cdot 5$. The number of 3-Sylow subgroups has the form $1 + 3k$ and is a divisor of 5. The only possibility is $k = 0$ and there is just one 3-Sylow subgroup. The number of 5-Sylow subgroups has the form $1 + 5k$ and is a divisor of 3^2. Again the only possibility is $k = 0$, so there is only one 5-Sylow subgroup. Thus, in a group of order 45, both the 2-Sylow subgroup and the 5-Sylow subgroup are normal subgroups. Let us show that a group of order 45 is necessarily abelian. All the elements of order a power of 3 lie in the unique 3-Sylow subgroup which has order 3^2. Because every group of order 3^2 is abelian (by Theorem 2.4), we see that any two elements of order a power of 3 must commute with each other. All elements of order 5 lie in the unique 5-Sylow subgroup which is a cyclic group of order 5 and hence is abelian. Now P_3 is a 3-Sylow subgroup and P_5 a 5-Sylow subgroup. For $x \in P_3$ and $y \in P_5$ we show $xy = yx$. The element xyx^{-1} is in $xP_5x^{-1} = P_5$ and so the element $c = (xyx^{-1})y^{-1}$ is in P_5. The element $yx^{-1}y^{-1}$ lies in $yP_3y^{-1} = P_3$ and so $c = x(yx^{-1}y^{-1})$ also lies in P_5. Thus, c lies in both P_3 and P_5. However, $P_3 \cap P_5 = \{e\}$ so that $e = c = (xyx^{-1})y^{-1}$. It follows that $xy = yx$. Hence, the group is abelian.

4.1 Groups of Order pq

We know that a group of prime order is cyclic. A group of order the square of a prime is abelian by Theorem 2.4. Next we consider groups of order equal to the product of two different primes. We will see that for certain pairs of primes, the groups are abelian, but in other cases some are not. We classify the nonabelian groups.

Theorem 4.1 *Let p and q be prime numbers with $p < q$. If G is a group of order pq, then either G is abelian or p divides $q-1$. If p divides $q-1$, there is one nonabelian group of order pq and any two nonabelian groups of order pq are isomorphic.*

Proof: Let G be a group of order pq. The number of q-Sylow subgroups of G is a number of the form $1 + qk$ and divides p, by the second Sylow theorem. Since $p < q$, the only possible choice of k is $k = 0$ because $k > 0$ implies $1 + kq > q > p$. Therefore, there is only one q-Sylow subgroup, call it Q, which is a normal subgroup of G. Now let P be a p-Sylow subgroup of G. Since $gQg^{-1} = Q$ for every $g \in G$, we may consider the action of P on the elements of Q by conjugation. We then have a p-group P acting on a group Q with q elements and Q is the disjoint union of the P-equivalence classes. One P-equivalence class consists of only the identity, e, as $xex^{-1} = e$ for $x \in P$. Suppose there is another equivalence class $[y]$ with only one element; then $xyx^{-1} = y$ for every x in P. Since Q is cyclic of prime order, every element of Q has the form y^i and we have

$$xy^ix^{-1} = (xyx^{-1})^i = y^i, \qquad x \in P.$$

Thus, every element of P commutes with every element of Q. The set PQ of all products with $x \in P$ and $y \in Q$ contains pq elements and so $PQ = G$. It follows that G is abelian. This conclusion is derived from the assumption that there is at least one nonidentity element of Q which forms a P-equivalence of one element. Now we may assume this is not the case.

Then each of the $q-1$ nonidentity elements of Q lies in a P-equivalence class with more than one element. The number of elements in a P-equivalence class is a divisor of the order of P, namely, a divisor of p. Since p is a prime, each P-equivalence class has p elements. Thus, the $q-1$ nonidentity elements of Q are partitioned into sets of p elements and so $q-1 = pt$ for some integer t. That is, p divides $q-1$.

The rest of the proof requires construction of some examples to show that the required nonabelian groups exist. Before giving the construction, we will derive some additional properties and essentially produce a multiplication table for G. Then we will show that there is only one group, up to isomorphism. Lastly, we give a construction of the group as a permutation group.

Let y be a nonidentity element of Q so that Q equals the cyclic group $\langle y \rangle$ of order q. Let x be a nonidentity element of P so that $P = \langle x \rangle$ has order p. Since Q is a normal subgroup of G, there is an integer r such that $1 < r < q$ and

$$\iota_x(y) = xyx^{-1} = y^r$$

since xyx^{-1} is one of the nonidentity elements of Q. Note that r is not 1 because G is nonabelian. We have written ι_x for the map $Q \to Q$ given by conjugation with x. We compute the powers of this map:

$$\iota_x^2(y) = \iota_x(\iota_x(y)) = \iota_x(y^r) = \iota_x(y)^r = (y^r)^r = y^{r^2}.$$

More generally, we find that

$$\iota_x^t(y) = y^{r^t} \qquad \text{or} \qquad x^t y = y^{r^t} x^t.$$

We know that $x^p = e$ and so this implies that

$$y = eye^{-1} = \iota_x^p(y) = y^{r^p}.$$

The order of y equals q so we conclude that $r^p - 1$ is divisible by q. Equivalently, we may say that $r^p \equiv 1 \pmod{q}$. This places a restriction on the integer r; there may be more than one integer r that satisfies this condition. If the group G exists, then there is at least one such integer r.

Given r, the integer elements r^i also satisfy the condition $(r^i)^p \equiv 1 \pmod{q}$ and the numbers $r, r^2, \ldots, r^{p-1}$ represent all the equivalence classes modulo q of nonidentity elements satisfying the equation. Each such integer may be obtained in the same way as r was obtained. At the start we selected x of order p; suppose we select $x_0 = x^i$ with $1 \le i < p$ as a different choice. Then x_0 still has order p and $x_0 y x_0^{-1} = y^s$ for some s. Let us compute this s:

$$x_0 y x_0^{-1} = x^i y x^{-i} = \iota_x^i(y) = y^{r^i} = y^s.$$

Thus, $s \equiv r^i \pmod{q}$.

Now let H be a nonabelian group of order pq. We want to show that G is isomorphic to H. We may select elements X and Y in H such that X has order p, Y has order q, and, because the q-Sylow subgroup must be normal in H, $XYX^{-1} = Y^s$ for some integer s that satisfies $s \not\equiv 1 \pmod{q}$ and $s^p \equiv 1 \pmod{q}$. Now we obtain an isomorphism from G to H by selecting x_0 and y in G so that $x_0 y x_0^{-1} = y^s$, as we previously showed was possible. Then map $x_0 \to X$ and $y \to Y$, or more generally,

$$x_0^i y^j \longrightarrow X^i Y^j.$$

With some effort, one may verify that this is an isomorphism from G onto H. Hence, there is only one nonabelian group, up to isomorphism, of order pq if p divides $q - 1$.

CONSTRUCTION BY PERMUTATIONS: We give a concrete construction of nonabelian groups of order pq with p and q primes and $q - 1 = pk$ for some integer k. The proof requires some facts proved earlier about finite fields.

Consider the field $\mathbb{Z}_q$ of q elements. Its multiplicative group is a cyclic group of order $q - 1$ and so it has a subgroup of order p because p divides $q-1$. A cyclic group of order p has $p-1$ generators so there are exactly $p-1$ elements $r \in \mathbb{Z}_q$ which satisfy $r \neq 1$ and $r^p = 1$. We let r stand for any one of these.

We construct a set of permutations of the set $\mathbb{Z}_q$. For $a, b \in \mathbb{Z}_q$ with $a \neq 0$, let

$$\tau_{a,b} : \mathbb{Z}_q \longrightarrow \mathbb{Z}_q, \qquad \tau_{a,b} : x \longrightarrow ax + b.$$

This is a permutation, as we now show. To see that $\tau_{a,b}$ is onto, let c be any element of $\mathbb{Z}_q$. The equation $ax + b = c$ has exactly one solution, $x = a^{-1}(c - b)$, so $\tau_{a,b}$ maps exactly one element to c, proving also that $\tau_{a,b}$ is one-to-one. Now let

$$\sigma = \tau_{r,0}, \qquad \gamma = \tau_{1,1}$$

so that for each $x \in \mathbb{Z}_q$ we have

$$\sigma(x) = rx, \qquad \gamma(x) = x + 1.$$

It follows that $\sigma^i(x) = r^i(x)$ and $\gamma^i(x) = x + i$. Thus, σ has order p because $r^p = 1$ and γ has order q. Moreover,

$$\sigma\gamma\sigma^{-1}(x) = \sigma(\gamma(r^{-1}x)) = \sigma(r^{-1}x + 1) = r(r^{-1}x + 1) = x + r.$$

It follows that $\sigma\gamma\sigma^{-1} = \gamma^r$. Thus, σ and γ are permutations that satisfy the relations as described previously. One uses this relation to show that the set of all products $\sigma^i\gamma^j$ with $0 \leq i < p$ and $0 \leq j < q$ is a group of order pq. It is nonabelian because we required that $r \not\equiv 1 \pmod{q}$ so that $\sigma\gamma \neq \gamma\sigma$. ■

EXERCISES

1. Suppose a group G has order q^2p with p and q different prime numbers and $p < q$. Suppose also that G has a cyclic q-Sylow subgroup Q. Prove that either G is abelian or p divides $q - 1$. [Hint: Show that Q is normal and then count the P-equivalence classes of Q that contain elements of order q^2 for a p-Sylow subgroup P.]
2. Let x be an element of finite order pq in some group G. Prove that $x = x_p x_q$, where x_p and x_q are elements of G such that x_p has order p, x_q has order q, and $x_p x_q = x_q x_p$. [Hint: Apply Sylow's theorem to the cyclic group generated by x.]
3. Make a list of the numbers n with $n < 30$ having the property that there might be a group of order n that is nonabelian.
4. Prove that if G is a finite group such that every p-Sylow subgroup is cyclic and normal (for every prime number p), then G is a cyclic group.
5. Let a and b be two elements in a finite group such that a has order n and b has order 2. Suppose also that $bab^{-1} = a^{-1}$. Prove that the set of elements of the form $a^i b^j$ with $0 \leq i < n$ and $0 \leq j < 2$ is a group of order $2n$. Show that the center of this group is the identity if n is odd and the center has order 2 if n is even. [This group is called the *dihedral group* of order $2n$.]
6. For a prime q, show that the permutations of the q element set $\mathbb{Z}_q$ defined by $\tau_{a,b} : x \to ax + b$, with $a, b \in \mathbb{Z}_q$ and $a \neq 0$ and all x in $\mathbb{Z}_q$ form a nonabelian group of order $q(q-1)$. Is there any connection between this exercise and Exercise 12 following Section 1?

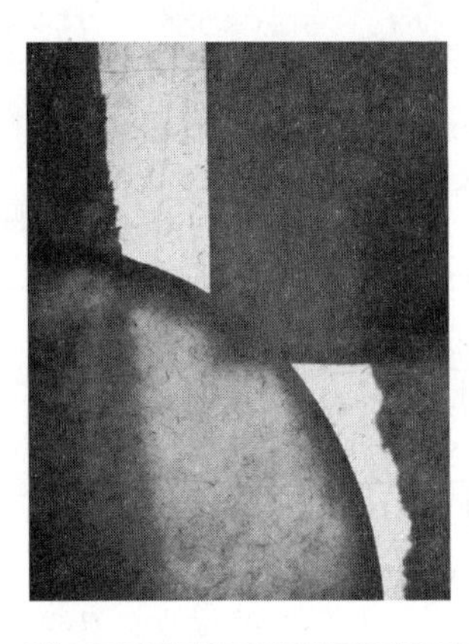

IX ADDITIONAL TOPICS ON RINGS AND FIELDS

In this chapter we discuss several topics that expand upon ideas studied previously. We have studied the factorization of elements of the ring of integers and of the ring of polynomials with coefficients in a field. In both cases we prove a unique factorization property. The rings $\mathbb{Z}$ and $F[x]$ share the common property of being integral domains in which every ideal is principal. We will show that any integral domain with every ideal principal has a unique factorization property. We also give examples to show that an integral domain may have unique factorization but does not have every ideal principal.

1 INTEGRAL DOMAINS: GENERAL PROPERTIES

Throughout this section, D denotes an integral domain; that is, D is a commutative ring with identity e, in which the product of nonzero elements is nonzero. The set of units of D is denoted by $U(D)$, or just U if D is clearly understood, and consists of all elements $u \in D$ such that $uv = e$ for some $v \in D$. The element v is written as u^{-1}. Thus, U is the multiplicative group of invertible elements of D.

Definition 1.1

If $a, b \in D$ with $b \neq 0$, we say that b is a factor of a (or that b divides a) if there is an element $c \in D$ such that $bc = a$. If b divides a, we may also say that b is a *divisor* of a and that a is a *multiple* of b, and we use the symbol $b|a$ to indicate that b divides a. The symbol $b \nmid a$ means b does not divide a.

Note that the units are divisors of every element; for any $a \in D$ and any $u \in U$ we have $a = u(u^{-1}a)$. In particular, every unit divides the identity e. Conversely, any divisor of the identity is a unit because $uv = e$ implies $v = u^{-1}$. If b is a divisor of a and if u is a unit, then bu is a divisor of a; this is almost immediate because the representation $a = bc$ implies $a = (bu)(u^{-1}c)$.

We will want to consider how an element a can be represented as a product of elements of D and we must take the unit multiples into account. We use the following terminology:

Definition 1.2 Two elements $a, b \in D$ are **associates** if there is a unit $u \in U(D)$ such that $a = bu$.

The phrase "is an associate of" defines an equivalence relation on the elements of D; a is an associate of itself because $a = ae$; if $a = bu$, then $b = au^{-1}$, so a is an associate of b implies b is an associate of a; $a = bu$ and $b = cv$ with $u, v \in U$ implies $a = c(uv)$ with $uv \in U$, so the transitive property holds.

The following is another way to determine when two elements are associates:

Lemma 1.1 *Two nonzero elements a and b of the integral domain D are associates if and only if $a|b$ and $b|a$.*

Proof: If $a|b$, then there is some $x \in D$ such that $ax = b$. Similarly, if $b|a$, then $a = by$ for some $y \in D$. When both of these hold, we may combine the equations to conclude $a = axy$ and $a(e - xy) = 0$. Since a is nonzero, it follows that $xy = e$, so x and y are units and a and b are associates. ∎

In the case $D = \mathbb{Z}$, the ring of integers, the units are ± 1 and the only associates of a nonzero integer m are the two integers $\pm m$. In the case of $D = F[x]$ with F a field, the units of D are the nonzero elements of F, that is, the polynomials of degree zero. If $f(x)$ is a nonzero polynomial, then the associates of $f(x)$ are the polynomials $cf(x)$ with $c \in F$ and $c \neq 0$.

Next we deal with the analogs in a general integral domain of prime numbers and irreducible polynomials.

Definition 1.3 An element a of an integral domain D is **irreducible** if it is not a unit and its only divisors are the units and the associates of a.

Definition 1.4 An element p of an integral domain D is **prime** if it has the following property: Whenever $a, b \in D$ and $p|(ab)$ then either $p|a$ or $p|b$.

When we discussed $\mathbb{Z}$ and $F[x]$, F a field, we saw that the notions of "prime" and "irreducible" were equivalent. That is, an element of $\mathbb{Z}$ or $F[x]$ is prime if and only if it is irreducible. This statement is not true if either $\mathbb{Z}$ or $F[x]$ is replaced by an arbitrary integral domain. There is a relationship that holds in general which we now prove.

Lemma 1.2 *A prime element of an integral domain is irreducible.*

Proof: Let p be a prime element of the integral domain D; we must show that it is irreducible. Therefore, let x be a divisor of p. Then there is some $y \in D$ such that $p = xy$. Since $p|(xy)$ and p is prime, it follows that either $p|x$

or $p|y$. If $p|x$, then we also have $x|p$, so p and x are associates by Lemma 1.1. If $p|y$, then $y = pt$ for some $t \in D$ and $p = xy = x(pt)$. We conclude that $e = xt$, so x and t are units. The divisor x has been shown to be either a unit or an associate of p. ■

We now give an example to show that there exist integral domains in which an irreducible element need not be prime. The example to follow has many other features to which we will refer later.

1.1 An Example of Nonunique Factorization

Let $D = \{a + b\sqrt{-5} : a, b \in \mathbb{Z}\}$ so that D is a subring of the complex field. It contains $\mathbb{Z}$ as a subring. If $x = a + b\sqrt{-5} \in D$, then its complex conjugate $\bar{x} = a - b\sqrt{-5}$ is also in D. Note that $x \to \bar{x}$ is an isomorphism of D with itself so that, in particular, $\overline{(xy)} = \bar{x}\bar{y}$ since this relation holds for any pair of complex numbers. The study of factorization in D is facilitated by using the *norm map*, $N(x)$, defined by

$$N(x) = x\bar{x}, \qquad x \in D.$$

Note that $x = a + b\sqrt{-5}$, $a, b \in Z$, implies

$$N(x) = (a + b\sqrt{-5})(a - b\sqrt{-5}) = a^2 + 5b^2.$$

Therefore, $N(x)$ is an integer when $x \in D$.

Let us determine the units of D. Suppose $a + b\sqrt{-5}$ is a unit of D; then there is some element $c + d\sqrt{-5} \in D$ such that

$$(a + b\sqrt{-5})(c + d\sqrt{-5}) = 1.$$

Take norms of each side to get

$$(a^2 + 5b^2)(c^2 + 5d^2) = N(1) = 1.$$

Since $a^2 + 5b^2$ and $c^2 + 5d^2$ are nonnegative integers, we must have $a^2 + 5b^2 = 1 = c^2 + 5d^2$. The only solutions to this equation are $b = 0$ and $a = \pm 1$ and $d = 0$ and $c = \pm 1$. If $a + b\sqrt{-5}$ is a unit, then $a + b\sqrt{-5} = \pm 1$. As an afterthought, we have shown that the units are the elements of norm 1.

Now we illustrate how one can show that certain elements are irreducible. The unique factorization of integers is used, along with the norm map from D to $\mathbb{Z}$, to draw conclusions. Let us prove that 2 is irreducible in D. Suppose on the contrary that $2 = (a + b\sqrt{-5})(c + d\sqrt{-5})$. Then take norms and use the fact that $N(2) = 4$ to conclude

$$4 = (a^2 + 5b^2)(c^2 + 5d^2).$$

We first argue that 2 cannot be the norm of an element of D. If $2 = x^2 + 5y^2$, then $y = 0$ and 2 is a square, which is an impossibility. Now 4 is written as the product of two norms, neither of which can be 2. Hence, one norm is 4 and the other is 1. The element of norm 1 is a unit and so one of the two factors of 2 is a unit; thus, 2 is irreducible. Similarly, there is no element of norm 3.

It might be tempting to guess that an irreducible element of $\mathbb{Z}$ is also an irreducible element of D. This is not true. The number 41 is a prime and so is irreducible in $\mathbb{Z}$. However, it is not irreducible in D in view of the factorization

$$41 = (6 + \sqrt{-5})(6 - \sqrt{-5}).$$

The situation is even more complicated. The following are factorizations of the number 6:

$$6 = 2 \cdot 3, \qquad 6 = (1 + \sqrt{-5})(1 - \sqrt{-5}).$$

It is clear that 2 does not divide either factor on the right because $\frac{1}{2} \pm \frac{1}{2}\sqrt{-5}$ is not an element of D. Furthermore $1 + \sqrt{-5}$ does not divide 2 because 2 is irreducible, as we have previously proved. Note that the element $1 + \sqrt{-5}$ is irreducible. We see this because its norm is 6 and if it were not irreducible there would be an element in D with norm 2. Thus, we have two different factorizations of 6 as a product of irreducibles, but the irreducibles in one factorization are not associates of the irreducibles in the second factorization. This is an example in which unique factorization fails.

Note also that none of the four elements $2, 3, 1+\sqrt{-5}, 1-\sqrt{-5}$ is prime. For example, 2 is not prime because 2 divides $(1 + \sqrt{-5})(1 - \sqrt{-5}) = 6$, but 2 does not divide either factor. Similar observations apply to show that the other three elements are not prime.

The factorization properties of this ring $\mathbb{Z}[\sqrt{-5}]$ are significantly different from those in $\mathbb{Z}$ or $F[x]$ (with F a field) where every nonzero element is a product of irreducible elements in a way that is unique, up to associates.

The following is a definition that incorporates the factorization properties that we hope to prove in certain situations:

Definition 1.5

An integral domain D is a **unique factorization domain** (UFD) if it has the following properties:

(i) Every nonzero element $a \in D$ is a unit, an irreducible element, or can be expressed as a product of a finite number of irreducible elements of D.

(ii) If $a = p_1 p_2 \cdots p_r$ and $a = q_1 q_2 \cdots q_s$ with all the p_i and q_j irreducible elements of D, then $r = s$ and, after a possible renumbering, p_i and q_i are associates for each i.

We have previously seen that $\mathbb{Z}$ and $F[x]$ are UFDs. We will show that both of these facts may be considered a special case of a more general result, to which we now turn.

2 PRINCIPAL IDEAL DOMAINS

Recall that in a commutative ring R with identity, an ideal of the form

$$aR = \{ar : r \in R\}$$

is called the **principal ideal** generated by a. For the ring of integers and for the polynomial ring over a field, every ideal is a principal ideal. This is not always true in a commutative ring. We will see later that the integral domain $\mathbb{Z}[\sqrt{-5}]$ has ideals that are not principal. We will discover that there is a connection between factorization of elements and principal ideals. We make a definition of the rings we wish to study:

Definition 2.1 An integral domain D is called a **principal ideal domain** (PID) if every ideal of D is principal.

Thus, we can say that $\mathbb{Z}$ and $F[x]$ are PIDs. The rest of this section will be devoted to proving the following important result:

Theorem 2.1 *A principal ideal domain is a unique factorization domain.*

In view of this result, the unique factorization in $\mathbb{Z}$ and $F[x]$ may both be considered a consequence of the fact that they are PIDs. The proof of this theorem will be given in a series of lemmas. The first of these is true even if D is not a PID.

Lemma 2.1 *If a and b are nonzero elements of the integral domain D, then $aD = bD$ if and only if a and b are associates.*

Proof: Since $a \in aD = bD$, there is some $x \in D$ such that $a = bx$. Similarly, there is a $y \in D$ such that $b = ay$. Thus, $a = bx = ayx$ and it follows that x and y are units. ∎

The following lemma plays an important part in the proof, although at first it seems to have nothing to do with factorization:

Lemma 2.2 *In a principal ideal domain D, an increasing chain of ideals*

$$a_1D \subseteq a_2D \subseteq a_3D \subseteq \cdots$$

has all its terms equal beyond some point; that is, for some index k, $a_kD = a_jD$ for all $j > k$.

Proof: Suppose that ideals are given such that

$$a_1D \subseteq a_2D \subseteq a_3D \subseteq \cdots.$$

We will show that one member of this chain contains all the others so the inclusions become equality beyond some point. Let A denote the union of all the ideals $a_j D$, $A = \cup_j a_j D$. We first show that A is an ideal. For any $a, b \in A$ we have $a \in a_i D$ and $b \in a_j D$ for some i and j. Suppose that $i \leq j$ so then $a_i D \subseteq a_j D$. Thus, both a and b are in $a_j D$ and thus the sum $a + b$ and difference $a - b$ are in $a_j D$ and therefore also in A. Similarly, $ar \in A$ for any $a \in A$ and $r \in D$. It follows that A is an ideal of D.

Since D is a PID, the ideal A has the form $A = xD$ for some $x \in A$. Since A is the union of the sets $a_i D$ and since $x \in A$, it follows that x is in some $a_k D$. For any index j greater than k we have $a_k D \subseteq a_j D$, but also for every index i we have

$$a_i D \subseteq A = xD \subseteq a_k D.$$

It follows that all of the ideals $a_j D$ for $j \geq k$ are equal and so the chain of ideals has only a finite number of different terms. ■

Now we discuss why this lemma is relevant to a discussion of factorization. If $aD \subseteq bD$, then $a = bx$ for some x in D. Moreover, if $aD \neq bD$, then the element x is not a unit. If $a_1 D \subset a_2 D \subset a_3 D$, then $a_1 = a_2 x$ with x not a unit and $a_2 = a_3 y$ with y not a unit and then $a_1 = a_3 yx$. Thus, chains of ideals produce factorizations of elements.

If it is true in a given ring that there cannot exist an infinite sequence of ideals such that each one is properly contained in the following one, we express this fact by saying that the **ascending chain condition** holds. The previous lemma shows that the ascending chain condition holds in a PID.

There is another equivalent condition that may sometimes be useful. Instead of dealing with an ascending chain of ideals, consider an arbitrary, but nonempty, set $\mathcal{S}$ of ideals. We say that A is a *maximal element* of $\mathcal{S}$ if $A \in \mathcal{S}$ and there is no ideal $B \in \mathcal{S}$ such that $A \subset B$ and $A \neq B$. In other words, A is a maximal element of $\mathcal{S}$ if it is a member of $\mathcal{S}$ but A is not contained in any larger ideal that is in the set $\mathcal{S}$. A commutative ring is said to satisfy the **maximum condition** if every nonempty set of ideals has a maximal element. The next result shows that one may use either the ascending chain condition or the maximum condition since they are equivalent.

Lemma 2.3 *Let R be a commutative ring. Then every nonempty set of ideals of R has a maximal element if and only if R satisfies the ascending chain condition.*

Proof: Assume the maximum condition holds and let there be given an ascending chain

$$A_1 \subseteq A_2 \subseteq A_3 \subseteq \cdots$$

of ideals A_i of R. Let $\mathcal{S} = \{A_i : i = 1, 2, 3 \cdots\}$. By the maximum condition, the set $\mathcal{S}$ must have a maximal element, call it B. Then B is one of the

A_j, by definition of $\mathcal{S}$. Suppose $B = A_k$. The definition of a maximal element implies no element of $\mathcal{S}$ properly contains B, yet by assumption $B = A_k \subseteq A_{k+t}$ for every $t \geq 0$. The only possibility consistent with these two statements is that $B = A_{k+t}$ for all $t \geq 0$ and so the chain has all terms equal to B after the kth term. Hence, the chain has only a finite number of distinct terms.

Now assume that the ascending chain condition holds for R and let $\mathcal{S}$ be a nonempty set of ideals of R. We must show that $\mathcal{S}$ contains a maximal element. Select any element A_1 in $\mathcal{S}$. If A_1 is a maximal element, we are done. Assume it is not maximal. Then there must be an element A_2 in $\mathcal{S}$ that properly contains A_1; that is, $A_1 \subset A_2$. If A_2 is a maximal element of $\mathcal{S}$, we are done. If not, then there is an $A_3 \in \mathcal{S}$ with

$$A_1 \subset A_2 \subset A_3.$$

If we continue this line of reasoning but never reach a maximal element of $\mathcal{S}$, then we produce an infinite ascending chain of ideals of R. Since no such chain exists by the ascending chain condition, it must happen that eventually we reach a maximal element of $\mathcal{S}$. ∎

The following is another idea that is useful:

Definition 2.2

An ideal M of a commutative ring R is a **maximal ideal** if $M \neq R$ and the only ideals N that satisfy $M \subseteq N \subseteq R$ are $N = M$ and $N = R$.

Thus, an ideal is a maximal ideal if it is not equal to R but is not contained in any ideal larger than itself, except R. In the case of a PID, a maximal ideal is one generated by an irreducible element. If $\mathcal{S}$ is the set of all ideals of R that do not equal R, then a maximal ideal is a maximal element of $\mathcal{S}$.

Now we may prove an important step of our main result.

Lemma 2.4

If a is a nonzero element of the PID D and a is not a unit, then $a = p_1 p_2 \cdots p_r$ for some $r \geq 1$ and irreducible elements $p_1, p_2, \ldots, p_r$ in D.

Proof: We give an argument by contradiction. Suppose that D contains some nonzero element a that is not a unit, is not irreducible, and is not equal to a product of irreducible elements. Let $\mathcal{S}$ be the collection of principal ideals aD with a such an element. Therefore, our assumption is that $\mathcal{S}$ is not the empty set. Since D is a PID, the maximum condition holds and $\mathcal{S}$ has a maximal element $A = bD$, where b is a nonzero element that is neither a unit nor an irreducible and it does not equal the product of irreducible elements. Since b is not irreducible it must have a factorization $b = xy$, where neither x nor y is a unit of D (by the definition of an irreducible element). Clearly, $bD \subseteq xD$ and $bD \subseteq yD$. Since neither x nor y is an associate of b we must have strict inclusions; that is, $bD \neq xD$ and $bD \neq yD$. In view of our choice of bD as a maximal element of $\mathcal{S}$, neither xD nor yD can be elements of $\mathcal{S}$.

Since x is a nonzero element but not a unit, it must either be irreducible or equal to a product of irreducible elements. The same is true of y. Both x and y are products of irreducible elements and so $b = xy$ also equals a product of irreducible elements. This is in conflict with the assumption that b was not such an element. Hence, there are no such elements and S is the empty set after all. ∎

We now know that every nonzero, nonunit of a PID is a product of irreducible elements. Next we work toward showing the uniqueness of such factorizations.

2.1 Greatest Common Divisors

We give the definition of a greatest common divisor (GCD) for a general PID. The language is precisely the same as the definition used for GCDs in $\mathbb{Z}$ or $F[x]$.

Definition 2.3 Let a, b be nonzero elements of a PID D. An element d of D is a **greatest common divisor** of a and b

(i) if $d|a$ and $d|b$ (so d is a common divisor of a and b);

(ii) if $c \in D$ and $c|a$ and $c|b$, then $c|d$ (d is divisible by every common divisor).

If d is a GCD of a and b, then any associate of d is also a GCD of a and b. In the case of the ring of integers, the only associates of d are $\pm d$ so we made a conventional choice that the GCD should always be positive, hence making the GCD unique. Similarly, for polynomials with coefficients in a field, we defined a unique GCD by requiring that it be a monic polynomial. However, for a general PID, there may be no obvious way to make a unique choice from the possible associates of a GCD, so we simply allow any of the associates of one GCD to be used. Just as in the two earlier cases, one must prove that the GCD of two elements exists. This will be seen as a consequence of the assumption that every ideal is principal.

Lemma 2.5 *Let a and b be two nonzero elements of the PID D and let $A = \{ax + by : x, y \in D\}$. Then A is an ideal of D and any element d such that $A = dD$ is a GCD of a and b. Moreover, there exist elements $u, v \in D$ such that $d = au + bv$.*

Proof: To see that A is an ideal we simply observe that A is the sum of the two ideals aD and bD, which is an ideal. If d is a generator for A, then $a \in dD$ and $b \in dD$ since a and b are in A. Thus, $d|a$ and $d|b$ so that d is a common divisor of a and b. Now let c be a common divisor of a and b. Since every element of A has the form $ax + by$ with $x, y \in D$ there must exist elements $u, v \in D$ with $d = au + bv$. Thus, the common divisor c of a and b

must also divide d, proving that d is a GCD of a and b. All parts of the lemma hold. ■

Now we can finally show the equivalence of prime and irreducible.

Lemma 2.6

In a PID, an irreducible element is a prime element.

Proof: Let c be an irreducible element such that c divides a product ab of nonzero elements $a, b \in D$. We must prove that either $c|a$ or $c|b$ (definition of prime). Suppose that c does not divide a. Then the elements a and c have a GCD which cannot be an associate of c because c does not divide a. The only factors of c are associates of c and units so the GCD of a and c is a unit. The principal ideal generated by a unit is D itself. In view of the preceding lemma we see that

$$D = \{ax + cy : x, y \in D\},$$

and, in particular, if e is the identity element of D then $e = ax_1 + cy_1$ for some $x_1, y_1 \in D$. Multiply this by b to obtain

$$(ab)x_1 + c(by_1) = b.$$

It is given that $c|(ab)$ and, of course, $c|c$, so that c divides the left side of this equation, and we conclude that $c|b$. Thus, c is prime. ■

This provides the final tool for the proof of our main theorem. We show that a nonzero, nonunit element a is a product of irreducible elements that are unique up to associates. Let

$$a = p_1 p_2 \cdots p_r = q_1 q_2 \cdots q_t$$

be two factorizations of the element a in a PID in which the elements p_i and q_j are all irreducible. Such a factorization exists by Lemma 2.4. Because an irreducible element is prime and because p_1 divides a, it follows that p_1 must divide the product $q_1 q_2 \cdots q_t$ and hence one of the q_j. Renumbering the qs if necessary, we may assume $p_1|q_1$. However, q_1 is irreducible so $q_1 = p_1 u$ for some unit u and thus p_1 and q_1 are associates. Cancel the factor p_1 from both sides and repeat the argument with p_2, then with p_3, and so on. We get that each p_i is an associate of q_i for $1 \leq i \leq r$. After canceling, the left side is e and the right side is a product of irreducibles starting at q_{r+1}. This is impossible so, in fact, $t = r$ and the uniqueness, up to associates, is established. We have proved that a PID is a unique factorization domain; in other words, a PID is a UFD.

This is not the complete story on unique factorization. There are UFDs that are not PIDs. We do not give a proof here, but the following is a general theorem that gives many examples of unique factorization domains:

Theorem 2.2 *If R is a unique factorization domain, then the polynomial ring $R[x]$ is also a unique factorization domain.*

Thus, the ring of polynomials $\mathbb{Z}[x]$ with integer coefficients is a UFD. An outline of the proof that $\mathbb{Z}[x]$ is a UFD is given in the exercises. It is not a PID because the ideal $A = 2\mathbb{Z}[x] + x\mathbb{Z}[x]$ is not principal, as we now show. Suppose d is a generator of A; then d is a common divisor of 2 and x that lies in $\mathbb{Z}[x]$. The divisors of 2 in $\mathbb{Z}[x]$ are the same as the divisors of 2 in $\mathbb{Z}$, namely, $\pm 1, \pm 2$. (Use a degree argument for this step.) Neither ± 2 divides x so the only possibility remaining is that $d = \pm 1$. If this holds then $1 = 2f(x) + xg(x)$ for some polynomials $f(x), g(x) \in \mathbb{Z}[x]$. Evaluate at $x = 0$ to conclude $1 = 2f(0)$. Since $f(x) \in \mathbb{Z}[x]$, $f(0)$ must be an integer and the equation $1 = 2f(0)$ is impossible. Thus, not all UFDs are PIDs.

We previously studied the ring $D = \mathbb{Z}[\sqrt{-5}]$ and saw that factorization was not unique. Thus, D cannot be a PID. To find an ideal that is not principal we consider two nonassociate factors of 6 and the ideal they generate. We had

$$6 = 2 \cdot 3 = (1 + \sqrt{-5})(1 - \sqrt{-5}),$$

where 2,3 and $1 \pm \sqrt{-5}$ are all irreducible and no two of them are associates. Consider the ideal

$$A = 2D + (1 + \sqrt{-5})D.$$

We argue that $A \neq dD$ for any $d \in D$. Suppose the contrary so that there exist such an element d; then there exist $a, b \in D$ such that $da = 2$ and $db = 1 + \sqrt{-5}$. We know 2 and $1 + \sqrt{-5}$ are irreducible elements. There are only two cases to consider. First, suppose that d is not a unit. Then both a and b are units and we conclude that 2 is an associate of d and $1 + \sqrt{-5}$ is an associate of d. It follows that 2 is an associate of $1 + \sqrt{-5}$, which we already knew was not the case. The only remaining possibility is that d is a unit. In this case, $dD = D$ so there exist $u, v \in D$ such that $1 = 2u + (1 + \sqrt{-5})v$. Multiply by $1 - \sqrt{-5}$ to get

$$(1 - \sqrt{-5}) = 2(1 - \sqrt{-5})u + 6v.$$

The right side is divisible by 2, whereas the left side is not divisible by 2. This case cannot occur either. The conclusion is that A is not a principal ideal.

EXERCISES

1. In $D = \mathbb{Z}[\sqrt{-5}]$, verify that $21 = 3 \cdot 7 = (1 + 2\sqrt{-5})(1 - 2\sqrt{-5})$ and that these are two different factorizations of 21 as a product of irreducible elements. Verify that the ideal $B = 3D + (1 + 2\sqrt{-5})D$ is not a principal ideal.

2. Prove that $\mathbb{Z}[x]$ is a unique factorization domain. The following is an outline of how this may be carried out:

(a) For $f(x) \in \mathbb{Z}[x]$ with $f(x) \neq 0$, let $c(f) =$ the GCD of the nonzero coefficients of $f(x)$. Then every nonzero polynomial has the factorization $f(x) = c(f) \cdot g(x)$ with $g(x) \in \mathbb{Z}[x]$ and $c(g) = 1$.

(b) The irreducible elements of $\mathbb{Z}[x]$ are the prime integers (irreducible elements of $\mathbb{Z}$) and the polynomials $f(x)$ in $\mathbb{Z}[x]$ which have $c(f) = 1$ and are irreducible polynomials when viewed as elements of $\mathbb{Q}[x]$.

(c) Complete the proof using unique factorization in $\mathbb{Q}[x]$ and Gauss's lemma (see Chapter V, Section 7.1).

3. An ideal aD of a PID D is a maximal ideal if and only if a is an irreducible element.

4. Let a be an element of a PID D such that $a = p_1 p_2 \cdots p_t$ is a factorization of a as a product of irreducible elements p_i. What are all the maximal ideals that contain a?

5. Let D be an integral domain whose only ideals are (0) and D. Show that D is a field.

6. Let M be a maximal ideal of a commutative ring R and let $a, b \in R$. Prove that $ab \in M$ if and only if either $a \in M$ or $b \in M$. [An ideal with this property is called a *prime ideal*. This exercise shows that maximal ideals are prime ideals.]

7. Let R be a commutative ring and M a maximal ideal of R. Prove that the factor ring R/M is a field. Prove conversely that if A is an ideal such that R/A is a field, then A is a maximal ideal.

8. Let p be a prime number and D the subset of the rational numbers consisting of all elements a/b, where $a, b \in \mathbb{Z}$ and $p \nmid b$. Prove that D is a subring of $\mathbb{Q}$ and has the following properties:

(a) The units of D are the elements a/b with $a, b \in \mathbb{Z}$ and $p \nmid a$.

(b) p is not a unit of D and the principal ideal pD is a maximal ideal. [Hint: All elements outside of pD are units.]

(c) The only nonzero ideals of D are the principal ideals $p^t D$ with $t \geq 0$.

(d) Conclude that D is a PID with only one maximal ideal.

9. For any positive integer n show that there is a PID that has exactly n different maximal ideals. [Hint: Let D_p be the ring constructed in the previous exercise. Consider the intersection of the rings D_{p_i} for n distinct prime numbers $p_1, \ldots, p_n$.]

10. Define the *least common multiple* (LCM) of two nonzero elements a, b in an integral domain D to be an element m with the following properties: (i) $a|m$ and $b|m$ and (ii) if x is any element such that $a|x$ and $b|x$, then also $m|x$. Show that the LCM of a and b exists and equals m if and only if the intersection $aD \cap bD$ is a principal ideal equal to mD.

11. Let D be a PID and A an ideal of D. Show that every ideal of the ring D/A is principal. Show by example that D/A need not be an integral domain. If D/A is an integral domain, what conclusion can you draw about a generator of A?

3 EUCLIDEAN DOMAINS

We have proved that a PID is a UFD. This is a very general theorem since the class of PIDs is broad. There may be a problem recognizing a PID when one makes an appearance. In the cases of the integers and polynomial rings over a field, we had a division algorithm that was instrumental in verifying that the ring was a PID and a UFD. Here, we extract the idea of a division algorithm and apply it to some more general situations.

Definition 3.1

An integral domain D is called a **Euclidean domain** if there is a function ϕ defined on the set of nonzero elements of D with values in the set of nonnegative integers such that the following property holds: If a and b are elements of D with $b \neq 0$, then there exists elements q and r in D such that $a = bq + r$ and either $r = 0$ or $\phi(r) < \phi(b)$.

Thus, a Euclidean domain has a division algorithm; the elements q and r are the quotient and remainder after division of a by b.

The function ϕ in the definition is called a *Euclidean function.* For the ring $\mathbb{Z}$ of integers, the function $\phi(m) = |m|$ is a Euclidean function. For the polynomial ring $F[x]$ with coefficients in a field F, the function $\phi(f(x)) = \deg(f(x))$ is a Euclidean function. We now show how the Euclidean function is used to show that the ring is a PID.

Theorem 3.1

A Euclidean domain is a principal ideal domain.

Proof: Let D be a Euclidean domain with ϕ its Euclidean function. Let A be an ideal of D. If $A = \{0\}$, then $A = 0D$ is principal. Therefore, now assume that A is not the ideal consisting of only the zero element. Then there are nonzero elements in A and the set of values $\{\phi(a) : a \in A, a \neq 0\}$ is a set of nonnegative integers which must have a smallest element. Select $a \in A$ such that $a \neq 0$ and $\phi(a) \leq \phi(b)$ for every $b \in A$ with $b \neq 0$. We will show that $A = aD$. Take any $x \in A$ and apply the condition in the definition of Euclidean domain to write

$$x = aq + r, \qquad q, r \in D,$$

where either $r = 0$ or $\phi(r) < \phi(a)$. If $r = 0$ then $x = aq$, so $x \in aD$. If $r \neq 0$, then $r = x - aq \in A$ because both x and a are elements of A. Thus, $r \in A$ but $\phi(r)$ is less than the smallest value $\phi(a)$ of ϕ at nonzero elements of A. This is an impossible situation, so $x = aq$ is the only possibility and $A = aD$. ■

We will apply this idea to the ring of **Gaussian integers** $G = \mathbb{Z}[i]$, where i is the complex number satisfying $i^2 = -1$. The elements of G are all the $a + bi$ with $a, b \in \mathbb{Z}$. We prove the following:

Theorem 3.2 *The function $\phi(a + ib) = a^2 + b^2$ is a Euclidean function for the ring G of Gaussian integers. Moreover, G is a PID.*

Proof: Let $x = a + bi$ and $y = u + vi$ be nonzero elements of G. We wish to show that $x = yq + r$ for some q and r in G with $\phi(r) < \phi(y)$. We begin by working in the field $\mathbb{Q}[i]$ and carry out the division in the form

$$\frac{a + bi}{u + vi} = \frac{(au + bv) + (bu - av)i}{u^2 + v^2}.$$

Any rational number p/q falls between two integers, $m \leq p/q \leq m + 1$, so that one of the distances $p/q - m$ or $m + 1 - p/q$ is at most $1/2$. Thus, it is possible to write any rational number in the form $p/q = m + h$ with m an integer and h a rational number satisfying $|h| \leq 1/2$. We apply this to the previous quotient to obtain

$$\frac{(au + bv) + (bu - av)i}{u^2 + v^2} = \frac{qu + bv}{u^2 + v^2} + \frac{bu - av}{u^2 + v^2}i = (m + ni) + (h + ki),$$

where m, n are integers and h, k are rational numbers with $|h| \leq 1/2$ and $|k| \leq 1/2$. Now clear the fractions to get

$$a + bi = (u + vi)(m + ni) + r, \quad r = (u + vi)(h + ki).$$

The element r must lie in G because $r = (a + bi) - (u + vi)(m + ni)$, which is in G. In order to compute $\phi(r)$ it is simplest to observe that for any element in $x = s + ti \in \mathbb{Q}[i]$ we have $x\bar{x} = s^2 + t^2$, where $\bar{x} = s - ti$ is the complex conjugate of x. Therefore, in particular we use this to show that $\phi(xy) = \phi(x)\phi(y)$ for any two elements of $\mathbb{Q}[i]$. Then we observe that

$$\phi(r) = \phi(u + vi)\phi(h + ki) = \phi(u + vi)(h^2 + k^2).$$

Since $|h|$ and $|k|$ are no more than $1/2$, $h^2 + k^2 \leq 1/4 + 1/4 = 1/2$. Thus, $\phi(r) \leq \phi(u + vi)/2 < \phi(u + vi)$. This shows that ϕ is a Euclidean function for G. Then we conclude G is a PID by Theorem 3.1. ■

There is a feature of this example of a Euclidean domain that did not appear in earlier examples. For $\mathbb{Z}$ and $F[x]$, we had a division algorithm for which we were able to show that the quotient and remainder were unique. This is not the case for Euclidean domains in general. We give an example using G. Take $a = 3 + 5i$ and $b = 2$. Then $\phi(2) = 4$, so we carry out a division of a by b and look for a remainder r with $\phi(r) < 4$. We have two equations:

$$\begin{aligned} a = 3 + 5i &= 2(2 + 2i) + (-1 + i) \\ &= 2(1 + 2i) + (1 + i). \end{aligned}$$

Note that $\phi(-1 + i) = \phi(1 + i) = 2 < 4$ so the conditions required of the division are satisfied by two quotients and remainders: $(q, r) = (2+2i, -1+i)$ and $(1 + 2i, 1 + i)$.

EXERCISES

1. For each pair $a, b \in G$, the ring of Gaussian integers, find the quotient and remainder after division of a by b:

a	$4 + 9i$	$4 + 9i$
b	3	$2 + i$

2. (An illustration to show that the Euclidean function is not unique.) Let G be the ring of Gaussian integers and let d be the Euclidean function defined for G by $d(a + bi) = a^2 + b^2$ for $a, b \in \mathbb{Z}$. Let δ be the function defined on G by the rule

$$\delta(x) = \begin{cases} d(x) & \text{if } x \neq 2; \\ 3 & \text{if } x = 2. \end{cases}$$

Show that δ is also a Euclidean function for G. [Hint: A value of d must be the sum of two square integers so the first few values are $0, 1, 2, 4, 5, 8, \ldots$. If some value of d is less than 4, it is also less than 3.]

3. Let R be a Euclidean domain with Euclidean function d. Show that the function $\delta(x) = d(x) + 1$ is also a Euclidean function for R.

4. Let R be a Euclidean domain with Euclidean function d. Show that the function θ defined by

$$\theta(x) = \min\{d(xy) : y \in R, \text{and } y \neq 0\}$$

is also a Euclidean function for R and that θ has the following properties:

(a) $\theta(a) \leq \theta(ab)$ for all nonzero $a, b \in R$;

(b) $\theta(u) = \theta(v)$ for all units u and v of R.

[Remark: Condition (a) is often included in the definition of a Euclidean function but, as this exercise shows, it is redundant to do so.]

4 THE FUNDAMENTAL THEOREM OF ALGEBRA

In this section we give a proof that every irreducible polynomial over the field of complex numbers has degree 1. This theorem, sometimes called the *fundamental theorem of algebra*, was first proved by Carl Friedrich Gauss (1777–1855). The level of this proof is considerably higher than that of most of the earlier material in this text. There is no elementary proof of the theorem. Perhaps, upon reflection, one should not expect an elementary proof. An equivalent statement of the theorem is that every polynomial with degree at least 1 over the complex field has a root in the complex field. This is surely not a true statement over the field of real numbers because the polynomial $x^2 + 1$ has no root in $\mathbb{R}$. Thus, the theorem depends on some subtle properties of the complex field that are not shared by all fields, in particular, some properties not shared by the field of real numbers. The proof we present makes use of properties of functions of two real variables that should be familiar to a student who has completed a course in calculus. Before stating this result, we provide is a definition that is needed.

Definition 4.1

A **real polynomial function** of the variables u and v is a function of the form

$$F(u, v) = a_{00} + a_{10}u + a_{01}v + \cdots + a_{ij}u^i v^j + \cdots + a_{mn}u^m v^n, \quad (9.1)$$

where the coefficients a_{ij} are real numbers.

For real polynomial functions we require the analog for functions of two variables of the familiar property of polynomial functions (or even continuous functions) of one variable: If $f(x)$ is a polynomial function with real coefficients, and $[a, b]$ is any closed interval on the real axis, then there is a point c on the interval at which $f(c) \leq f(r)$ for all r on the interval; that is, f takes on its minimum value on a closed interval. For two variable functions, the interval $[a, b]$ is replaced by a disc $C(R)$, where

$$C(R) = \{(x, y) : x, y \in \mathbb{R}, \text{ and } x^2 + y^2 \leq R^2\}.$$

We require the following property:

Theorem 4.1

If $F(u, v)$ is a real polynomial function, and if $C(R)$ is the disc centered at the origin having radius R, then there is a point (u_0, v_0) in $C(R)$ such that $F(u_0, v_0) \leq F(u, v)$ holds for all $(u, v) \in C(R)$.

To paraphrase this theorem, we could say that the set of values, $F(u, v)$ for $(u, v) \in C(R)$, has a greatest lower bound which is equal to one of those values.

This idea is applied to study polynomials with complex coefficients as follows. Recall that the absolute value of a complex number $z = a + bi$, with $a, b \in \mathbb{R}$, is defined as

$$|z| = |a + bi| = \sqrt{a^2 + b^2}$$

and the conjugate of z is $\bar{z} = a - bi$. Thus, $|z|^2 = z\bar{z}$.

Now suppose we are given a polynomial $f(x)$ of positive degree and having coefficients in $\mathbb{C}$. We obtain a real polynomial of two variables u and v in the form $F(u, v) = |f(u + vi)|^2$. We state this more formally. First, we introduce some common terminology. For any complex number $\zeta = a + bi$, with $a, b \in \mathbb{R}$, we call a the *real part* of ζ and b the *imaginary part* of ζ.

Lemma 4.1 *Let $f(x) \in \mathbb{C}[x]$ be a polynomial with complex coefficients. If we write a complex number z as $z = u + vi$ with $u, v \in \mathbb{R}$, then the function*

$$F(u, v) = |f(u + vi)|^2$$

is a real polynomial function of the two variables u and v.

Proof: Write $f(x) = c_0 + \cdots + c_j x^j + \cdots + c_n x^n$ with the $c_j \in \mathbb{C}$. Substitute $z = u + iv$ for x to get

$$f(u + iv) = c_0 + \cdots + c_j (u + vi)^j + \cdots + c_n (u + vi)^n.$$

Expand each power $(u+iv)^j$ using the binomial theorem to get a sum of terms $u^r v^s$ for various r and s multiplied by complex numbers depending on the c_k. After all the expansion is complete, we may separate the real and imaginary parts of $f(u + iv)$ to obtain

$$f(u + iv) = a(u, v) + b(u, v)i,$$

where both $a(u, v)$ and $b(u, v)$ are real polynomial functions of u and v.

Then $|f(z)|^2$ may be written as $f(z)\overline{f(z)}$ to obtain

$$|f(u + iv)|^2 = a(u, v)^2 + b(u, v)^2,$$

which is the sum of two real polynomials and is itself a real polynomial. ∎

As preparation for the proof of the fundamental theorem, we remind the reader of some geometric facts about complex numbers as discussed earlier. We think of the complex number $z = a + bi$ to be associated with the point (a, b) in the xy-plane. The distance from the complex number z to the number w is $|z - w|$ and an inequality of the form $|w - q| \leq r$ means that the complex number w lies within the circle of radius r with center at q.

Now we discuss the fundamental theorem.

Theorem 4.2 THE FUNDAMENTAL THEOREM OF ALGEBRA. *If $f(x)$ is a polynomial of positive degree with complex coefficients, then there is a complex number z with $f(z) = 0$.*

Proof: We assume the result is not true and will eventually reach a contradiction. Thus, we assume the complex polynomial $f(x)$ has degree $n > 0$ and there is no complex number z such that $f(z) = 0$. Then $F(u, v) = |f(u+iv)|^2$ is a real polynomial function of u and v and this function is nonzero for every choice of u and v. Our proof requires that we find an element $z_0 = u_0 + iv_0$ such that the inequality $F(u_0, v_0) \leq F(u, v)$ holds for *all* u and v. Theorem 4.1 ensures that there is a disc $C(R)$ such that it holds for all (u, v) inside the disc. As our first step, we show that we can select R so that the inequality $F(0, 0) \leq F(u, v)$ holds for all (u, v) outside the disc.

Let

$$f(x) = a_0 + a_1 x + \cdots + a_n x^n, \quad a_n \neq 0.$$

Select a real number R large enough so that all of the inequalities

$$|a_j| \leq \frac{R^{n-j}|a_n|}{2n}, \quad j = 0, 1, \ldots, n-1$$

hold. This is possible because the exponent $n - j$ on R is never 0. Now suppose that z is a complex number with $|z| > R$ [so if $z = u+iv$, then (u, v) is outside the disc $C(R)$]. We have the computation

$$\frac{f(z)}{z^n} - a_n = \frac{a_0}{z^n} + \frac{a_1}{z^{n-1}} + \cdots + \frac{a_{n-1}}{z}$$

and, by the triangle inequality for complex numbers (see Exercises), we obtain

$$\left|\frac{f(z)}{z^n} - a_n\right| \leq \left|\frac{a_0}{z^n}\right| + \left|\frac{a_1}{z^{n-1}}\right| + \cdots + \left|\frac{a_{n-1}}{z}\right|.$$

Now use the inequalities $1/|z| < 1/R$ and $|a_j|/|z|^{n-j} \leq |a_n|/2n$ to conclude

$$\left|\frac{f(z)}{z^n} - a_n\right| \leq \frac{|a_n|}{2}.$$

This inequality is interpreted as follows: For each complex number z with $|z| > R$, the complex number $f(z)/z^n$ lies inside the circle of radius $|a_n|/2$ centered at a_n. In particular, the complex number $f(z)/z^n$ cannot lie inside the circle of radius $|a_n|/2$ centered at the origin; that is, $|f(z)/z^n| \geq |a_n|/2$ provided $|z| > R$. We now obtain a string of inequalities:

$$|f(z)| \geq |z^n||a_n|/2 \geq R^n|a_n|/2 \geq n|a_0| \geq |a_0| = |f(0)|.$$

This implies, for $z = u + iv$,

$$F(u, v) = |f(z)|^2 \geq |f(0)|^2 = F(0, 0)$$

for all (u, v) outside $C(R)$. Now we consider points *inside* $C(R)$. There is a point $z_0 = u_0+iv_0$ such that (u_0, v_0) is inside $C(R)$ and such that $F(u_0, v_0) \leq$

$F(u, v)$ for all (u, v) inside $C(R)$ (by Theorem 4.1). By definition $(0, 0)$ is inside $C(R)$ so, in particular, $F(0, 0) \geq F(u_0, v_0)$ and so for every (u, v) *outside* $C(R)$ we have

$$F(u, v) \geq F(0, 0) \geq F(u_0, v_0).$$

It follows that the inequality $F(u, v) \geq F(u_0, v_0)$ holds for *all* points (u, v). This completes the first step of the proof.

Next we will modify the original polynomial, $f(x)$, to obtain a new polynomial of the form $cf(px + q)$ for certain constant complex numbers c, p, q to be selected so as to force the polynomial to satisfy certain conditions. We take $c = 1/f(z_0)$ so that the polynomial $|cf(u + iv)|^2$ has minimum value 1 at $z_0 = u_0 + iv_0$. We arrange this minimum to occur at the origin by replacing the variable x with $px + z_0$; set $h(x) = cf(px + z_0)$, where p is a nonzero complex number still to be selected. Now we have $h(0) = 1$ and $|h(z)| \geq 1$ for all complex numbers z. We write the polynomial $h(x)$ as

$$h(x) = cf(px + z_0) = 1 + b_k p^k x^k + \cdots + b_n p^n x^n,$$

where the b_i are complex constants and the two numbers b_k and b_n are nonzero. We assume that k is the smallest positive integer such that x^k has a nonzero coefficient. We now select the number p so that $b_k p^k = -1$. This is possible because the equation $p^k = -1/b_k$ has a solution $p \in \mathbb{C}$. With these choices we now have

$$h(x) = 1 - x^k q(x),$$

where $q(x)$ is a polynomial in $\mathbb{C}[x]$ that satisfies $q(0) = 1$. The inequality satisfied by $h(z)$ can be written as

$$1 \leq |h(z)|^2 = |1 - z^k q(z)|^2.$$

This may be interpreted geometrically as saying that the distance between the complex numbers $z^k q(z)$ and 1 is never less than 1, for any choice of complex number z. In other words, for every complex number z, the point $z^k q(z)$ lies outside the circle C of radius 1 with center at 1. Now let us consider small, real values of the variable; let $0 \leq u \leq 1$. From the fact that $q(0) = 1$ we see that $u^k q(u)$ is a complex number that lies on the line segment between $q(u)$ and 1 because the angle of the complex number $q(u)$ is the same as the angle of $u^k q(u)$. Since $q(u)$ is close to 1, $u^k q(u)$ is close to u^k, which is approaching 0 as u gets closer to 0. Now use the continuity of polynomials to conclude that for very small, real values of u, $u^k q(u)$ can be made as close to 0 as desired, e.g., at most a distance $1/2$ from 0. This conflicts with the assertion that $u^k q(u)$ is always at least 1 unit away from 1. In other words, we have reached a contradictory situation caused by the assumption that $f(x)$ has no root in $\mathbb{C}$. ∎

Corollary 4.1 *If $f(x)$ is a polynomial of positive degree n in $\mathbb{C}[x]$, then there exist complex numbers $z_1, \ldots, z_n$ and a constant a such that*

$$f(x) = a(x - z_1)(x - z_2) \cdots (x - z_n).$$

Proof: By the fundamental theorem, there is a number z_1 such that $f(x) = (x - z_1)f_1(x)$ for some polynomial $f_1(x)$ of degree $n - 1$. Apply the same argument to the polynomial $f_1(x)$ if the degree $n - 1$ is positive. Eventually, we reach the stage where the polynomial remaining is a constant a of degree 0. ∎

We obtain information about irreducible polynomials over the field of real numbers from the fundamental theorem.

Theorem 4.3 *Let $g(x)$ be an irreducible polynomial in $\mathbb{R}[x]$. Then $g(x)$ has either degree 1 or degree 2. If $g(x) = ax^2 + bx + c$ has degree 2, then $b^2 - 4ac < 0$. Conversely, every such degree 2 polynomial is irreducible.*

Proof: Let $g(x)$ be a polynomial with real coefficients that is irreducible in $\mathbb{R}[x]$. By the fundamental theorem, there is a complex number $z = p + qi$, $p, q \in \mathbb{R}$, that is a root of $g(x)$. Consider the polynomial

$$h(x) = (x - (p + qi))(x - (p - qi)) = x^2 - 2px + (p^2 + q^2).$$

This is a polynomial in $\mathbb{R}[x]$ and also has z as a root. Since $g(x)$ was assumed to be irreducible, it follows that $h(x)$ is divisible by $g(x)$ (Theorem 4.1) and so the degree of $g(x)$ is at most 2. If the degree is 1, then $g(x) = a(x - b)$ for some real numbers a and b; if the degree is 2, then $g(x) = ah(x) = a(x^2 - 2px + (p^2 + q^2))$ for some real numbers a, p, q. Note in this case that $b^2 - 4ac = 4p^2a^2 - 4a^2(p^2 + q^2) = -4a^2q^2 < 0$. The quadratic formula assures us that the polynomial $ax^2 + bx + c$ is reducible when $b^2 - 4ac \geq 0$. ∎

Another application of this line of reasoning can be used to show that certain reducible polynomials must have a real root.

Corollary 4.2 *A polynomial with real coefficients and of odd degree must have a real root.*

Proof: Let $f(x)$ be a polynomial in $\mathbb{R}[x]$ having odd degree. By the unique factorization theorem for polynomials, $f(x)$ may be expressed as a product of irreducible polynomials in $\mathbb{R}[x]$. Since the degree of $f(x)$ is odd, not all these irreducible factors can have degree 2. The only other possibility is that $f(x)$ has a factor of degree 1. If $x - r$ is a factor of degree 1, then r is a root of $f(x)$. ∎

The general facts about the existence of roots of polynomials over the real or complex fields are of great theoretical importance but they do not give any computational clues as to how one might find the roots in a given instance. Approximation methods, such a Newton's method, provide a means for finding approximate roots of polynomial equations with real coefficients. Newton's method is described in the following exercises. Other numerical methods are also possible. Modern symbolic manipulation computer programs make finding approximate solutions of equations a relatively easy task. However, many of these approximation methods ultimately rely on the theoretical results that assert solutions must exist.

EXERCISES

1. Let $f(x)$ be a polynomial with real coefficients. Suppose that z is a nonreal complex number such that $f(z) = 0$. Prove that complex conjugate $\bar{z}$ also satisfies $f(\bar{z}) = 0$.

2. Prove the triangle inequality for complex numbers which asserts

$$|z_1 + z_2 + \cdots + z_n| \leq |z_1| + |z_2| + \cdots + |z_n|.$$

[Hint: Prove this for $n = 2$ and then use induction.]

3. Newton's method for finding approximate solutions of the equation $f(x) = 0$, when $f(x)$ is a polynomial with real coefficients, is as follows. Let x_0 be an initial "guess" at a solution of $f(x) = 0$. If $f(x_0) \neq 0$, then improve this first guess by a sequence of successive approximations $x_1, x_2, x_3, \ldots$ defined by setting

$$x_{n+1} = x_n - \frac{f(x_n)}{f'(x_n)},$$

where $f'(x)$ is the first derivative of $f(x)$ with respect to x. If the sequence $x_1, x_2, x_3, \ldots$ converges to a limit l, then $f(l) = 0$ and l is a root of $f(x)$. (The sequence fails to converge only rarely, if x_1 is "near" a true solution.) Different initial guesses x_1 may lead to different roots when there is more than one root. Use this idea to find four decimal approximations of the solutions to the following equations starting with the first guess x_1:

(a) $x^3 - 6 = 0$, $x_1 = 1$; repeat with $x_1 = 2$.

(b) $x^4 - 2 = 0$, $x_1 = 1$; repeat with $x_1 = -1$.

REFERENCES

G. Birkhoff and S. MacLane, *A Survey of Modern Algebra,* 3rd ed., Macmillain, New York, 1965.

B. L. van der Waerden, *Modern Algebra,* Fredrick Ungar, New York, 1953.

I. N. Herstein, *Topics in Algebra,* Blaisdell, New York, 1964.

M. Artin, *Algebra,* Prentice Hall, Englewood Cliffs, 1991.

T. W. Hungerford, *Algebra,* Springer-Verlag, New York, 1974.

N. Jacobson, *Basic Algebra, I and II,* W. H. Freeman, San Francisco, 1980.

The books listed above represent a small sample of the many texts currently available and contain bibliographies that will supplement this limited one. The first three are classic introductory texts on abstract algebra. Those remaining are more advanced texts suitable for a second course in algebra.

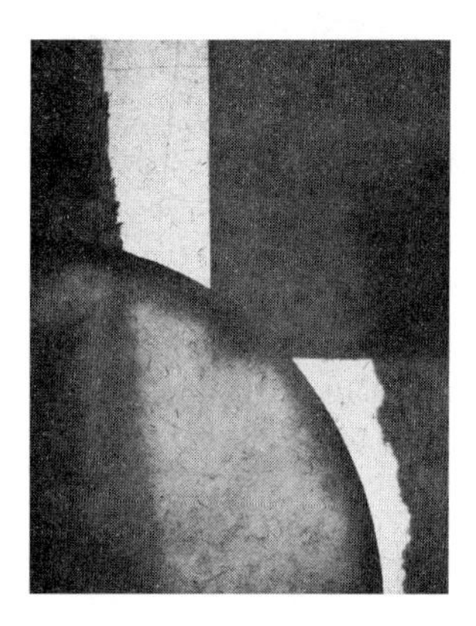

INDEX